Impact Dynamics

Impact Dynamics: A Numerical Approach enables safer aerospace and mechanical designs by providing state-of-the-art techniques and methodologies for impact simulation and crashworthiness analysis of critical structural components. It investigates the shock type of dynamic loading situations as a part of mechanical design discussions.

Classical approaches to impact problems rely on purely empirical methods that correlate large amounts of experimental data; this approach, while useful for specific problems, often fails when extrapolating to design conditions outside the tested velocity ranges. In contrast, this book focuses on numerical simulations using current models to explain the underlying physics and demonstrate how this physics-based approach provides better insight into impact events and enables successful extrapolation of test data into new design spaces for handling complex industrial problems.

This book will be useful to aviation safety professionals, researchers, and practicing aerospace and mechanical design engineers.

Sunil K. Sinha is an independent technical consultant based in the USA. In 1977, he earned his Ph.D. in Theoretical and Applied Mechanics from the University of Illinois at Urbana-Champaign, USA, and has since been active in various applications of fragment containment, target penetration, and nonlinear dynamics, especially numerical simulations of contact-impact events in various aerospace applications in a propulsion system. He retired from GE Aerospace after 31 years, where he specialized in analytical predictions of flight-safety and engine survivability during catastrophic events and extreme conditions. During his tenure at GE, he also worked on several programs and projects with Livermore Software Technology Corporation (LS-DYNA®), Boeing, Safran, the FAA, and the US Air Force. He has published over 35 papers in the area of impact dynamics and structural vibrations and serves as a technical reviewer for several leading publications.

Impact Dynamics
A Numerical Approach

Sunil K. Sinha

CRC Press
Taylor & Francis Group
Boca Raton London New York

CRC Press is an imprint of the
Taylor & Francis Group, an **informa** business

First edition published 2026
by CRC Press
2385 NW Executive Center Drive, Suite 320, Boca Raton FL 33431

and by CRC Press
4 Park Square, Milton Park, Abingdon, Oxon, OX14 4RN

CRC Press is an imprint of Taylor & Francis Group, LLC

ISBN: 978-1-041-26275-6 (hbk)
ISBN: 978-1-041-26276-3 (pbk)
ISBN: 978-1-003-74975-2 (ebk)

DOI: 10.1201/9781003749752

Typeset in Times
by codeMantra

Contents

Acknowledgments

This book is dedicated to my late parents Mr. Awadh Bihari Sinha and Mrs. Sumitra Sinha. Their memory inspired me to write this book, who always taught me that advancement of knowledge should be our lifetime goal. And, knowledge advances only when it is shared with your peers and contemporaries and then passed on to the next generation of researchers. My interest in the problems related to impact dynamics goes to my childhood, which led to my groundbreaking contributions in my doctoral work at the University of Illinois at Urbana-Champaign. I will take this opportunity to thank my doctoral thesis adviser late Professor George Costello.

In the course of my later professional work, I had the opportunity to interact with many people who helped and mentored me to advance my career. At the top of that list is Dr. John Hallquist, the original developer and author of codes DYNA3D and later LS-DYNA through his company Livermore Software Technology Corporation (LSTC). I had a great opportunity to work with him and helped in early developments on the theoretical and coding aspects of today's LS-DYNA.

I would also like to thank my many colleagues such as Mr. Jan Schilling, retired as a Chief Engineer at GE Aerospace in Cincinnati, and other peers, namely Mr. Gary Wollenweber, Mr. Rolf Hetico, Mr. Pat Niskode, and Mr. Greg Czarnecki.

I would also like to remember my late wife Mrs. Suman Sinha. I would like to acknowledge the support and encouragement of my daughters Neena Sinha, Dr. Mrs. Anjuli Sinha Campbell and her family in Menlo Park, California, in writing this book.

Dr. Sunil K. Sinha
November 2025

1 Fundamentals of Impact Dynamics

1.1 INTRODUCTION

Impact scenarios occur and happen to develop in our daily routine activities quite frequently and sometimes unexpectedly. It is a common observation that when two moving bodies come in the path of each other, they would quickly come into contact and the collision takes place. Fortunately, most of these impact events are usually benign without any observable damage, and as such they go unnoticed. In all impact events as soon as the crash happens, an audible banging sound is created due to the pressure pulse traveling through the air medium. During a collision, some of the particles' initial kinetic energy will be lost in the form of heat, sound, or due to localized deformation. The two surfaces that come into contact deform locally, which could cause large or small deformation depending upon the magnitude and direction of the relative velocity vector of the two colliding bodies moving in the path of each other. The dynamic deformation can be elastic deformation, which is fully recoverable after the impact event is over. In certain cases, depending upon the material ductility, it can also be elastic–plastic deformation with some permanent set taking place both in the moving impactor and in the crash barrier in its path. In a typical mechanical engineering curriculum, the design of machinery components subjected to impact, shock, or crash types of dynamic loading is usually treated as simply an extension of vibration design issues and procedures [see References 1, 2]. In 1982, Zukas and his other co-authors were the first ones to publish a textbook titled *Impact Dynamics* [3], which treated all impact-related mechanical design and analysis issues in a comprehensive manner. During the last 40+ years, a lot of advancements have been made in this field of specialization, driven by the availability of faster computers and analytical techniques such as direct time-integration numerical schemes using explicit method formulation of such fast-moving high-velocity dynamic events. This book covers this particular discipline of computational aspects of *Impact Dynamics* with a special emphasis on aerospace applications. Starting with the fundamental equations of motion related to the impact dynamics of the two colliding objects, we will have an in-depth discussion of the analytical formulation and numerical approach in a high-speed computing environment. The technical contents of this chapter are completely based upon the author's own lecture notes as well as his other original publications and presentations [see References 4–6].

To gain an insight into the dynamics of any typical simple impact event involving two objects, one starts with the concept of (a) "projectile" and (b) "target". In a very simple analytical model, the target is usually considered stationary and the projectile of mass "M" with an incoming impact velocity vector $\boldsymbol{V}$ comes into contact and hits the target. Normally, in "Ballistic impact", a low mass projectile moving with

DOI: 10.1201/9781003749752-1

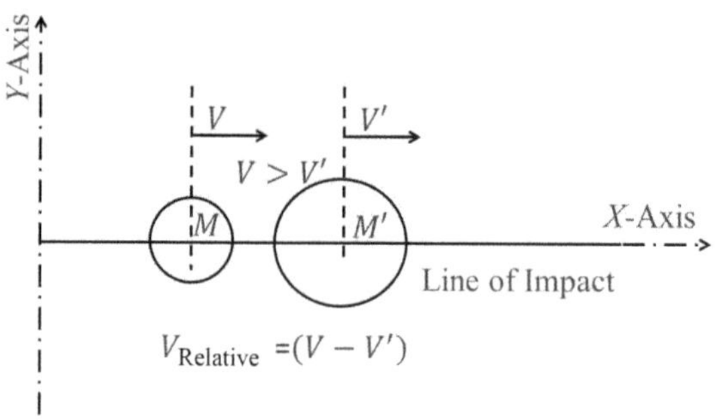

FIGURE 1.1 Direct impact of two bodies on a collision course simulated as point masses M and M'.

high-velocity impacts a stationary target and causes significant damage. The direction at which two objects hit each other will determine the trajectory of their path. The line of impact is the trajectory of the velocity vector of the moving projectile with respect to the target. The line of impact becomes the common normal line between the contacting areas of the particles during impact, and thus it is always perpendicular to the plane of impact in a normal impact scenario (Figure 1.1). In two moving point masses in contact, the line of impact is a line through the particles' mass centers. In the case of direct impact, the relative velocity vector of the mass center between the two impacting bodies coincides with the line of impact. The velocity of impact V_{Impact} is also the relative velocity V_{Relative} of the mass M being considered like a moving projectile.

In a general case of a two-body collision scenario of an oblique impact situation, the angle between the "line of impact" and the "plane of impact" is also known as "angle of incidence". Similarly, the angle between the relative velocity vector and the line of impact is called the "angle of attack θ". Thus, the sum of the "angle of incidence" and the "angle of attack" is always 90°. The common tangent plane between two impacting bodies at the point of initial contact is known as the "plane of impact". In the case of deformable impacting bodies, the plane of impact is not stationary and can change very rapidly as the impact event progresses.

1.2 DYNAMICS OF COLLISION DURING TWO BODIES IN CONTACT

From Newton's first law of motion, a moving body will keep on traveling on a straight line path with constant velocity in the absence of any external forces. Usually, these external forces could be considered nodal point loads concentrated at certain discrete locations at the boundary or can be distributed on some outside surface areas of the structure as pressure loads. An object can also experience body forces such as its own weight due to gravity or other inertia forces when subjected to either acceleration or deceleration due to external dynamic loading situations. Other body forces could be centrifugal loading types, when mass is spinning with an angular velocity due to its rotation about an axis. Impact loads of equal and opposite magnitude develop as transient dynamic loads on the moving object or even on a stationary object at the contact location as the two objects come in contact with each other during a collision or a crash.

Depending upon the relative flexibilities of two impacting objects in collision at the point of contact and its immediate vicinity, some impact events can be analyzed under the simplifying assumptions of (a) rigid body dynamic equations versus (b) conservation of momentum equations involving the coefficient of restitution e. It should be noted that in nature there is no such material like rigid material, and every structural material will deform elastically within the domain of small deformations by the time dynamic stresses σ are less than its yield limit σ_y, i.e., $\left(\sigma \leq \sigma_y\right)$.

1.2.1 Rigid Body Dynamic Equations

Under the rigid body dynamic assumptions (see Figure 1.2), the mass M of the entire object is considered concentrated as a point mass located at its center of gravity (CG).

Any external force vector on the object can be replaced by a combination of a force and moment about its CG using the following equations.

For any point P on the rigid body, located at the position vector $\boldsymbol{r}(x, y)$ translational velocity $\boldsymbol{V} = \boldsymbol{V_{CG}} + \omega X(\boldsymbol{r} - \boldsymbol{r_{CG}})$.
Linear momentum due to $\boldsymbol{V_{CG}}$ translational velocity of mass center $= M\left[\left(\boldsymbol{V_{CG}}\right)_{Initial}\right]$
Angular momentum due to rotational velocity $= I_{CG}\omega$
Change in the linear momentum due to impulse $\boldsymbol{p} = M\left(\boldsymbol{V_{Final}} - \boldsymbol{V_{Initial}}\right)$
Centrifugal forces due to rotational velocity of ω radians/s $= M\omega X\left[\omega X\left(\boldsymbol{r} - \boldsymbol{r_{CG}}\right)\right]$
Moment about the rigid body CG due to the external force vector $\boldsymbol{F_C}$ acting at the contact point C, $= \left(\boldsymbol{r_C} - \boldsymbol{r_{CG}}\right)X\boldsymbol{F_C}$

It should be noted that an impulsive external force changes the linear momentum of the projectile and the impulsive moment affects its angular momentum at the end of its impact event. In a rigid body impact scenario, the freely moving projectile always

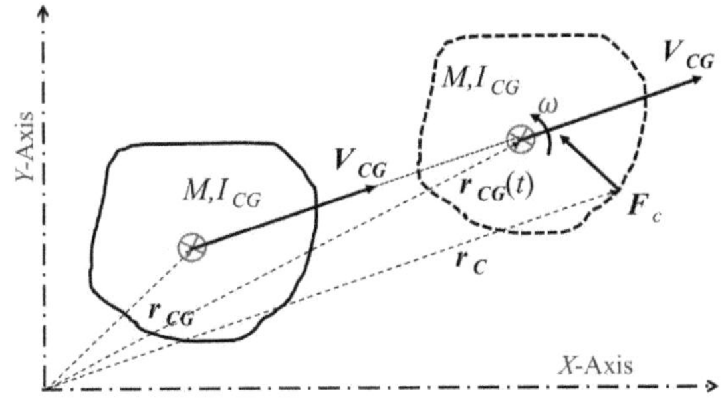

FIGURE 1.2 Dynamics of a rigid body with initial state and after time t.

makes contact with the target at the farthest corner from its mass center (CG). The following equations are additional energy equations used in the rigid body dynamics:

$$\text{Kinetic energy (KE) due to translational motion } (\text{KE})_{\text{Translational}} = \frac{1}{2} M (\boldsymbol{V} \circ \boldsymbol{V}) \tag{1.1}$$

$$\text{Kinetic energy (KE) due to rotational motion } (\text{KE})_{\text{Rotational}} = \frac{1}{2} I_{CG} (\omega \circ \omega) \tag{1.2}$$

1.2.2 Impact of Deformable Bodies and Coefficient of Restitution

It may be convenient to consider a typical collision as consisting of a (i) deformation phase, during which the shape of each body is deformed and a (ii) restitution phase, during which the shape of each body is completely or partially restored (see Figure 1.3). Newton's law follows from the supposition that, for each body, the impulse during restitution is *e* times the impulse during deformation. Coefficient of restitution is a parameter associated with the deformation of a moving projectile during a collision. In any typical collision, the impulse-momentum equation is expressed by the following equation:

$$\text{Momentum at the end of impact event} = MV_{\text{Final}} = MV_{\text{Initial}} + \int F(t)\,dt \tag{1.3}$$

Since the impulse $\int F(t)\,dt$ is integral of contact force time history $F(t)$, which is always compressive in nature, one can write the total impulse on the moving projectile ball as a combination of two impulses such as

$$\int F(t)\,dt = -\left[\int F_D dt + \int F_R dt\right] \tag{1.4}$$

$$\text{Momentum at the end of deformation phase} = MV_{\text{deformation}} = MV_{\text{Initial}} - \int F_D dt$$

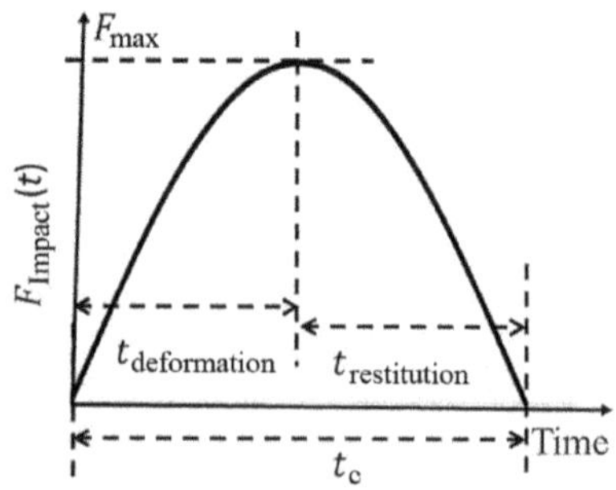

FIGURE 1.3 A typical force time history during impact.

$$\text{Final momentum at the end of impact event} = MV_{\text{Final}} = MV_{\text{deformation}} - \int F_R dt$$

The coefficient of restitution is a ratio of the magnitudes of impulse between the period of restoration and the period of deformation and is defined as

$$e = \frac{\int F_R dt}{\int F_D dt} \tag{1.5}$$

In an impact involving two balls, the forces will act in the opposite directions. Suppose that two spherical balls of equal masses are traveling in the same straight line path with velocities V_{Initial} and V_{initial}', respectively, such that $V_{\text{Initial}} > V_{\text{Initial}}'$. Then, in the impact dynamics terminology, the first ball can be considered the moving "projectile" with respect to the second ball as the "target" being impacted by a relative velocity of collision with a magnitude of $(V_{\text{Initial}} - V_{\text{Initial}}')$. In other words, the two moving objects have initial velocities of V_{Initial} and V_{Initial}', before the collision, and velocities V_{Final} and V_{Final}', after the collision. If the coefficient of restitution is e, then the relative velocities after the impact $(V_{\text{Final}} - V_{\text{Final}}')$ and before the impact $(V_{\text{Initial}} - V_{\text{Initial}}')$ are related as

$$(V_{\text{Final}} - V_{\text{Final}}') = -e(V_{\text{Initial}} - V_{\text{Initial}}') \tag{1.6}$$

This formula is Newton's law of restitution. The coefficient of restitution always satisfies $0 \le e \le 1$. When $e = 0$, it is a plastic impact and the balls remain in contact after the collision. When $e = 1$, the collision is fully elastic: there is no loss of kinetic energy. In two identical moving spheres coming into collision contact with each other, one can write

$$\text{Coefficient of restitution } e = -\frac{\text{Final relative velocity}}{\text{Initial relative velocity}} = -\frac{(V_{\text{Final}} - V_{\text{Final}}')}{(V_{\text{Initial}} - V_{\text{Initial}}')} \tag{1.7}$$

The above is the definition of the coefficient of restitution e. This term is especially useful in determining the resulting velocities of inelastic collisions, i.e., $e \neq 1$. The characteristics of a collision can be examined with this ratio, which are as follows:

a. Elastic collision ($e = 1$)
b. Inelastic (elastic–plastic) collision ($0 < e < 1$)
c. Perfectly inelastic (plastic) collision ($e = 0$).

It should be noted that the conservation of momentum equation for two dissimilar point masses M and M' yields

$$(MV_{\text{Final}} + M'V_{\text{Final}}{}') = (MV_{\text{Initial}} + M'V_{\text{Initial}}{}') \tag{1.8}$$

Combining the two equations (1.8) and (1.9) results in

$$V_{\text{Final}} = \frac{(M + eM')V_{\text{Initial}} + (1 - e)M'V_{\text{Initial}}{}'}{(M + M')} \tag{1.9}$$

and

$$V_{\text{Final}}{}' = V_{\text{Final}} + e(V_{\text{Initial}} - V_{\text{Initial}}{}') \tag{1.10}$$

Thus, the postimpact velocities of both the point masses V_{Final} and $V_{\text{Final}}{}'$ can easily be determined for any value of coefficient of restitution e between the two colliding objects.

In short, the coefficient of restitution of $e = 1$ describes an elastic collision and a value of zero results in a perfectly inelastic collision where the deformation never recovers, and after the impact the relative velocity is equal to zero. In a typical mathematical representation involving a global Lagrangian frame of reference, the observer is sitting on the deformable target with its origin attached to a clamping point of the structure. The normal duration of an impact event in structural design is relatively very short period, which usually lasts for 2–5 ms. During the impact event, the projectile decelerates and the target accelerates until the relative velocity between the two objects becomes zero. The duration for which the two bodies stay in contact with each other is an important parameter and is usually represented by "t_c". The peak dynamic contact force "$F_{\max}$" developed between two impacting bodies can be extremely large. Depending upon the material capabilities of both the projectile and the target, such an impact event can result in either some minor damage or significant fragmentation and penetration of the target.

1.2.3 Simple Estimation of Peak Dynamic Force During Impact

In the dynamic analysis of impact problems, one of the most difficult parameters, on which major design decisions depend, is the dynamic force developed during the impact. It is a transient force that reaches a certain peak value and goes to zero at the end of the impact. This quantity is extremely difficult to determine either by analytical means or even by experimental method. In the analytical approach, the problem arises due to several unknowns such as the coefficient of restitution, the damping of impacting bodies, the local deformation of surfaces during the impact, and the change of material behavior during impact from elastic to plastic. There is no convenient method for making a quick estimate of dynamic design loads, which are usually needed for the preliminary design. In the experimental approach, the problem arises due to severe limitations of experimental technique in accurate measurements of deformations, loads, stresses, etc. during the very short contact period, which normally is of the order of a fraction of milliseconds depending upon the stiffness and masses of the impacting bodies. One approach, which has been used

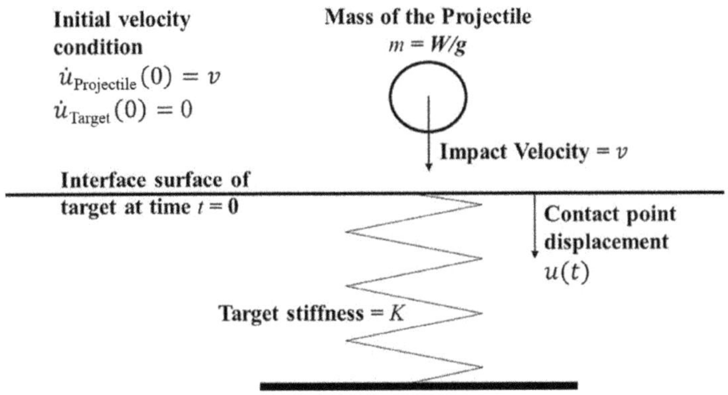

FIGURE 1.4 A rigid projectile of mass "m" moving with velocity v impacting an elastically deformable target of linear stiffness K.

in the past quite extensively, is to determine the acceleration experimentally on the existing baseline design by some accelerometer in terms of g's and then to apply it as a static load on the structure to improve or optimize the design. The common mistake, which originates from static world reasoning, is that if a design is not working, then making it stiffer will enable it to withstand the load. This may appear to be a conservative design approach; however, it is prone to yield wrong results. The reason is that the dynamic forces are functions of stiffness and as the structure or a component is stiffened, the dynamic loads go up as well. Many times, the static method of design becomes a losing battle in dynamic situations, such as increasing the fan case thickness to contain the blade. It may be counterintuitive, but sometimes it is better to make the structure more flexible and allow it to deform so that it can absorb the kinetic energy of the projectile using the ductility and plastic deformation characteristic of the target.

In a classic impact scenario shown in Figure 1.4, a moving projectile of mass "m" traveling with initial velocity v hits an elastically deformable barrier of linear stiffness K sitting in its path of motion. In a simple example, the impactor can be a point mass with its velocity vector perpendicular to the surface of the target, which can be considered a thin plate clamped at its edges. As soon as the projectile comes into contact in its path with the target, a dynamic contact force $F_{\text{contact}}(t)$ develops at the interface, which is always compressive in nature. An equal and opposite time-dependent contact force $F_{\text{contact}}(t)$ acts in such a way that it decelerates the projectile and tries to accelerate the target surface in the direction of motion. Figure 1.5 shows the free-body diagram of the dynamic force equilibrium at any time t after the projectile makes contact with the target surface at time $t = 0$.

Thus, it should be noted that during the impact loading phase, the sign of $\ddot{u}$ for a node at the centroid of the projectile will be negative (-) and for a node at the target contact point, it will be with a positive (+) sign. Referring to Figure 1.5 and shown as $t_{\text{deformation}}$, this loading period of the target continues until the relative velocity between the projectile and the target has come down to zero. At this point, the transient dynamic impact force reaches its peak value shown as $F_{\max}$. If the target is

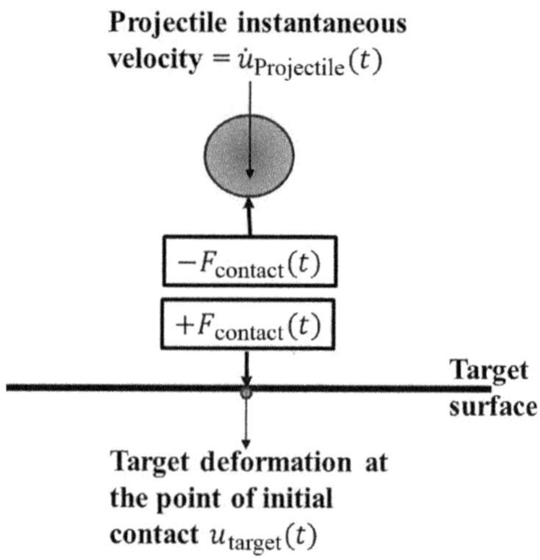

FIGURE 1.5 Schematic free-body dynamic force equilibrium diagram for the projectile and the target at any typical loading time t after the initial contact at time $t=0$.

elastic–plastic, then the dynamic force related to the impact event ends at that instant. However, if the target is perfectly elastic, then the rebound phase shown as $t_{\text{restitution}}$ starts, where the plate deformation recovers and starts pushing back the projectile "m" in such a way that after time $t_c/2$, the projectile separates and bounces back with the return velocity magnitude $-v$, which is acting in the opposite direction. One must take into account the dynamic characteristics of the structure, which is responsible for the change of momentum into impulse. In other words, the kinetic energy of the impactor gets converted into the strain energy of the target. The actual forces may be higher or lower depending upon the stiffnesses and mass of the impacting bodies. In the following section, a relatively simple method is presented for the analytical determination of dynamic forces. To represent the transient dynamics of a contact-impact event for a simple spring-mass system shown in Figure 1.5 in the differential equation form, we can write

a. Equation for the projectile : $m_{\text{Projectile}}\ \ddot{u} = -\ F_{\text{contact}}(t)$ (1.11)

Similarly, for a target of mass m_{target} with its spring stiffness K, we have,

b. Equation for the target : $m_{\text{target}}\ \ddot{u} + Ku = +F_{\text{contact}}(t)$ (1.12)

It is obvious that for a massless target, the above equations would simplify as

$$Ku = F_{\text{contact}}(t) = -m_{\text{Projectile}}\ \ddot{u} \tag{1.13}$$

In the current simple problem description of a rigid point mass projectile hitting an elastically deformable target during the impact event, our goal is to obtain a reasonably accurate estimate of the peak impact force F_{max}, which is as follows from the impulse-momentum considerations.

$$\int_0^{t_c/2} F_{contact}(t)\,dt = m_{Projectile}\; v \tag{1.14}$$

A set of formulas have been developed for the determination of crucial parameters such as peak dynamic forces, contact time, and peak deformation of the structure under dynamic loads F_{max}. Depending upon the nonlinearities involved in an impact event, the actual load-time history during the impact event will be extremely complex, but for simplicity could be approximated as having a sinusoidal distribution to capture the shock loading characteristics as shown in Figure 1.6a. Here, Figure 1.6b shows the corresponding time history of the velocity of the projectile during the same impact event like a rigid point mass hitting an elastically deformable target.

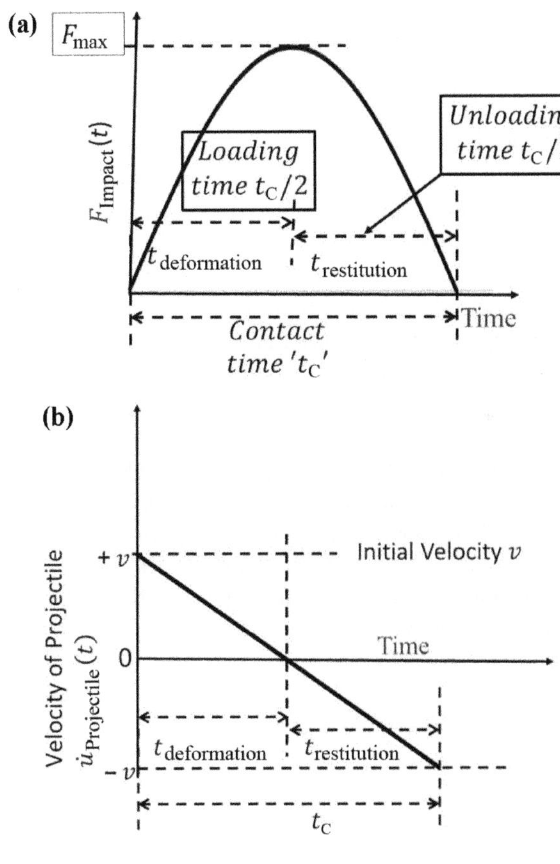

FIGURE 1.6 Typical dynamic response time history of a deformable target surface being impacted by a rigid projectile. (a) Impact force time-history loading applied on the target. (b) Velocity time history of the projectile decelerating during its contact with the target surface.

TABLE 1.1
Simple Design Parameters for Rigid Impactor Hitting an Elastically Deformable Target

Design Parameters	Half-Sine Wave Function	Saw-Tooth Linear Impulse Function
1. Loading time = $t_c/2$	$t_L = \frac{t_c}{2} = \frac{\pi}{2}\sqrt{\frac{m}{K}}$	$t_L = \frac{t_c}{2} = \frac{3}{2}\sqrt{\frac{m}{K}}$
2. Maximum force F_{max}	$F_{max} = v\sqrt{Km}$	$F_{max} = v\sqrt{Km}$
3. Maximum acceleration	$A_{max} = v\sqrt{K/m}$	$A_{max} = v\sqrt{K/m}$
4. Maximum elastic deflection of target δ_{max}	$\delta_{max} = v\sqrt{m/K}$	$\delta_{max} = v\sqrt{m/K}$

The typical dynamic parameters commonly useful in the design of mechanical components subjected to impact loading are shown in Table 1.1 with $m = W/g$.

One should note some of the differences between the dynamic and static characteristics of the structure, which are as follows:

a. In a static loading situation, the deflection and resulting strains in the structure can be reduced directly in proportion to the increasing stiffness, whereas in the dynamic situations they are only inversely proportional to the square root of the stiffness.
b. Similarly, in a static loading situation, the externally applied load to the structure is considered an independent quantity, whereas in the dynamic situations keeping parameters other than stiffness constant, the peak load F_{max} actually increases in proportion to the square root of the stiffness.

A classical design approach for dynamic load situations is to design the structure for static load capacity of $F_{max}/2$. Under this approach, a factor called impact factor is introduced, and the dynamic deflection is assumed to be twice the static deflection for a given load F_{max}. It should be noted that this approach does not take into account the impactor velocity, and the peak load is assumed to be known.

With the elastic Young's modulus "E", the typical stiffness "K" for simple loading conditions of the most commonly used beam-type structural components is as follows:

a. Pure longitudinal axis loading in a beam-column (Length 'L'): $K = (EA)/L$ (1.15)

b. Beam bending (cantilever span "L", Impact loading at the end): (1.16)

$$K = (3EI)/L^3$$

c. Beam bending (simply supported ends with span "L", impact in the middle): $K=(24EI)/L^3$. (1.17)

d. Beam bending (clamped ends with span "L", impact in the middle) : $K=(192EI)/L^3$ (1.18)

For example, for beams of rectangular cross-sectional area $A=(bh)$, such that width $= b$, and thickness $= h$, we can write $I=(1/12)bh^3$.

Using the above equations for dynamic loads, the corresponding stresses for the bending moment (BM) can be determined by using the traditional strength of material formulas:

$$\sigma=\left[\frac{(BM)h}{2I}\right] \text{[where } h = \text{thickness of the beam]} \tag{1.19}$$

Thus, for example, in the case of a cantilever beam impact problem, the dynamic peak bending moment capability can be determined as follows:

$$\sigma=\left[\frac{(BM)h}{2I}\right]=\left[\frac{(F_{\max}L)h}{2I}\right] \tag{1.20}$$

After some algebraic manipulation, one can obtain the peak dynamic strain $\epsilon_{\max}$ in the cantilever beam due to its impact by a projectile of mass "m" traveling with velocity "v" and hitting at its free end of span $= L$:

$$\epsilon_{\max}=\left(h\sqrt{\frac{3(k.e.)}{2EIL}}\right) \text{[where, } (k.e.) = \text{Kinetic Energy of the projectile]} \tag{1.21}$$

1.3 TARGET FAILURE, PENETRATION, PERFORATION, AND PROJECTILE FRAGMENTATION

The best reference on this topic is a classic paper of 1978 authored by Backman and Goldsmith [see Reference 7]. For any detailed understanding and an in-depth discussion of target penetration by a moving projectile, this reference is highly recommended by all readers. The amount of damage to the target in any crash-impact situation is heavily dependent upon the magnitude of the relative impact velocity between the two colliding objects. The following criteria are provided to classify the type of impact based upon the impact velocity and their specific application for the purpose of numerical modeling and simulation. These finite-element numerical analysis techniques are highly valuable tools to design the structures to withstand and enhance their damage-tolerance capability for varying impact velocities ranging from low-velocity applications like dynamic loading on passengers sitting in an aircraft seat during a crash landing to very high-velocity impact due to a disk burst in a running engine.

a. *Hypersonic velocity impact*: 1500 m/s and higher (for example: cruise missile impact).
b. *Supersonic velocity impact*: 350–1500 m/s and higher (for example: ballistic missile impact).
c. *Very high-velocity impact*: 100–300 m/s (aerospace applications, such as bird strike, hail ingestion, fan blade out, disk burst, and containment structure design).
d. *High-velocity impact*: 50–100 m/s (automotive crash simulations, airbag analysis).
e. *Medium-velocity impact*: 20–50 m/s (automotive passenger injury)
f. *Low-velocity impact*: 5–20 m/s (loading on a dummy during the crash landing of aircraft).
g. *Very low-velocity impact*: Less than 5 m/s (injury to humans due to falling on the floor, usually during slipping in the bathroom or on ice).

To summarize the above criteria, it should be noted that 1500 m/s is usually considered the upper limit of ballistic impact velocity. During a normal impact as the projectile starts deforming the target, the local strain in the vicinity of the impact location increases, and so does the stress. Depending upon the material characteristics such as yield strength, failure strain, ductility, and strain rate behavior, the target may crack and other impact-related damage may occur. To understand the different aspects of target failure subjected to impact loading, we will consider the same case of a classic example of a rigid mass projectile impacting the deformable target (see Figure 1.4) presented in Section 1.2 and extend our discussion further in the light of material properties such as its stress–strain characteristics. As previously mentioned, under the assumptions of a rigid point mass impactor and an elastically deformable target, collision is a simplified example of a very complex dynamic impact event. In the majority of impact and crash situations, there are all kinds of nonlinearities present such as elastic–plastic behavior, structural failure, and other damages observed both in the projectile and in the target.

Referring to Figure 1.7, after the projectile traveling with velocity v makes initial contact with the target surface at time $t = 0$ (see Figure 1.7a), the target starts being pushed in the direction of motion. In the transient dynamic process of locally deforming the target, the contact forces $F_{contact}(t)$ at the interface between the projectile and the target keep on increasing (see Figure 1.7a) to reach a peak value of contact load shown as F_{max}. The interface contact forces between the projectile and the target are inherently compressive in nature and are self-balancing at the point of contact acting in the normal direction at the contact surface. The direction of loading is in such a way that external loading on the target is in the direction of motion, and simultaneously the projectile encounters an equal and opposite reaction load acting as a resistance against its direction of motion. The external loading on the target during the loading duration of $t_{deformation}$ produces a local strain in the target, and as

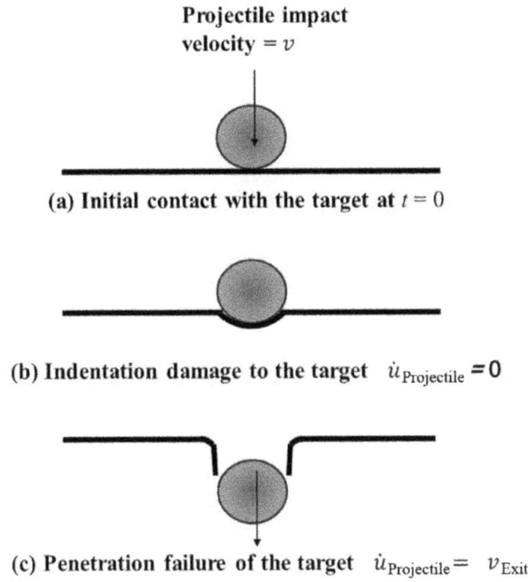

FIGURE 1.7 Schematic representation of indentation and penetration type damage to the target caused by a rigid projectile traveling with initial velocity v.

the deflection increases so does the strain and the corresponding elastic stresses in the target as well. Since most of the target materials are metals exhibiting an elastic–plastic material characteristics, as soon as the dynamic stresses in the target reach a yield strength of the material called σ_{yield}, the contact surface in the target forms a local plastic deformation, appearing like an indentation in the shape of the projectile (see Figure 1.7b). If the relative velocity of the projectile and the target becomes zero, then the impact event is over and there is no more loading of the target. At that point, the indentation in the target becomes permanent as a plastic deformation and the projectile comes to a rest with $\dot{u}_{\text{Projectile}} = 0$, and the projectile remains on contact with the target from that point onwards. This type of permanent plastic deformation in the target after the impact event is usually observed in a low and medium impact velocity regime. In aerospace applications, such permanent deformation is observed as a "cusp" at the leading edge of a titanium fan blade after bird strike.

However, if the impact velocity is in the range of high to very high and ballistic impact range, then target damage after the impact event will not be limited to indentation and local plastic deformation only. Rather in such cases, the target can experience structural failure during an impact event with rupture and a penetration of the target can also be expected. A typical such failure of the target is shown schematically (see Figure 1.7c) with a rigid impactor puncturing the target with full penetration and coming out of it with an exit velocity shown as v_{Exit}. During the penetration failure of the target, it can also rupture and the liberated debris can form another ricochet, which sometimes can travel even faster than the original projectile.

Such liberated small masses in the form of multiple ricochets can act like a sling shot, which then can again be pushed by the moving projectile mass.

The energy-absorbing capability of the target plate is measured in terms of toughness. Toughness is known as energy-absorption capacity and is generally defined as the area under load–deflection curve of a flexural or stress–strain curve of a compressive test of a specimen. We will discuss the "Fracture toughness".

The "Toughness" is measured as the strain energy absorbed for an unnotched specimen to be loaded in a tensile testing machine until failure. As opposed to it, the "Fracture toughness" is the amount of energy absorbed during crack propagation to a complete failure for a fatigued specimen with a very small microscopic-sized initial crack. The fracture toughness term is usually used in a low-cycle fatigue failure context, whereas "Impact toughness" is determined for the kinetic energy to be absorbed by a ductile metal usually measured in a Charpy impact test. Usually in an impact scenario, higher yield strength metal alloys have low ductility, hence low "Fracture toughness" capability. In impact event, low yield strength metal alloys exhibit much better energy-absorbing capability by large deformation of the target due to their inherent higher ductility characteristics. At the other extreme, brittle materials such as carbon and glass fibers exhibit very high failure limit, but with no ductility have very poor impact resistance capability. The impact resistance capability of a containment-type structure can also be scaled using a coupon test kinetic energy absorption data obtained in a Charpy impact testing machine. For high-velocity impact test data, the reader is referred to an excellent set of results compiled by the Watertown arsenal test laboratory during mid-1950s and early 1960s time-frame. Empirically, it has been observed that the projectile kinetic energy-absorption capability of a containment structure is proportional to the square-to-cube power of the thickness ratio.

It should be noted that until now we have assumed the projectile as a rigid mass. However, most of the projectiles are made up of metals in the shape of a spear-like front end. During the penetration process, they can also fragment into smaller debris and pieces, which can again release other sets of ricochets. This process of the projectile breaking into smaller pieces is called fragmentation.

In the classic impact problem of a moving projectile hitting a flat plate target, the fragmentation process is shown schematically in Figure 1.8. As soon as the projectile

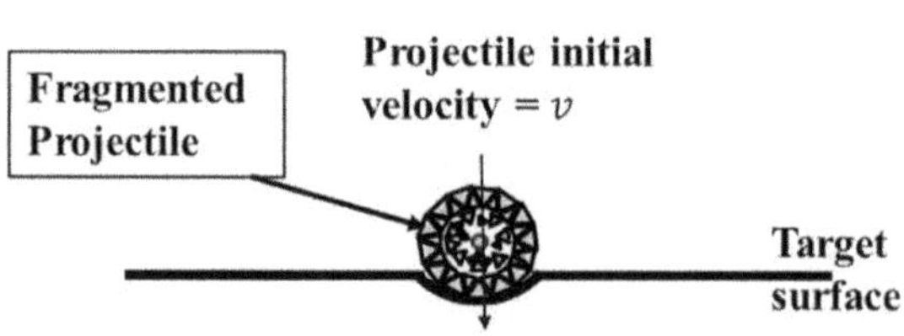

FIGURE 1.8 Schematic representation of "Fragmentation" type failure of the deformable projectile caused by transient dynamic strain in the impactor in excess of its material capability during the forward piercing movement into the target after making initial plastic indentation ($0 < \dot{u}_{\text{Projectile}}(t) < v$).

traveling with initial velocity v makes contact with the target surface, a dynamic compressive contact force of magnitude $F_{contact}(t)$ creates a stress in the impactor. If the equivalent von Mises exceeds the material yield strength of σ_{yield}, first it will plastically deform and if it crosses the failure limit of the projectile material $\sigma_{Failure}$, it will fragment the projectile and that will release some of the strain energy stored in the projectile material. In short, the fragmentation failure will dissipate a lot of kinetic and strain energies in the projectile, and eventually it will exhibit its effect in reducing the damage or failure to the target material. Sometimes at very high strain rates, the projectile material can fail like a brittle material, which is commonly observed in ceramics.

1.4 PROJECTILE VELOCITY VECTOR AND TYPES OF IMPACT

Depending upon the line of impact trajectory and its orientation angle with respect to the plane of contact, one can classify different impact scenarios under the following major categories:

a. Central impact
b. Normal impact
c. Oblique impact
d. Glancing impact

The above-mentioned types of impact play an important role in determining the post-impact trajectory of the projectile or the two or multiple bodies coming in contact with each other. The angle θ which the translational velocity vector of the mass center makes with the line of impact for the moving body is also known as "angle of attack". Central impact (see Figure 1.9) happens when the traveling velocity vectors of the two approaching objects coincide with the line of impact. In the basic simple form of normal impact (see Figure 1.10), the projectile hits a rigid flat wall or a plane barrier with velocity of impact $V_{Impact} = V$ exactly perpendicular to the plane of impact. Conceptually, the classical definition of terms such as "Projectile" and

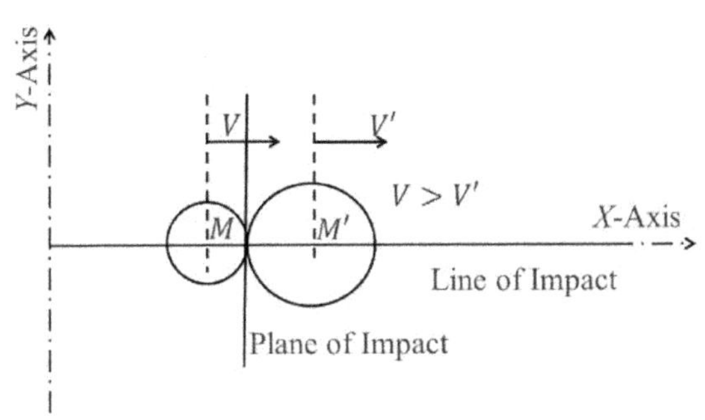

FIGURE 1.9 Central impact with the relative velocity of impact $V_{Impact} = V - V'$.

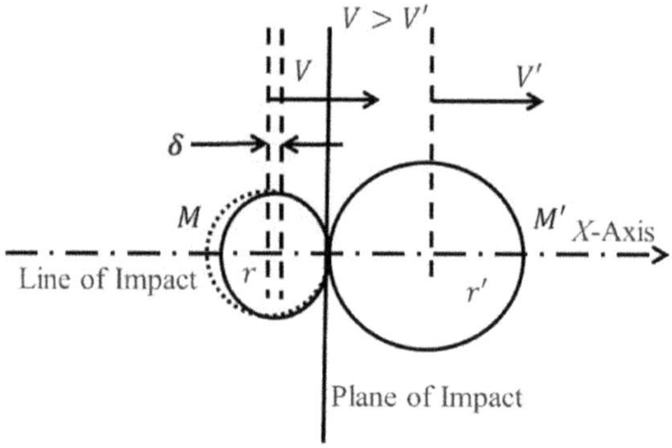

FIGURE 1.10 Elastically deformable mass M impacting a slow-moving rigid mass M' in its path with a relative velocity $V_{\text{Relative}} = (V - V')$. Deformation on only one sphere $= \delta$.

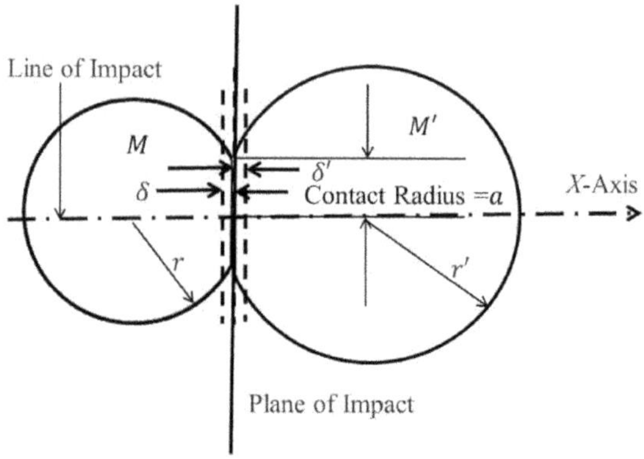

FIGURE 1.11 Normal impact on a flat target.

"Target" in an impact dynamics scenario is highlighted very well in the context of "Normal impact" as illustrated in Figure 1.11.

1.4.1 Example of Central Impact

As shown earlier in Eqs. (1.9) and (1.10), the postimpact velocities of the two masses in a central impact scenario are

$$V_{\text{Final}} = \frac{(M + eM')V_{\text{Initial}} + (1 - e)M'V_{\text{Initial}}'}{(M + M')} \tag{1.22}$$

and,

$$V_{\text{Final}}' = V_{\text{Final}} + e(V_{\text{Initial}} - V_{\text{Initial}}'). \tag{1.23}$$

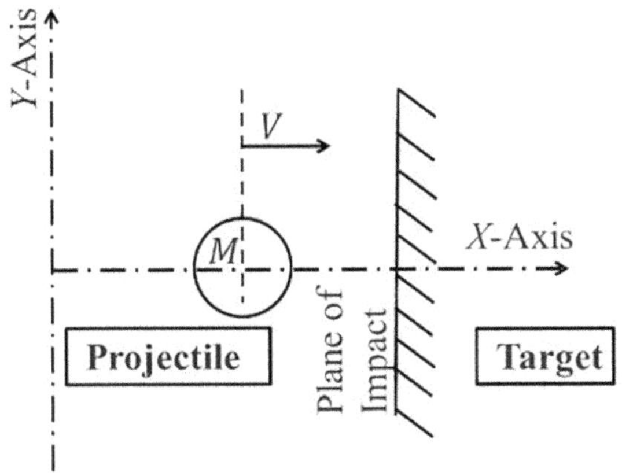

FIGURE 1.12 Elastically deformable mass M impacting another elastically deformable mass M' in its path with a relative velocity $V_{\text{Relative}} = (V - V')$, Total deformation $= \delta - \delta'$.

1.4.2 Two Deformable Spheres in Hertzian Contact

Now, we will consider two spheres of radii r and r' coming in contact with each other. Here, we assume either one (see Figure 1.10) or both the spheres (see Figure 1.12) approaching each other are deforming under contact-impact load of $F_c = F_{\text{Impact}}(t)$. The dynamic impact load equation can be written as

$$\text{Contact force during impact event on the first mass} = M\frac{dV}{dt} = -F(t) = -F_{\text{Impact}}(t) \tag{1.24}$$

$$\text{Contact force during impact event on the second mass} = M\frac{'dV'}{dt} = F(t) = F_{\text{Impact}}(t) \tag{1.25}$$

$$\text{Thus, the acceleration during the contact-impact event} = -\frac{F_{\text{Impact}}(t)}{M_{\text{Eqv}}} \tag{1.26}$$

$$\text{Stress wave velocity } c = \sqrt{\frac{E}{\rho(1-\nu^2)}} \tag{1.27}$$

$$E_{\text{Eqv}} = \frac{1}{\left\{\frac{1-\nu^2}{E} + \frac{1-\nu'^2}{E'}\right\}} = \frac{EE'}{\left[(1-\nu^2)E' + (1-\nu'^2)E\right]} \tag{1.28}$$

$$r_{\text{Eqv}} = \frac{1}{\frac{1}{r} + \frac{1}{r'}} = \frac{rr'}{(r+r')} \tag{1.29}$$

$$M_{\text{Eqv}} = \frac{1}{\frac{1}{M} + \frac{1}{M'}} = \frac{MM'}{M+M'} \tag{1.30}$$

and the yield strength,

$$\sigma_y = \min\left[\sigma_y, \sigma_y'\right]. \tag{1.31}$$

The elastic Hertzian contact force between the two colliding spheres is determined as

$$\text{Contact-impact force} = F_{\text{Impact}}(t) = F_c = \frac{4}{3} E_{\text{Eqv}} \sqrt{r_{\text{Eqv}}} (\delta^{3/2}) \tag{1.32}$$

and the contact radius

$$a = \sqrt{r_{\text{Eqv}} \delta}. \tag{1.33}$$

The elastic regime during impact (see Figures 1.10 and 1.12) is defined from the instant the contact starts $(t = 0,\ \delta = 0)$ until the local yielding with stress reaching up to σ_y in the vicinity of contact starts. Depending upon the material characteristics, the yielding can start either on the projectile and the target or only on one.

1.4.3 Example of Normal Impact Against a Rigid Wall

In case of a normal impact against a rigid flat barrier (see Figure 1.11), the final post-impact velocity is determined as $V_{\text{Final}} = -eV_{\text{Initial}}$.

1.4.4 Example of Oblique Impact Against a Rigid Wall

Similarly, in an oblique impact event (see Figure 1.13) against a rigid wall, the velocity vector V of the projectile makes an attack angle θ with the normal to the plane of impact, such that $V_{\text{Impact}} = V\cos\theta$. Thus, the component of the velocity vector V in the plane of impact becomes $V_{\text{Tangent}} = V\sin\theta$.

In case of an oblique impact by a point mass M against a rigid wall (see Figure 1.13), the final postimpact velocity vector magnitude and directions are determined using the following two parameters:

$$\theta_{\text{Final}} = \tan^{-1}\left[-\frac{1}{e}\tan\theta_{\text{Initial}}\right] \tag{1.34}$$

and

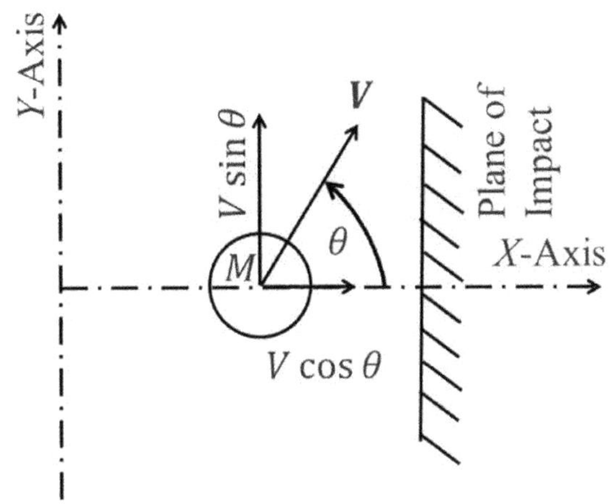

FIGURE 1.13 Oblique impact on a rigid wall.

$$V_{\text{Final}} = \left[\frac{\sin\theta_{\text{Initial}}}{\sin\theta_{\text{Final}}}\right] V_{\text{Initial}} \tag{1.35}$$

1.4.5 Example of Oblique Collision of Two Spheres

Since conservation of momentum is a vector equation in a general 3-D dynamical system, the (X, Y, Z) components of the momentum must be balanced separately for both the initial and final momentum of all the masses before and after the impact. As a simple planar example of a general oblique impact situation, we consider two masses M and M' moving along two different velocity vectors $\boldsymbol{V}$ and $\boldsymbol{V'}$ (see Figure 1.14). The respective translational velocity vectors $\boldsymbol{V}$ and $\boldsymbol{V'}$ of the two colliding masses before the impact are expressed in terms of their X and Y direction components as

$$\boldsymbol{V} = V_{\text{Initial}}\left(\cos(\theta_{\text{Initial}})\hat{\imath} + \sin(\theta_{\text{Initial}})\hat{\jmath}\right) = V\left(\cos\theta\hat{\imath} + \sin\theta\hat{\jmath}\right) \tag{1.36}$$

$$\boldsymbol{V'} = V_{\text{Initial}}'\left(\cos(\theta_{\text{Initial}}')\,\hat{\imath} + \sin(\theta_{\text{Initial}}')\hat{\jmath}\right) = V'\left(\cos\theta'\,\hat{\imath} + \sin\theta'\hat{\jmath}\right) \tag{1.37}$$

$$\text{Here, } V_{\text{Impact}} = V\cos\theta - V'\cos\theta' \tag{1.38}$$

Figure 1.14 shows a more general case of oblique impact involving two moving masses M and M' coming in contact with each other but traveling in two different directions, such that the initial and final postimpact velocities have both X and Y direction components. Thus, the corresponding momentum balance equations result in

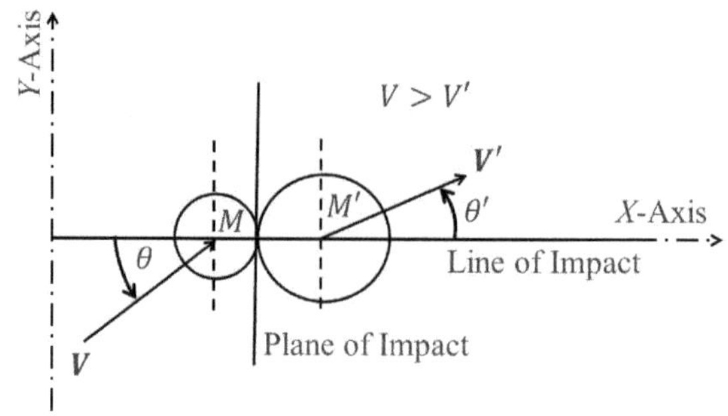

FIGURE 1.14 Oblique Impact of two spheres with the relative velocity of initial contact along the line of impact.

$$\left(M\ V_{\text{Final}}\cos(\theta_{\text{Final}})+M'\ V_{\text{Final}}{}'\cos(\theta_{\text{Final}}{}')\right)$$

$$=\left(M\ V_{\text{Initial}}\cos(\theta_{\text{Initial}})+M'V_{\text{Initial}}{}'\cos(\theta_{\text{Initial}}{}')\right) \tag{1.39}$$

$$\left(M\ V_{\text{Final}}\sin(\theta_{\text{Final}})+M'\ V_{\text{Final}}{}'\sin(\theta_{\text{Final}}{}')\right)=$$

$$\left(M\ V_{\text{Initial}}\sin(\theta_{\text{Initial}})+M'\ V_{\text{Initial}}{}'\sin(\theta_{\text{Initial}}{}')\right) \tag{1.40}$$

In the above discussion section, it should be noted that for brevity, we have used V for V_{Initial} and V' for $V_{\text{Initial}}{}'$.

$$(V_X)_{\text{Final}}=\frac{(M+eM')V\cos\theta+(1-e)M'V'\cos\theta'}{(M+M')} \tag{1.41}$$

$$\left(V_X{}'\right)_{\text{Final}}=(V_X)_{\text{Final}}+e(V\cos\theta-V'\cos\theta') \tag{1.42}$$

$$(V_Y)_{\text{Final}}=\frac{(M+eM')V\sin\theta+(1-e)M'V'\sin\theta'}{(M+M')} \tag{1.43}$$

and

$$\left(V_Y{}'\right)_{\text{Final}}=(V_Y)_{\text{Final}}+e(V\sin\theta-V'\sin\theta') \tag{1.44}$$

We also have

$$V_{\text{Final}}=\sqrt{(V_X)_{\text{Final}}^2+(V_Y)_{\text{Final}}^2} \tag{1.45}$$

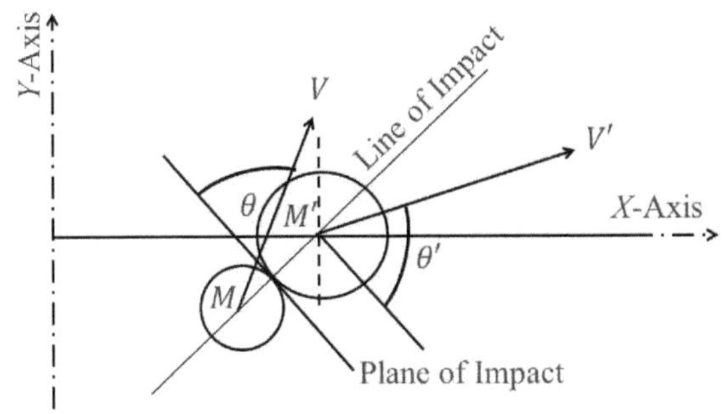

FIGURE 1.15 Glancing impact and angle of incidence of the relative velocity vector with respect to the plane of impact <3°.

$$V_{\text{Final}}' = \sqrt{\left(V_X'\right)_{\text{Final}}^2 + \left(V_Y'\right)_{\text{Final}}^2} \tag{1.46}$$

$$\theta_{\text{Final}} = \tan^{-1}\left[\frac{(V_Y)_{\text{Final}}}{(V_X)_{\text{Final}}}\right] \tag{1.47}$$

$$\theta_{\text{Final}}' = \tan^{-1}\left[\frac{\left(V_Y'\right)_{\text{Final}}}{\left(V_X'\right)_{\text{Final}}}\right] \tag{1.48}$$

Next, Figure 1.15 illustrates a special case of oblique impact known as "glancing impact". In this type of impact, the relative angle of incidence is very shallow and is usually less than 3°.

For such a very shallow angle of impact, the damage to the colliding bodies is relatively minimal, but can produce a very significant diversion in the trajectory of the two objects. When there is a large difference in the masses of M and M' with $M \ll M'$, the smaller mass M, as shown in Figure 1.15, can gain a significant spin about its mass center, after it is impacted by the larger mass M'.

1.5 CONTACT DYNAMICS MODEL FOR A FALLING ROD

First, we will consider a free-falling rigid beam of weight W and length L in an inclined orientation from height h and hitting the floor behaving like a rigid barrier (see Figure 1.16). We will assume that the beam touches the ground with an impact velocity of V_{Initial} such that $V_{\text{Initial}} = \sqrt{2gh}$.

As soon as the lower end of the beam touches the floor, the ground reacts with a vertical upward load of static weight W, such that $W = (\rho ALg)$. The ground contact point acts like a hinge and becomes the instantaneous center of rotation of the beam.

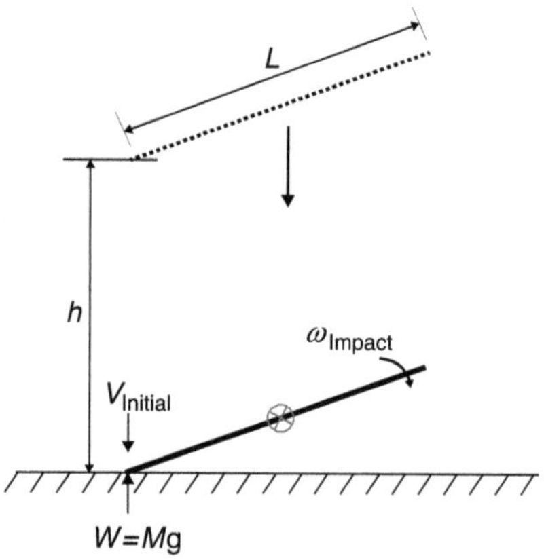

FIGURE 1.16 A 1-D beam falling on the ground in an inclined orientation.

Immediately after the contact with the ground, the beam starts rotating clockwise with an angular velocity of ω about the instantaneous center being the point of initial contact, which can be determined using the kinetic energy balance equation.

Kinetic energy $(KE)_{\text{Iniiial}}$ of the beam just before the ground contact $= \dfrac{1}{2}\left[\dfrac{W}{g}\right]V_{\text{Initial}}{}^2$.

Kinetic energy $(KE)_{\text{Final}}$ of the beam immediately after the impact $= \dfrac{1}{2}\left[I_{CG} + \dfrac{WL^2}{4g}\right]\omega^2$

which yields the angular velocity of beam about the point of contact as $\omega = \sqrt{3}\,\dfrac{V_{\text{Initial}}}{L}$.

Thus, the angular velocity of impact by the broad side of the beam along its entire length on the floor is expressed as

$$\omega_{\text{Impact}} = \frac{\sqrt{6gh}}{L} \tag{1.49}$$

1.5.1 1-D Beam Hitting the Floor with Friction

Now we will consider a moving rigid beam hitting the floor with friction behaving like a rigid wall (see Figure 1.17).

The governing equations of motion for the CG of the falling beam are

$$M\ddot{x} = \mu F_y \tag{1.50}$$

$$M\ddot{y} = F_y - Mg \tag{1.51}$$

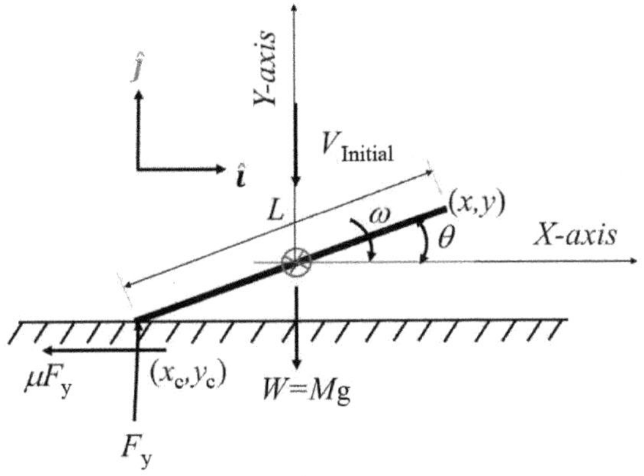

FIGURE 1.17 A 1-D beam falling on the ground (rigid wall) with friction μF_y.

$$I_{CG}\ddot{\theta} = \frac{L}{2}[\mu \sin\theta - \cos\theta]F_y \tag{1.52}$$

From the impulse-momentum equation, we obtain the change in the vertical velocity of the normal contact load acting on the beam, at the point of contact with the ground.

$$\Delta\dot{y} = \frac{1}{M}\int_{t}^{t+\Delta t} F_y dt \tag{1.53}$$

1.5.2 Details of Impulse-Momentum Equation

With respect to the mass center (CG) of the beam as the origin, we have the left-end corner of the beam coordinates contacting with the floor as

$$x_c = -\frac{L}{2}\cos\theta \tag{1.54}$$

$$y_c = -\frac{L}{2}\sin\theta \tag{1.55}$$

The vector components (p_x, p_y) of the impulse vector are expressed as

$$\boldsymbol{p} = p_x\hat{\boldsymbol{i}} + p_y\hat{\boldsymbol{j}} \tag{1.56}$$

$$M\left[(V_y)_{\text{Final}} - (V_y)_{\text{Initial}}\right] = p_y \tag{1.57}$$

$$M\left[(V_x)_{\text{Final}} - (V_x)_{\text{Initial}}\right] = p_x = \mu p_y \tag{1.58}$$

We have $I_{CG} = \frac{1}{12}ML^2$, which enables us to write the angular momentum balance equation for the beam as

$$I_{CG}\left[\omega_{\text{Final}} - \omega_{\text{Initial}}\right] = p_x y_c - p_y x_c = \left(\mu p_y y_c\right) - p_y x_c \tag{1.59}$$

The coefficient of restitution equation yields

$$\left(V_y\right)_{\text{Final}} - \omega_{\text{Final}} x_c = -e\left[\left(V_y\right)_{\text{Initial}} - \omega_{\text{Initial}} x_c\right] \tag{1.60}$$

The simultaneous solution of the above equations yields the numerical values of $(V_x)_{\text{Final}}, \left(V_y\right)_{\text{Final}}, \omega_{\text{Final}}$, and impulse p_y.

When in the special case of $\omega_{\text{Initial}} = 0$, we have the following sets of equations:

$$M\left[\left(V_y\right)_{\text{Final}} - \left(V_y\right)_{\text{Initial}}\right] = p_y \tag{1.61}$$

$$M\left[(V_x)_{\text{Final}} - (V_x)_{\text{Initial}}\right] = p_x = \mu p_y \tag{1.62}$$

$$\left(V_y\right)_{\text{Final}} = -e\left[\left(V_y\right)_{\text{Initial}}\right] \tag{1.63}$$

1.6 2-D PLANAR ANALYTICAL MODEL FOR RIGID IMPACT

Let rigid bodies B and B' with masses M and M' collide at contact point C where at least one body has a surface with continuous curvature. At the contact point, there is a common tangent plane and perpendicular to this plane, a common normal direction $\hat{\boldsymbol{j}}$. The perpendicular direction $\hat{\boldsymbol{i}}$ lies in the tangent plane and is the direction of sliding at incidence. Position vectors, $\boldsymbol{r}$ and $\boldsymbol{r}'$ from the centers of mass, G and G', to the contact point C define the configuration of the system, Figure 1.18. At the contact point C bodies B and B' have velocities $\vec{V}$ and $\vec{V}'$, respectively, so that there is a relative velocity at C, $\vec{v} = \vec{V} - \vec{V}'$ where $= \hat{\boldsymbol{i}} + \hat{\boldsymbol{j}}$. We assume that the period of collision is very brief and the contact forces are very large so that there is negligible displacement or change in configuration during impact.

In case of two moving rigid masses M and M' coming in contact with each other, the conservation of momentum equation yields

$$M\boldsymbol{V}_{\text{Initial}} + M'\boldsymbol{V}_{\text{Initial}}' = M\boldsymbol{V}_{\text{Final}} + M'\boldsymbol{V}_{\text{Final}}' \tag{1.64}$$

In case of these two elastically deformable masses M and M' moving with linear initial velocities of V_{Initial} and V_{Initial}', respectively, such that $V_{\text{Initial}} > V_{\text{Initial}}'$ coming into contact with each other.

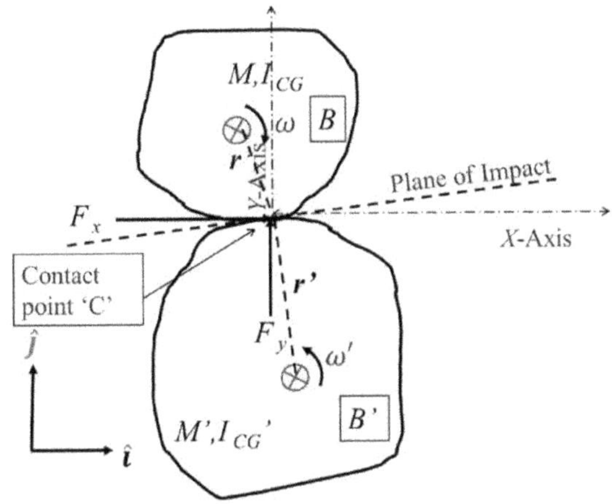

FIGURE 1.18 2-D in-plane collision of two rigid bodies.

$$\text{Momentum of 'Projectile' mass } M \text{ after impact event} = M\boldsymbol{V}_{\text{Final}} = M\boldsymbol{V}_{\text{Initial}} - \int_0^{t_c} \boldsymbol{F}(t)\,dt \tag{1.65}$$

$$\text{Momentum of 'Target' mass } M' \text{ after impact event} = M'\boldsymbol{V}_{\text{Final}}{}' = M'\boldsymbol{V}_{\text{Initial}}{}' + \int_0^{t_c} \boldsymbol{F}(t)\,dt \tag{1.66}$$

In a golf game, the mass of the club is M with the impact velocity that V_{Initial} and the mass of the golf ball is M' with the initial velocity of $V_{\text{Initial}}{}' = 0$. Thus, after being hit by the club, the velocity of ball $V_{\text{Final}}{}'$ is determined as

$$V_{\text{Final}}{}' = \left[\frac{M}{M'}\right] V_{\text{Initial}} \tag{1.67}$$

1.6.1 2-D In-Plane Collision of Two Deformable Bodies

A collision occurs when the bodies come into contact with some normal relative velocity that is negative $\left[V_y(0) - V_y{}'(0)\right] < 0$ at the point where the bodies are coincident, C (see Figure 1.19). During collision, equal but opposite impulses $\boldsymbol{p}$ and $\boldsymbol{p}'$ act at C on bodies B and B', respectively, where $\boldsymbol{p} = \left(p_x\hat{\boldsymbol{i}} + p_y\hat{\boldsymbol{j}}\right)$. The normal component of impulse $p_y \equiv \int_0^t F_y(t)\,dt$ acts in a direction that opposes interpenetration of the bodies. Because the normal contact force is compressive throughout the brief period of contact, during impact the normal impulse p_y is a monotonically increasing function of time $dp_y = F_y(t)\,dt > 0$.

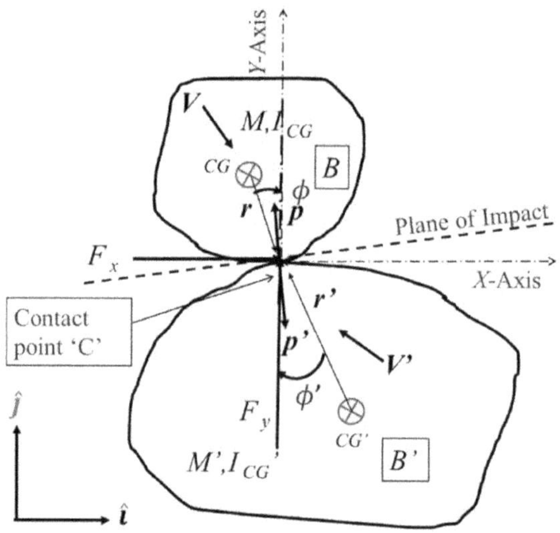

FIGURE 1.19 2-D in-plane oblique collision of two bodies with relative impact velocity $\left(V_y - V_y'\right)$.

Thus, the total normal impulse imparted during impact can be separated into an initial impulse period of compression p_c where $v_3(p) \le 0$ and a subsequent period of restitution $p_f - p_c$ where $v_3(p) > 0$. Likewise, if $p_s < p_f$, the tangential impulse can be separated into an initial impulse period of positive sliding and a subsequent impulse period of either reverse sliding or stick. If $p_s \ge p_f$, initial sliding continues throughout the impact, a condition known as continuous or gross slip.

1.7 PROBLEMS OF STRUCTURAL DYNAMICS AND MECHANICAL SYSTEMS

In this section, we will briefly review the classical equations of structural dynamics with which the readers are already familiar from their previous vibrational background. To this end, we start from the following governing equations of motion in Lagrangian systems for the dynamic response of elastically deformable structures or mechanical components. In matrix form with standard notations, these systems of simultaneous second-order ordinary differential equations are as follows:

$$[\boldsymbol{M}]\{\ddot{\boldsymbol{u}}\}+[\boldsymbol{C}]\{\dot{\boldsymbol{u}}\}+[\boldsymbol{K}]\{\boldsymbol{u}\}=\{\boldsymbol{F}\}_{external} \tag{1.68}$$

Where

$[\boldsymbol{M}]$ Mass matrix (symmetrical). It contains inertia-related terms.
$[\boldsymbol{C}]$ Damping matrix. It contains velocity-dependent terms.

$[K]$ Stiffness matrix (symmetrical). It contains deformation and stiffness-related terms.

$\{F\}_{external}$ External force column vector. It contains all external forces acting on the dynamical system. In a contact-impact problem, this also contains all the contact forces acting. In the dynamical system, the external forces are time-dependent.

$\{\ddot{u}\}$ Acceleration column vector. It is the response of the dynamical system due to external forces acting on it. The terms of the column vector are time-dependent.

$\{\dot{u}\}$ Velocity column vector terms are also time-dependent. It is also the response of the dynamical system and results in a change in momentum due to impulse.

$\{u\}$ Displacement or deformation column vector terms are also time-dependent, which travel through the structure like a stress wave. Since it is the dynamic response of the system, it results in deformation of the structure or mechanical components. It creates strain in the structural components, which generates internal stresses to counter the effect of external forces.

The solution of the matrix equation outlined in Eq. (1.68) is usually obtained in two steps. In the elementary vibration textbook terminology, they are known as follows:

a. Free vibration solution, which is also called "Eigenvalue solution", and the numerical results are referred to as "Natural Frequencies of Vibrational Modes".

$$[M]\{\ddot{u}\}+[C]\{\dot{u}\}+[K]\{u\}=0 \tag{1.69}$$

b. Forced vibrational excitation, which is also called "Harmonic response" of the dynamical system subjected to some external periodic excitation. In the context of transient dynamic analysis, sometimes this state of dynamic response is also called "Steady-state response". Suppose the external force column vector's peak magnitude is represented by $\{F_{max}\}$ such that its excitation frequency is Ω radians/s, which enables us to express Eq. (1.68) as follows.

$$[M]\{\ddot{u}\}+[C]\{\dot{u}\}+[K]\{u\} = \{F_{max}\}e^{i\Omega t} \tag{1.70}$$

1.7.1 Eigenvalue Solution of Damped Dynamical System

In this section, we will seek the solution of Eq. (1.69) as an exponential function $e^{\lambda t}$ with a characteristic exponent λ such that we can write the following relationships among various related parameters of velocity and acceleration column vector terms:

$$\{u\} = \{u(t)\} = \{U_{max}\}e^{\lambda t} \tag{1.71}$$

$$\{\dot{u}\} = \{\dot{u}(t)\} = \lambda\{U_{max}\}e^{\lambda t} \tag{1.72}$$

$$\{\ddot{u}\} = \{\ddot{u}(t)\} = \lambda^2\{U_{max}\}e^{\lambda t} \tag{1.73}$$

On substituting Eqs. (1.71), (1.72), and (1.73) in Eq. (1.69), one obtains the following equation:

$$\left[[K] + \lambda[C] + \lambda^2[M]\right]\{U\} = \{0\} \tag{1.74}$$

In the above equation, the magnitude of terms in the displacement column vector known as "Mode shape", which are arbitrary and are usually normalized with respect to the largest number such that the peak value is always unity. The eigenvalue problem of Eq. (1.74) can be solved easily by first converting the equation into a "state" form by introducing a column vector $\{V\} = \lambda\{U\}$ such that a new set of matrix equation is obtained as follows:

$$\begin{bmatrix} 0 & [I] \\ -[M]^{-1}[K] & -[M]^{-1}[C] \end{bmatrix} \begin{Bmatrix} \{U\} \\ \{V\} \end{Bmatrix} = \lambda \begin{Bmatrix} \{U\} \\ \{V\} \end{Bmatrix} \tag{1.75}$$

where $[I]$ is an identity or unit matrix. The general solution of this set of equations (Eq. 1.75) results in eigenvalues λ with complex roots, such that one can write $\lambda = \lambda_r \pm i\lambda_i$, where $i = \sqrt{-1}$. The positive and negative signs of the real part of eigenvalues (λ_r) indicate the unstable and stable behaviors of the dynamical system. In dynamic stability problems, the usual interest is in identifying the unstable zones, i.e., to determine the boundaries of the principal instability regions in the frequency domain, where the parametric resonance is most likely to occur. The parametric resonance takes place in those types of structural components or mechanical systems, where part of the stiffness matrix represented by matrix term $[B]$ becomes periodic time-dependent function of the excitation frequency like rotational angular velocity of Ω in a bent rotor shaft and whirling about the engine axis.

$$[M]\{\ddot{u}\} + [C]\{\dot{u}\} + \left[[K] - [B]\cos\Omega t\right][K]\{u\} = \{0\} \tag{1.76}$$

The boundary frequency equation (roots of Ω) for the dynamic system typically governed by the Mathieu–Hill equation of Eq. (1.76) is written as follows:

$$\text{Determinant}\begin{vmatrix} \left[[K] + \frac{1}{2}[B] - \frac{\Omega^2}{4}[M]\right] & -\frac{\Omega}{2}[C] \\ \frac{\Omega}{2}[C] & \left[[K] - \frac{1}{2}[B] - \frac{\Omega^2}{4}[M]\right] \end{vmatrix} = 0 \tag{1.77}$$

After some lengthy algebraic manipulations, Eq. (1.77) reduces to finding the eigenvalues of the following matrix:

$$\begin{bmatrix} 0 & 0 & 2[\boldsymbol{I}] & 0 \\ 0 & 0 & 0 & 2[\boldsymbol{I}] \\ [\boldsymbol{M}]^{-1}\left[2[\boldsymbol{K}]+[\boldsymbol{B}]\right] & 0 & 0 & -2[\boldsymbol{M}]^{-1}[\boldsymbol{C}] \\ 0 & [\boldsymbol{M}]^{-1}\left[2[\boldsymbol{K}]-[\boldsymbol{B}]\right] & 2[\boldsymbol{M}]^{-1}[\boldsymbol{C}] & 0 \end{bmatrix} \tag{1.78}$$

Since the matrices $[\boldsymbol{M}]$, $[\boldsymbol{C}]$, $[\boldsymbol{K}]$, and $[\boldsymbol{B}]$ are fully defined and known, either of the complex eigenvalue problem described by Eqs. (1.75) and (1.78), respectively, is easily solved by using the IMSL-EISPACK routine. The form in Eq. (1.75) is more suitable for determining the stability for a known value of the periodic load P_i and excitation frequency Ω. On the other hand, the form of Eq. (1.78) is more convenient at the design stage, where the attempt is to identify the instability zones a priori and, if possible, to avoid them altogether. After determining the roots of λ in Eq. (1.75), one can also quickly determine the effective logarithmic decrement by using the following equation:

$$\text{Logarithmic decrement } \delta = -2\pi\left[\frac{\lambda_r}{|\lambda_i|}\right] \tag{1,79}$$

The dynamic system will be stable only when the effective logarithmic decrement δ for each mode is either zero or positive. The parametric resonance condition can be determined by the imaginary part (λ_i) of the eigenvalue. Simple parametric resonance in the neighborhood of twice the natural frequency of the system would always be observed. The imaginary roots always yield a very important piece of data, i.e., the critical speed of the rotor.

1.7.2 Forced Vibratory Response Under Harmonic Excitation

In the field of classical vibration analysis, this is the most common textbook problem of forced response of a dynamical system subjected to a periodic external force excitation. When the excitation frequency is close to the natural frequency of the mechanical structure, it can go into resonance, which can be very destructive in the case of either zero or a very low negligible value of system damping. Even in a time marching forward numerical scheme for direct time integration used in all transient dynamic problems, the initial condition of steady state is first established by the implicit method solution of forced harmonic response by stress initialization. A spinning rotor shaft with nominal unbalance is a very good example of such an application. In this section, we will consider one such example of a turbofan rotor to demonstrate the analytical methodology and its application in a numerical approach. To that end, suppose we have a fan rotor shaft with a mass moment unbalance of m_r, spinning at a constant angular velocity of Ω radians/s, The units of this term m_r are mass times radial distance from the engine axis of rotation (dimension: mass × length units). Hence, the unbalance force of magnitude F_U is generated in such a way that it becomes a rotating load vector acting radially outward with respect to the engine axis in the circumferential location of the unbalance mass. Thus,

$$F_U = m_r\Omega^2 \tag{1.80}$$

If "w" is the deformation of the rotor in a Lagrangian frame of reference attached to the fan rotor, then in the global fixed frame of reference. If (u, v) represent the components of displacement in the $(X - Y)$ plane corresponding to $(X\text{-axis})$ and $(Y\text{-axis})$, respectively, as the horizontal and vertical directions. Thus, using the usual notations of equations of this dynamical system in the Lagrangian frame of reference equation (Eq. 1.70) attached to the rotating fan rotor shaft can be written as shown below:

$$[\boldsymbol{M}]\{\ddot{\boldsymbol{w}}\}+[\boldsymbol{C}]\{\dot{\boldsymbol{w}}\}+[\boldsymbol{K}]\{\boldsymbol{w}\} = \{\boldsymbol{F}_U\}e^{i\Omega t} = \left(m_r\Omega^2\right)\left[\cos(\Omega t) + i\,\sin(\Omega t)\right] \tag{1.81}$$

Thus, in the complex number notation, we can use the following relationships. Here, the external force components as their real and imaginary parts are as follows:

$$\{\boldsymbol{F}_U\}_r = \left(m_r\Omega^2\right)\cos(\Omega t) \text{ and, } \{\boldsymbol{F}_U\}_i = \left(m_r\Omega^2\right)\sin(\Omega t) \tag{1.82}$$

If the resulting magnitude of maximum radial deflection is δ_{max}, then in Eq. (1.81), we can express $\{\boldsymbol{w}\}$ as a complex number with its real part $\{\boldsymbol{w}_r\}$ and imaginary part $\{\boldsymbol{w}_i\}$, in the form of a periodic function of time "t" as follows:

$$\{\boldsymbol{w}(t)\} = \{\boldsymbol{w}\} = \{\boldsymbol{w}_r\} + i\,\{\boldsymbol{w}_i\} = \delta_{\text{max}}\,e^{i\Omega t} = \delta_{\text{max}}\left[\cos(\Omega t) + i\,\sin(\Omega t)\right] \tag{1.83}$$

Then, in the matrix equation form, the harmonic response of displacement column vector $\{\boldsymbol{w}\}$ for an unbalance force vector $\{\boldsymbol{F}_U\}e^{i\Omega t}$ can be solved as follows:

$$\begin{bmatrix} \left[-\Omega^2[\boldsymbol{M}]+[\boldsymbol{K}]\right] & -\Omega[\boldsymbol{C}] \\ \Omega[\boldsymbol{C}] & \left[-\Omega^2[\boldsymbol{M}]+[\boldsymbol{K}]\right] \end{bmatrix} \begin{Bmatrix} \{\boldsymbol{w}_r\} \\ \{\boldsymbol{w}_i\} \end{Bmatrix} = \begin{Bmatrix} \{m_r\Omega^2\} \\ \{-m_r\Omega^2\} \end{Bmatrix} \tag{1.84}$$

It should be noted that the matrix equation Eq. (1.76) will yield as many eigenvectors or the mode shapes as the dimension of the column vector. In the forced harmonic response, only the mode that is closest to the excitation frequency Ω will contribute to determine the steady-state deformed shape of the structure. The participation of other modes away from the excitation frequency of Ω will essentially be zero. Thus, the phase difference for peak dynamic response of each mode with respect to the rotating unbalance load vector "ψ" is written as follows:

$$\psi = \tan^{-1}\left(\frac{\boldsymbol{w}_i}{\boldsymbol{w}_r}\right) \tag{1.85}$$

1.8 3-D ELASTICITY AND STRESS WAVE EQUATION

In the earlier section, we have discussed the rigid impact scenario, where the two bodies coming in contact are considered essentially rigid. Even in the deformable bodies with the coefficient of restitution, it is inherently assumed that the deformation is confined to the immediate vicinity of the contact location. Until now, we have neglected the effect of internal vibrations inside the deformable body, which generate stress waves that become the main mode of kinetic energy transfer from the point of initial contact to the rest of the body. During the impulsive loading, the peak magnitude of stress waves can become extremely large, creating catastrophic damage to the structure. The contact-impact loads between two colliding bodies are always compressive in nature, and hence they produce compressive stress waves which travel from the point of initial impact to other ends of the deformable structure, and then they reflect back. Thus, during the impact duration, the stress wave can travel back and forth many times, whose magnitude may decay depending upon the material and joint damping present as boundary conditions in the mechanical system. The stress waves at the clamped ends reflect back with the same sign, and thus instantly double in magnitude at that location. Hence, the compressive wave will reflect back as a compressive wave of identical magnitude and so does the tensile wave. Similarly, at the free boundary surface, the stress wave would rebound with opposite signs, thus turning a compressive stress wave into a tensile stress wave and vice versa. Compressive waves and extensional waves are also known as "dilatational waves". A dilatational wave causes a change in the volume and introduces volumetric strain in the material as the wave travels through the medium. On the other hand, the shear waves cause distortion of the element and introduce shear strain in the object as the wave travels through the solid medium. The general stress wave equation in a 3-D elastically deformable object (see Figure 1.20) can be determined by the following set of dynamic equilibrium equations in structural mechanics:

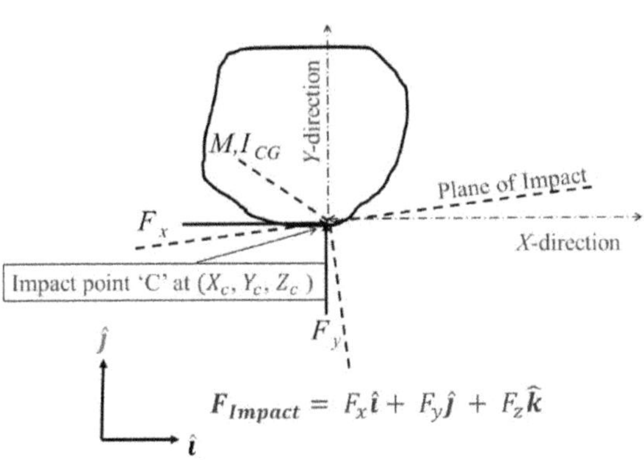

FIGURE 1.20 3-D impact loading at a contact point "C" at (X_c, Y_c, Z_c) in a 3-D elastically deformable body looking from the Z-direction.

$$\rho\ddot{u}-\left[\frac{\partial\sigma_{xx}}{\partial x}+\frac{\partial\tau_{xy}}{\partial y}+\frac{\partial\tau_{xz}}{\partial z}\right]=q_x(x,y,z)=F_x\left(\delta(x-X_c),\delta(y-Y_c),\delta(z-Z_c)\right) \quad (1.86)$$

$$\rho\ddot{v}-\left[\frac{\partial\tau_{yx}}{\partial x}+\frac{\partial\sigma_{yy}}{\partial y}+\frac{\partial\tau_{yz}}{\partial z}\right]=q_y(x,y,z)=F_y\left(\delta(x-X_c),\delta(y-Y_c),\delta(z-Z_c)\right) \quad (1.87)$$

$$\rho\ddot{w}-\left[\frac{\partial\tau_{zx}}{\partial x}+\frac{\partial\tau_{zy}}{\partial y}+\frac{\partial\sigma_{zz}}{\partial z}\right]=q_z(x,y,z)=F_z\left(\delta(x-X_c),\delta(y-Y_c),\delta(z-Z_c)\right) \quad (1.88)$$

In the above set of equations, (u,v,w) are the vector components of the deformation, $(\sigma_{xx},\sigma_{yy},\sigma_{zz},\tau_{xy},\tau_{yz},\tau_{xz})$ are the tensor components of the stress field, and (q_x,q_y,q_z) are the vector components of the body forces per unit volume. It should be noted that vector components (q_x,q_y,q_z) also include all types of external forces acting at the centroid of the infinitesimal element in terms of per unit volume, which is represented by a Dirac delta function acting at a point located at (X_c,Y_c,Z_c) such that $\boldsymbol{F}_{Impact} = F_x\hat{\boldsymbol{i}} + F_y\hat{\boldsymbol{j}} + F_z\hat{\boldsymbol{k}}$. In an impact situation, the vector components of external forces.

Thus, the general form of 3-D elastic stress wave equations in terms of deformation vector terms (u,v,w) becomes as follows:

$$\rho\ddot{u}-\frac{E}{2(1+\nu)}\left[\left(\frac{2(1-\nu)}{(1-2\nu)}\frac{\partial^2 u}{\partial x^2}+\frac{\partial^2 u}{\partial y^2}+\frac{\partial^2 u}{\partial z^2}\right)+\frac{\partial^2 v}{\partial x\partial y}+\frac{\partial^2 w}{\partial x\partial z}\right]$$
$$=F_x\left(\delta(x-X_c),\delta(y-Y_c),\delta(z-Z_c)\right). \quad (1.89)$$

$$\rho\ddot{v}-\frac{E}{2(1+\nu)}\left[\frac{\partial^2 u}{\partial x\partial y}+\left(\frac{\partial^2 v}{\partial x^2}+\frac{2(1-\nu)}{(1-2\nu)}\frac{\partial^2 v}{\partial y^2}+\frac{\partial^2 v}{\partial z^2}\right)+\frac{\partial^2 w}{\partial y\partial z}\right]$$
$$=F_y\left(\delta(x-X_c),\delta(y-Y_c),\delta(z-Z_c)\right). \quad (1.90)$$

$$\rho\ddot{w}-\frac{E}{2(1+\nu)}\left[\frac{\partial^2 u}{\partial x\partial z}+\frac{\partial^2 v}{\partial y\partial z}+\left(\frac{\partial^2 w}{\partial x^2}+\frac{\partial^2 w}{\partial y^2}+\frac{2(1-\nu)}{(1-2\nu)}\frac{\partial^2 w}{\partial z^2}\right)\right].$$
$$=F_z\left(\delta(x-X_c),\delta(y-Y_c),\delta(z-Z_c)\right) \quad (1.91)$$

It should be noted that the general form of 3-D elastic equilibrium equations may be good for theoretical discussions; however, they do not have much use in real engineering design applications. For this reason, we will focus our stress wave derivations on orthotropic materials applications only.

1.8.1 2-D Simplification of a 3-D Elastic Stress Wave Equation

The following section shows the application of the above set of equilibrium in a general 3-D elastically deformable body for a specific case of an anisotropic elastic material in two dimensions. For such a type of anisotropic thin plane stress problems, the corresponding stress–strain constitutive equations are written as follows. For composite materials made up of laminated multilayered plies, the 2-D orthotropic formulations with terms like $E_{11}, E_{22}, \nu_{12}, \nu_{21}$, G_{12} etc. are sufficient, which are as follows:

$$\sigma_{xx} = \frac{E_{11}}{(1-\nu_{12}\nu_{21})}\left[\varepsilon_{xx} + \nu_{21}\varepsilon_{yy}\right] = \frac{E_{11}}{(1-\nu_{12}\nu_{21})}\left[\frac{\partial u}{\partial x} + \nu_{21}\frac{\partial v}{\partial y}\right] \tag{1.92}$$

$$\sigma_{yy} = \frac{E_{22}}{(1-\nu_{12}\nu_{21})}\left[\varepsilon_{yy} + \nu_{12}\varepsilon_{xx}\right] = \frac{E_{22}}{(1-\nu_{12}\nu_{21})}\left[\frac{\partial v}{\partial y} + \nu_{12}\frac{\partial u}{\partial x}\right] \tag{1.93}$$

$$\tau_{xy} = G_{12}\gamma_{xy} = G_{12}\left[\frac{\partial u}{\partial y} + \frac{\partial v}{\partial x}\right] \tag{1.94}$$

On substituting the above relationships in the equations of equilibrium, we obtain the ply-level stress wave equations as follows:

$$\frac{E_{11}}{(1-\nu_{12}\nu_{21})}\left[\frac{\partial^2 u}{\partial x^2} + \nu_{21}\frac{\partial^2 v}{\partial x\,\partial y}\right] + G_{12}\left[\frac{\partial^2 u}{\partial y^2} + \frac{\partial^2 v}{\partial x\,\partial y}\right] + q_x = \rho\ddot{u} \tag{1.95}$$

$$G_{12}\left[\frac{\partial^2 v}{\partial x^2} + \frac{\partial^2 u}{\partial x\,\partial y}\right] + \frac{E_{22}}{(1-\nu_{12}\nu_{21})}\left[\frac{\partial^2 v}{\partial y^2} + \nu_{12}\frac{\partial^2 u}{\partial x\,\partial y}\right] + q_y = \rho\ddot{v} \tag{1.96}$$

The above coupled set of typical stress wave equations in 2-D stress field can be simplified for various classes of problems usually encountered in structural dynamics. For example, in a free vibration problem formulation, in the absence of any external forcing function the terms q_x and q_y can easily be dropped, which enables us to simplify the ply-level coupled stress wave equation in composites as

$$\frac{E_{11}}{(1-\nu_{12}\nu_{21})}\frac{\partial^2 u}{\partial x^2} + G_{12}\frac{\partial^2 u}{\partial y^2} + \left[\frac{\nu_{21}E_{11}}{(1-\nu_{12}\nu_{21})} + G_{12}\right]\frac{\partial^2 v}{\partial x\,\partial y} = \rho\ddot{u} \tag{1.97}$$

$$\frac{E_{22}}{(1-\nu_{12}\nu_{21})}\frac{\partial^2 v}{\partial y^2} + G_{12}\frac{\partial^2 v}{\partial x^2} + \left[\frac{\nu_{12}E_{22}}{(1-\nu_{12}\nu_{21})} + G_{12}\right]\frac{\partial^2 u}{\partial x\,\partial y} = \rho\ddot{v} \tag{1.98}$$

In some structural design problems, it is possible to neglect the in-plane shear modulus term G_{12}, which allows us to simplify the full set of coupled partial differential equations in an impact problem along with external dynamic forcing function the terms $q_x(x,y,t)$ and $q_y(x,y,t)$ as follows:

$$\frac{E_{11}}{(1-\nu_{12}\nu_{21})}\frac{\partial^2 u}{\partial x^2}+\frac{\nu_{21}E_{11}}{(1-\nu_{12}\nu_{21})}\frac{\partial^2 v}{\partial x\,\partial y}+q_x(x,y,t)=\rho\ddot{u} \tag{1.99}$$

$$\frac{\nu_{12}E_{22}}{(1-\nu_{12}\nu_{21})}\frac{\partial^2 u}{\partial x\,\partial y}+\frac{E_{22}}{(1-\nu_{12}\nu_{21})}\frac{\partial^2 v}{\partial y^2}+q_y(x,y,t)=\rho\ddot{v} \tag{1.100}$$

The above set of equations for isotropic thin 2-D membrane-type in-plane deformation problems is simplified as follows. Here, we have

$$E\equiv E_{11}=E_{22} \tag{1.101}$$

$$\nu\equiv\nu_{21}=\nu_{12} \tag{1.102}$$

$$G\equiv G_{12}=\frac{E}{2(1+\nu)} \tag{1.103}$$

$$\frac{E}{(1-\nu^2)}\frac{\partial^2 u}{\partial x^2}+\frac{E}{2(1+\nu)}\frac{\partial^2 u}{\partial y^2}+\frac{E}{2(1-\nu)}\frac{\partial^2 v}{\partial x\,\partial y}+q_x(x,y,t)=\rho\ddot{u} \tag{1.104}$$

$$\frac{E}{2(1+\nu)}\frac{\partial^2 v}{\partial x^2}+\frac{E}{(1-\nu^2)}\frac{\partial^2 v}{\partial y^2}+\frac{E}{2(1-\nu)}\frac{\partial^2 u}{\partial x\,\partial y}+q_y(x,y,t)=\rho\ddot{v} \tag{1.105}$$

1.8.2 Stress Wave Equation in an Euler–Bernoulli's Beam-Column Equation

Similarly, for the Euler–Bernoulli's beam-column equation, the dynamical response is governed by the following equation. Here, $u(x,t)$ is the axial displacement in the longitudinal direction (x-axis) and $w(x,t)$ is the lateral displacement in the transverse direction (z-axis) of the beam. In the following equation, we consider a beam of cross-sectional area A with its flexural rigidity EI and the beam cross section oriented at an angle β with respect to the axis of rotation. In a centrifugal force field generated membrane stress σ_{cf} by the angular velocity Ω radian/s, the dynamic equilibrium equations are coupled through Coriolis forces, which are as follows:

$$-EI\frac{\partial^4 w}{\partial x^4}+\frac{\partial}{\partial x}\left(A\sigma_{cf}\frac{\partial w}{\partial x}\right)=\rho A\begin{pmatrix}\ddot{w}-(x+u)\cos\beta\dot{\Omega}-\Omega^2\cos^2\beta w\\-2\dot{u}\Omega\cos\beta\end{pmatrix}-q_z(x,t) \tag{1.106}$$

$$\frac{EA}{\left(1-\nu^2\right)}\frac{\partial^2 u}{\partial x^2}+\rho A\Omega^2 u=\rho A\left(\ddot{u}\text{-}\Omega^2 x+w\cos\beta\dot{\Omega}+2\dot{w}\Omega\cos\beta\right)-q_x(x,t) \quad (1.107)$$

1.8.3 Stress Wave Equation in Timoshenko's Beam-Column Equation

In the more advanced beam 2-D formulation of the Timoshenko beam, there are three dynamic response parameters called cross-sectional rotation $\psi(x,t)$, lateral displacement $w(x,t)$, and axial extension $u(x,t)$, which define the movement of any point at the beam centerline. In the following equation, we consider a Timoshenko beam of cross-sectional area A with its shear coefficient κ and the beam cross section oriented at an angle β with respect to the axis of rotation. The corresponding equations are as follows:

$$\frac{\partial}{\partial x}\left(EI\frac{\partial\psi}{\partial x}\right)+\kappa AG\left(\frac{\partial w}{\partial x}-\psi\right)=\rho I\left(\ddot{\psi}-\cos\beta\dot{\Omega}-\Omega^2\psi\right) \quad (1.108)$$

$$\frac{\partial}{\partial x}\left(\left[F_a+A\sigma_{cf}\right]\frac{\partial w}{\partial x}+\kappa AG\left[\frac{\partial w}{\partial x}-\psi\right]\right)=\rho A\begin{pmatrix}\ddot{w}-(x+u)\cos\beta\dot{\Omega}\\ -\Omega^2\cos^2\beta w-2\dot{u}\Omega\cos\beta\end{pmatrix}-q_z(x,t) \quad (1.109)$$

$$\frac{EA}{\left(1-\nu^2\right)}\frac{\partial^2 u}{\partial x^2}+\rho A\Omega^2 u=\rho A\left(\ddot{u}-\Omega^2 x+w\cos\beta\dot{\Omega}+2\dot{w}\Omega\cos\beta\right)-q_x(x,t) \quad (1.110)$$

1.8.4 Stress Waves in a Thin Plate Subjected to Impact Loading

We will consider the flexural bending stress waves in a thin plate loaded with in-plane mid-surface membrane stresses subjected to impact shock loading. The corresponding flat rectangular plate equations per Kirchhoff–Levy's thin plate theory with its mid-surface being in the $(x-y)$ plane are as follows:

$$\begin{aligned}&-D\left[\frac{\partial^4 w}{\partial x^4}+2\frac{\partial^4 w}{\partial x^2\partial y^2}+\frac{\partial^4 w}{\partial y^4}\right]+\frac{\partial}{\partial x}\left(h\sigma_{cf}\frac{\partial w}{\partial x}\right)+\rho h\Omega^2\cos^2\beta w-2\rho h\Omega x\frac{\partial\dot{w}}{\partial x}\\&=\rho h\left[\ \ddot{w}-(x+w)\cos\beta\ \dot{\Omega}-\Omega^2\cos^2\beta w-2\dot{w}\ \Omega\cos\beta-2\dot{w}\ \Omega\sin\beta\ \right]\\&-\rho\frac{h^2}{12}\left[\frac{\partial^2\ddot{w}}{\partial x^2}+\frac{\partial^2\ddot{w}}{\partial y^2}\right]-\rho h(\beta' x)^2\left[\ddot{w}-\nu\left(\frac{Cy}{L}\right)^2\frac{\partial^2\ddot{w}}{\partial y^2}\right]-q_z(x,y,t)\end{aligned} \quad (1.111)$$

$$\frac{Eh}{\left(1-\nu^2\right)}\frac{\partial^2 u}{\partial x^2}+\rho h\Omega^2 u=\rho h\left(\ddot{u}\text{-}\Omega^2 x+w\cos\beta\dot{\Omega}+2\dot{w}\Omega\cos\beta\right)-q_x\left(x,y,t\right) \tag{1.112}$$

Similar equations for pretwisted plate about the x-axis with the twist rate of β' are as follows:

$$\begin{aligned}
&D\left[\frac{\partial^4 w}{\partial x^4}+2\frac{\partial^4 w}{\partial x^2\partial y^2}+\frac{\partial^4 w}{\partial y^4}\right]+D(\beta')^2\left[3\frac{\partial^2 w}{\partial x^2}+\frac{\partial^2 w}{\partial y^2}\right]\\
&-D(\beta')^4\left[x^2\frac{\partial^2 w}{\partial x^2}+\frac{(1-\nu)}{2}x^2\frac{\partial^2 w}{\partial y^2}+2x\frac{\partial w}{\partial x}\right]-h\frac{\partial(\sigma_{cf})}{\partial x}\frac{\partial w}{\partial x}-h\left(\sigma_{cf}\right)\frac{\partial^2 w}{\partial x^2}\\
&-\rho h\Omega^2\cos^2\beta w+2\rho h\Omega x\frac{\partial\dot{w}}{\partial x}=-\rho h\begin{bmatrix}\ddot{w}-(x+w)\cos\beta\ \dot{\Omega}-\Omega^2\cos^2\\ \beta w-2\dot{w}\ \Omega\cos\beta-2\dot{w}\ \Omega\sin\beta\end{bmatrix}\\
&+\rho\frac{h^2}{12}\left[\frac{\partial^2\ddot{w}}{\partial x^2}+\frac{\partial^2\ddot{w}}{\partial y^2}\right]-\rho h(\beta' x)^2\left[\ddot{w}-\nu\left(\frac{Cy}{L}\right)^2\frac{\partial^2\ddot{w}}{\partial y^2}\right]+q(x,y,t)
\end{aligned} \tag{1.113}$$

Where

$$D=\frac{Eh^3}{12\left(1-\nu^2\right)}. \tag{1.114}$$

Similar dynamic equilibrium equation using the classical Kirchhoff–Levy's thin plate bending formulation with $q(x,y,t)$ acting as a lateral pressure load on the plane surface at $z=-h/2$ is written as follows:

$$D\nabla^2\left(\nabla^2 w\right)=-\rho h\frac{\partial^2 w}{\partial t^2}+q(x,y,t) \tag{1.115}$$

Where

$$\nabla^2 w\equiv\frac{\partial^2 w}{\partial x^2}+\frac{\partial^2 w}{\partial y^2}+\frac{\partial^2 w}{\partial z^2}\quad\text{in 3-D rectilinear cartesian coordinate}\left(x,y,z\right)\text{ system} \tag{1.116}$$

$$\nabla^2 w\equiv\frac{1}{r}\frac{\partial}{\partial r}\left(r\frac{\partial w}{\partial r}\right)+\frac{1}{r^2}\frac{\partial^2 w}{\partial\theta^2}+\frac{\partial^2 w}{\partial z^2}\quad\text{in 3-D cylindrical circular coordinate }(r,\theta,z)\text{ system} \tag{1.117}$$

$$
D\left[\frac{\partial^4 w}{\partial \xi^4}+2\frac{\partial^4 w}{\partial \xi^2\partial \eta^2}+\frac{\partial^4 w}{\partial \eta^4}\right]+D(\beta')^2\left[3\frac{\partial^2 w}{\partial \xi^2}+\frac{\partial^2 w}{\partial \eta^2}\right]-D(\beta')^4
$$

$$
\left[\xi^2\frac{\partial^2 w}{\partial \xi^2}+\frac{(1-\nu)}{2}\xi^2\frac{\partial^2 w}{\partial \eta^2}+2\xi\frac{\partial w}{\partial \xi}\right]-h\frac{\partial(\sigma_{cf})}{\partial \xi}\frac{\partial w}{\partial \xi}-h(\sigma_{cf})\frac{\partial^2 w}{\partial \xi^2}
$$

$$
-\frac{F_{\text{rub}}}{C}\frac{\partial^2 w}{\partial \xi^2}-\rho_b h\Omega^2 cos^2\beta w+2\rho_b h\Omega\xi\frac{\partial \dot{w}}{\partial \xi}
$$

$$
=-\rho_b h\begin{bmatrix}\ddot{w}-(r+\xi+w)\cos\beta\,\dfrac{d\Omega}{dt}-\Omega^2\cos^2\beta w-2\dot{w}\,\Omega\cos\beta\\ -2\dot{w}\,\Omega\sin\beta\end{bmatrix} \tag{1.118}
$$

$$
+\rho_b\frac{h^2}{12}\left[\frac{\partial^2 \ddot{w}}{\partial \xi^2}+\frac{\partial^2 \ddot{w}}{\partial \eta^2}\right]-\rho_b h(\beta'\xi)^2\left[\ddot{w}-\nu\left(\frac{C\eta}{L}\right)^2\frac{\partial^2 \ddot{w}}{\partial \eta^2}\right]+q(\xi,\eta,t)
$$

1.8.5 Stress Waves in a Thin Rotating Composite Airfoil Subjected to Impact Loading

For an orthotropic airfoil with the mid-surface at $\zeta = 0$ coinciding with the reference (ξ,η) curvilinear surface, we have the following equation:

$$
\frac{E_{11}h}{(1-\nu_{12}\nu_{21})}\frac{\partial^2 u}{\partial \xi^2}+\rho h\Omega^2 u=\rho h\left(\ddot{u}-\Omega^2\xi+w\cos\beta\dot{\Omega}+2\dot{w}\Omega\cos\beta\right)-q_\xi(\xi,\eta,t) \tag{1.119}
$$

$$
\left[D_{11}\frac{\partial^4 w}{\partial \xi^4}+2(2D_{66}+D_{12})\frac{\partial^4 w}{\partial \xi^2\partial \eta^2}+D_{22}\begin{bmatrix}\dfrac{\partial^4 w}{\partial \eta^4}-\kappa^4\xi^2\dfrac{\partial^2 w}{\partial \xi^2}-2\kappa^4\xi\dfrac{\partial w}{\partial \xi}-\\ \kappa^2\dfrac{\partial^2 w}{\partial \eta^2}\end{bmatrix}\right]
$$

$$
-3D_{12}\kappa^2\frac{\partial^2 w}{\partial \xi^2}-(\beta')^4\left[D_{11}\xi^2\frac{\partial^2 w}{\partial \xi^2}+2D_{11}\xi\frac{\partial w}{\partial \xi}+D_{66}\xi^2\frac{\partial^2 w}{\partial \eta^2}\right]
$$

$$
+(\beta')^2\left[3D_{11}\frac{\partial^2 w}{\partial \xi^2}+(2D_{66}+D_{12})\frac{\partial^2 w}{\partial \eta^2}-2D_{66}\left[3\frac{\partial^2 w}{\partial \eta^2}+\xi\frac{\partial^3 w}{\partial \xi\partial \eta^2}\right]\right]
$$

$$
-h\frac{\partial(\sigma_{cf})_\eta}{\partial \eta}\frac{\partial w}{\partial \eta}-h\frac{\partial(\sigma_{cf})_\xi}{\partial \xi}\frac{\partial w}{\partial \xi}-h(\sigma_{cf})_\xi\frac{\partial^2 w}{\partial \xi^2}-\rho h\Omega^2\cos^2\beta w \tag{1.120}
$$

$$
+2\rho h\Omega\xi\frac{\partial \dot{w}}{\partial \xi}=-\rho h\left[\ddot{w}-(\xi+w)\cos\beta\,\dot{\Omega}-\Omega^2\cos^2\beta w-2\,\dot{u}\Omega\cos\beta\right]
$$

$$
+\rho\frac{h^2}{12}\left[\frac{\partial^2 \ddot{w}}{\partial \xi^2}+\frac{\partial^2 \ddot{w}}{\partial \eta^2}\right]-\rho h(\beta'\xi)^2\left[\ddot{w}-\nu_{12}\left(\frac{C\eta}{L}\right)^2\frac{\partial^2 \ddot{w}}{\partial \eta^2}\right]+q_\zeta(\xi,\eta,t)
$$

Where

$$\left(\sigma_{cf}\right)_{\xi} = \frac{\rho_b \Omega^2}{2}\left[\ 2RL - L^2 - 2(R-L)\xi - \xi^2\right] = \tag{1.121}$$

$$\text{Curvature } \kappa = \frac{1}{\text{Camber Radius}} \tag{1.122}$$

$$D_{11} = \frac{E_{11}h^3}{12(1-\nu_{12}\nu_{21})} \tag{1.123}$$

$$D_{22} = \frac{E_{22}h^3}{12(1-\nu_{12}\nu_{21})} \tag{1.124}$$

$$D_{12} = \frac{\nu_{21}E_{11}h^3}{12(1-\nu_{12}\nu_{21})} = \frac{\nu_{12}E_{22}h^3}{12(1-\nu_{12}\nu_{21})} \tag{1.125}$$

$$D_{66} = \frac{G_{12}h^3}{12} \tag{1.126}$$

$$D_{16} = \frac{G_{13}h^3}{12} \tag{1.127}$$

$$D_{26} = \frac{G_{23}h^3}{12} \tag{1.128}$$

$$\frac{\nu_{21}}{E_{22}} = \frac{\nu_{12}}{E_{11}} \tag{1.129}$$

And

$$\left(\sigma_{cf}\right)_{\eta} = \rho_b \Omega^2 r_{\text{camber}}{}^2\ \sin^2\beta \int_{+\vartheta/2}^{0} \varphi d\varphi = -\ \frac{\rho_b \Omega^2 r_{\text{camber}}{}^2\ \vartheta^2\ \sin^2\beta}{8} \tag{1.130}$$

Thus,

$$\frac{\partial\left(\sigma_{cf}\right)_{\xi}}{\partial\xi} = -\rho_b \Omega^2\left[r + \xi\ \right] \tag{1.131}$$

$$\frac{\partial\left(\sigma_{cf}\right)_{\eta}}{\partial\eta} = -\frac{\rho_b \Omega^2}{4}\ r_{\text{camber}}\vartheta\ \sin^2\beta \tag{1.132}$$

On substituting the above relationships in the equations of equilibrium, we obtain the ply-level stress wave equations.

In this chapter, we have presented all the relevant governing mathematical equations for various types of structural elements used in the design of various components from some very simple to other complex mechanical systems subjected to impact and shock loads.

REFERENCES

1. S. Timoshenko, D. H. Young and W. Weaver, Jr. (1974) *Vibration Problems in Engineering*, Fourth Edition, John Wiley & Sons, New York.
2. C. M. Harris and C. E. Crede (1976) *Shock and Vibration Handbook*, Second Edition, McGraw-Hill Book Company, New York.
3. J. A. Zukas, T. Nicholas, H. F. Swift, L. B. Greszczuk and D. R. Curran (1982) *Impact Dynamics*, Wiley-Interscience Publication, John Wiley & Sons, New York.
4. S. K. Sinha (1983) "A Mathematical Technique to Optimize a Mechanical Design with Many Parameters", *Mathematical Modeling in Science and Technology, The 4th International Conference*, Zurich, Switzerland, August 1983 (Pergamon Press, New York), pp. 224–227. https://doi.org/10.1016/B978-0-08-030156-3.50046-1
5. S. K. Sinha (1988) "On a General Method of Vibration Analysis in Curvilinear Coordinates", *AIAA Journal*, Vol. 26, No. 5, pp. 629–631. https://doi.org/1010.2514/3.9947
6. S. K. Sinha (1987) "Determination of Natural Frequencies of a Thick Spinning Annular Disk Using a Numerical Rayleigh-Ritz's Trial Function", *The Journal of the Acoustical Society of America*, Vol. 81, No. 2, pp. 357–369. https://doi.org/1010.1121/1.394901
7. M. E. Backman and W. Goldsmith (1978) "The Mechanics of Penetration of Projectiles into Targets", *International Journal of Engineering Science*, Vol. 16, No. 1, pp. 1–99. https://doi.org/10.1016/0020-7225(78)90002-2

2 Numerical Methods for Solving Stress and Shock Wave Problems

2.1 INTRODUCTION

The governing differential equations for stress and shock waves traveling through multiple different structural components of a mechanical system due to the impact event are invariably nonlinear in nature and have been a big challenge to solve them analytically in any mathematical technique. As such, various types of numerical schemes have been proposed and each of which usually has its own limitations due to the requirements of very small time-step size Δt. Here, it should be noted that the majority of stress wave partial differential equations are hyperbolic in nature. The technical contents of this chapter are completely based upon the author's own doctoral thesis [1] as well as his other original publications and presentations [2–5].

In the direct-time integration numerical schemes of equations of motion, the commonly used methodologies in the finite-element analysis are classified under two categories: (a) implicit method and (b) explicit method. In the implicit method numerical scheme, one sets up the coupled equations of motion for the entire structure in a matrix form generating terms for stiffness matrix $[\boldsymbol{K}]$ and mass matrix $[\boldsymbol{M}]$. Depending upon the nonlinearities such as large deflection, material elastic–plastic behavior, and contact separation, the stiffness matrix $[\boldsymbol{K}]$ needs to be updated, and a converged solution for each time step $(t+\Delta t)$ is computed using the Newton–Raphson method. The main drawback of this method is that during computation, it needs a very large memory space, and the solution may not converge at all due to the complexity of contact situation, even after using higher numerical values of the convergence tolerance limit. Additionally, this method is not capable of handling stress-wave propagation problems during high-velocity impact simulation. The implicit method scheme is the most commonly used in commercial codes like ANSYS® and NASTRAN®.

In comparison, on the other hand, in the explicit method numerical formulation, there is no need to form the traditional mass $[\boldsymbol{M}]$ and stiffness matrices $[\boldsymbol{K}]$ at all, rather one solves the uncoupled nodal equations of motion for each node separately. In an explicit method, it is possible to uncouple the equations of motion due to the inherent assumption of the time step Δt being so small that the information about the transient deformation does not travel through the structure more than the

 DOI: 10.1201/9781003749752-2

characteristic length, which is the size of the smallest element in the entire model. The commercial codes like LSDYNA®, PAMCRASH®, and DYTRAN® use an explicit formulation scheme in their solver algorithm.

2.1.1 Implicit Method Scheme

In this method, the entire structural matrix is solved simultaneously at each time step. In this method, one can use a relatively large time-step size. The numerical scheme commonly used in this methodology is called Newmark's $-\beta$ method. The general equation of dynamic equilibrium in the matrix form of the structure is written as

$$[\boldsymbol{M}]\{\ddot{u}\}+[\boldsymbol{C}]\{\dot{u}\}+[\boldsymbol{K}]\{\boldsymbol{u}\}=\{\boldsymbol{F}\}_{external} \tag{2.1}$$

Where the typical damping matrix $[\boldsymbol{C}]$ is expressed as a linear combination of mass matrix $[\boldsymbol{M}]$ and stiffness matrix $[\boldsymbol{K}]$ in the following form:

$$[\boldsymbol{C}]=\alpha\,[\boldsymbol{M}]+\beta\,[\boldsymbol{K}] \tag{2.2}$$

Here, it should be noted that the coefficient term β used in the above equation is a completely different parameter from the one used in the next paragraph. The terms of the nodal acceleration column vector at the new time step represented as $\{\ddot{u}\}_{t+\Delta t}$ are computed by solving the matrix equation of motion shown in Eq. (2.1) as follows:

$$\{\ddot{u}\}_{t+\Delta t}=[\boldsymbol{M}]^{-1}\left[\left[\{\boldsymbol{F}\}_{external}\right]_{t+\Delta t}-[\boldsymbol{C}]\{\dot{u}\}_t-[\boldsymbol{K}]\{\boldsymbol{u}\}_t\right] \tag{2.3}$$

From the unconditional numerical stability criterion, Newmark's method with $\gamma=0.5$ unconditional and $\beta=0.25$ is a good choice. The basic equations for determining the nodal velocity $\{\dot{u}\}_{t+\Delta t}$ and displacement vector terms $\{\boldsymbol{u}\}_{t+\Delta t}$ at the new time step of $(t+\Delta t)$ in Newmark's method are

$$\{\dot{u}\}_{t+\Delta t}=\{\dot{u}\}_t+(1-\gamma)(\Delta t)\{\ddot{u}\}_t+\gamma(\Delta t)\{\ddot{u}\}_{t+\Delta t} \tag{2.4}$$

$$\{\boldsymbol{u}\}_{t+\Delta t}=\{\boldsymbol{u}\}_t+(\Delta t)\{\dot{u}\}_t+\left(\frac{1}{2}-\beta\right)(\Delta t)^2\{\ddot{u}\}_t+\beta(\Delta t)^2\{\ddot{u}\}_{t+\Delta t} \tag{2.5}$$

Artificial positive damping is introduced if $\gamma>0.5$ and artificial negative damping if $\gamma<0.5$. The method is unconditionally stable if $\beta\geq\frac{(2\gamma+1^2)}{16}$.

2.1.2 Explicit Method Scheme

In this numerical FEA (Finite Element Analysis) integration scheme, one does not form the mass, damping, and structural matrix at all, rather one solves each node is solved individually at each time step. In an explicit method, it is possible to uncouple the equations of motion due to inherent assumption of the time-step size being so small that the information about the deformation does not travel through the structure more than the characteristic length, which is the size of the smallest element in the entire model.

$$m\ddot{u} = \boldsymbol{F}_{external} + \boldsymbol{F}_{contact} - \boldsymbol{F}_{nodal-stiffness-forces} \tag{2.6}$$

In the explicit method time marching forward integration scheme, the vector components of the nodal acceleration terms such as $(\ddot{\boldsymbol{u}})_{t+\Delta t}$ in the global (X,Y,Z) directions are computed as follows:

$$(\ddot{\boldsymbol{u}})_{t+\Delta t} = \frac{1}{m}\left[\left(\boldsymbol{F}_{external} + \boldsymbol{F}_{contact}\right)_{t+\Delta t} - \left(\boldsymbol{F}_{nodal-stiffness-forces}\right)_{t}\right] \tag{2.7}$$

In this chapter, we will discuss in detail the methodologies and challenges for determining the numerical values of each of the three terms in the right-hand side of equation (2.7) representing the nodal forces viz. $(\boldsymbol{F}_{external})_{t+\Delta t}$, $(\boldsymbol{F}_{contact})_{t+\Delta t}$, and $(\boldsymbol{F}_{\text{nodal-stiffness-forces}})_{t}$.

a. The global directional components of nodal force $(\boldsymbol{F}_{external})_{t+\Delta t}$ terms are determined from the time history of the imposed forced boundary conditions on the structural component being analyzed in the dynamic model. It should be noted that one should always avoid applying any concentrated external load on a single node. Every external load on the structure must be distributed over multiple nodes in the vicinity of load application zone.
b. The global directional components of nodal force $(\boldsymbol{F}_{contact})_{t+\Delta t}$ terms are numerically determined by a penalty method. If the node under consideration due to its computed displacement $(\boldsymbol{u})_t$ would have physically penetrated any surrounding structure in the previous time step "t", then the element being penetrated will apply a reaction load on that node in the opposite direction. These are the components of reaction forces encountered by the node under consideration being generated by the surrounding brick element to push the node back on its surface in order to stop being penetrated. Thus, the magnitude of this force is computed as follows:

$$(\boldsymbol{F}_{contact})_{t+\Delta t} = -[\text{Stiffness of the Element}]*[\text{Penetration depth}]_t \tag{2.8}$$

c. The global directional components of $\left(F_{nodal-stiffness-forces}\right)_t$ terms at the current time step of "t" are computed by integrating the components of respective stress tensor terms (σ_{XX}, σ_{YY}, σ_{ZZ}, τ_{XY}, τ_{YZ}, τ_{ZX}) in global directions for all the surrounding elements connected to that node under consideration. The following section describes the step-by-step procedure for integrating the stress field in a typical eight-noded hexahedron solid element.

2.2 COMPUTATION OF NODAL STIFFNESS FORCES IN A BRICK ELEMENT MESH

This section describes the full details of the procedure for integrating the stress field in a typical eight-noded hexahedron solid element (with node numbers N1, N2, N3, N4, N5, N6, N7, and N8, respectively) commonly used in almost all explicit method finite-element analysis codes such as LS-DYNA and DYTRAN (refer to Figure 2.1).

The following equation (Eq. 2.8) shows the relationship between nodal stiffness forces and the elemental stress forces:

$$\left(F_{\text{nodal-stiffness-forces}}\right)_t = \sum_{\text{element}\,\#1}^{\text{All elements connected to NODE}} \frac{\left(F_{\text{elemental-stiffness-forces}}\right)_t}{8} \tag{2.9}$$

The global directional components of the elemental stiffness forces at a typical time step "t" in the (X, Y, Z) directions, respectively, called (F_X, F_Y, F_Z) acting at the element centroid (CG) of that particular solid brick are computed as follows:

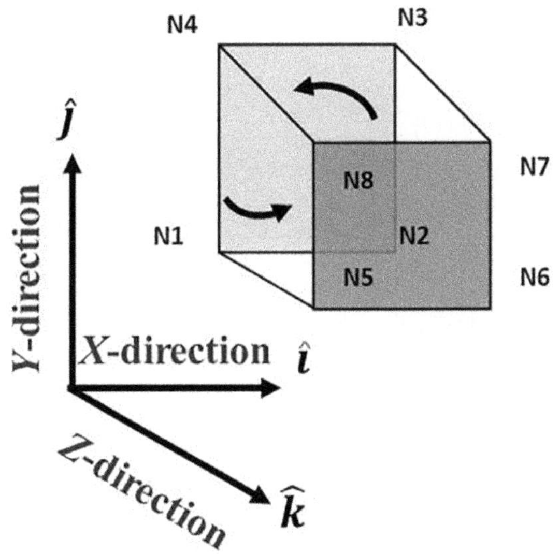

FIGURE 2.1 Eight-noded hexahedron solid brick element with a single-point integration at the element centroid.

$$[F_X]_{\text{elemental-stiffness-forces}} = (\sigma_{XX})A_X + (\tau_{XY})A_Y + (\tau_{XZ})A_Z$$

$$= (\sigma_{XX})\left[\begin{array}{l}\dfrac{1}{16}\left[\begin{array}{l}(Y_{N5}-Y_{N1})+(Y_{N6}-Y_{N2})+(Y_{N7}-Y_{N3})\\ +(Y_{N8}-Y_{N4})\end{array}\right]\\ *\left[\begin{array}{l}(Z_{N5}-Z_{N1})+(Z_{N6}-Z_{N2})+(Z_{N7}-Z_{N3})\\ +(Z_{N8}-Z_{N4})\end{array}\right]\end{array}\right]$$

$$+(\tau_{XY})\left[\begin{array}{l}\dfrac{1}{16}\left[\begin{array}{l}(Z_{N5}-Z_{N1})+(Z_{N6}-Z_{N2})+(Z_{N7}-Z_{N3})\\ +(Z_{N8}-Z_{N4})\end{array}\right]\\ *\left[\begin{array}{l}(X_{N5}-X_{N1})+(X_{N6}-X_{N2})+(X_{N7}-X_{N3})\\ +(X_{N8}-X_{N4})\end{array}\right]\end{array}\right]$$

$$+(\tau_{XZ})\left[\begin{array}{l}\dfrac{1}{16}\left[\begin{array}{l}(X_{N5}-X_{N1})+(X_{N6}-X_{N2})+(X_{N7}-X_{N3})\\ +(X_{N8}-X_{N4})\end{array}\right]\\ *\left[\begin{array}{l}(Y_{N5}-Y_{N1})+(Y_{N6}-Y_{N2})+(Y_{N7}-Y_{N3})\\ +(Y_{N8}-Y_{N4})\end{array}\right]\end{array}\right] \tag{2.10}$$

$$[F_Y]_{\text{elemental-stiffness-forces}} = (\tau_{XY})A_X + (\sigma_{YY})A_Y + (\tau_{YZ})A_Z$$

$$= (\tau_{XY})\left[\begin{array}{l}\dfrac{1}{16}\left[(Y_{N5}-Y_{N1})+(Y_{N6}-Y_{N2})+(Y_{N7}-Y_{N3})+(Y_{N8}-Y_{N4})\right]\\ *\left[(Z_{N5}-Z_{N1})+(Z_{N6}-Z_{N2})+(Z_{N7}-Z_{N3})+(Z_{N8}-Z_{N4})\right]\end{array}\right]$$

$$+(\sigma_{YY})\left[\begin{array}{l}\dfrac{1}{16}\left[\begin{array}{l}(Z_{N5}-Z_{N1})+(Z_{N6}-Z_{N2})+(Z_{N7}-Z_{N3})\\ +(Z_{N8}-Z_{N4})\end{array}\right]\\ *\left[\begin{array}{l}(X_{N5}-X_{N1})+(X_{N6}-X_{N2})+(X_{N7}-X_{N3})\\ +(X_{N8}-X_{N4})\end{array}\right]\end{array}\right]$$

$$+(\tau_{YZ})\left[\begin{array}{l}\dfrac{1}{16}\left[\begin{array}{l}(X_{N5}-X_{N1})+(X_{N6}-X_{N2})+(X_{N7}-X_{N3})\\ +(X_{N8}-X_{N4})\end{array}\right]\\ *\left[(Y_{N5}-Y_{N1})+(Y_{N6}-Y_{N2})+(Y_{N7}-Y_{N3})+(Y_{N8}-Y_{N4})\right]\end{array}\right] \tag{2.11}$$

$$[F_Z]_{\text{elemental-stiffness-forces}} = (\tau_{XZ})A_X + (\tau_{YZ})A_Y + (\sigma_{ZZ})A_Z$$

$$= (\tau_{XZ})\left[\frac{1}{16}\begin{bmatrix}(Y_{N5}-Y_{N1})+(Y_{N6}-Y_{N2})+(Y_{N7}-Y_{N3})\\ +(Y_{N8}-Y_{N4})\end{bmatrix} * \begin{bmatrix}(Z_{N5}-Z_{N1})+(Z_{N6}-Z_{N2})+(Z_{N7}-Z_{N3})\\ +(Z_{N8}-Z_{N4})\end{bmatrix}\right]$$

$$+(\tau_{YZ})\left[\frac{1}{16}\begin{bmatrix}(Z_{N5}-Z_{N1})+(Z_{N6}-Z_{N2})+(Z_{N7}-Z_{N3})\\ +(Z_{N8}-Z_{N4})\end{bmatrix} * \begin{bmatrix}(X_{N5}-X_{N1})+(X_{N6}-X_{N2})+(X_{N7}-X_{N3})\\ +(X_{N8}-X_{N4})\end{bmatrix}\right]$$

$$+(\sigma_{ZZ})\left[\frac{1}{16}\begin{bmatrix}(X_{N5}-X_{N1})+(X_{N6}-X_{N2})+(X_{N7}-X_{N3})\\ +(X_{N8}-X_{N4})\end{bmatrix} * \begin{bmatrix}(Y_{N5}-Y_{N1})+(Y_{N6}-Y_{N2})+(Y_{N7}-Y_{N3})\\ +(Y_{N8}-Y_{N4})\end{bmatrix}\right] \tag{2.12}$$

Thus, the global directional components of elemental stress-field terms called $(F_X, F_Y, F_Z)_{\text{elemental-stiffness-force}}$ are computed as shown above, which enables us to determine the respective nodal components called $\left(\boldsymbol{F}_{\textit{nodal-stiffness-forces}}\right)_t$ and used in Eqs. (2.7) and (2.8).

2.2.1 Computation of Contact Forces in the Penalty Method

In the explicit method using the penalty method, the nodes and the respective elements are allowed to penetrate with the adjoining interface surfaces. However, in the next time step, the penetrating nodes are forced back in by the amount of penetration $\delta_{\text{Penetration}}$. It should be noted that in the penalty method, the contact forces are actually reaction forces accounted in the next time-step calculation as an external force acting on that node. The global directional components of the penalty contact forces at a typical time step $(t+\Delta t)$ in the global (X,Y,Z) directions, respectively, called $\left([F_X]_{\text{contact}}, [F_Y]_{\text{contact}}, [F_Z]_{\text{contact}}\right)$ acting at the node under consideration are computed as follows. For brevity of nomenclature, we will introduce two geometrical parameters related to the contact interface surface called P and Q as follows:

$$P = \frac{(X_{N4} - X_{N1})(Z_{N3} - Z_{N1}) - (X_{N3} - X_{N1})(Z_{N4} - Z_{N1})}{(X_{N3} - X_{N1})(Y_{N4} - Y_{N1}) - (X_{N4} - X_{N1})(Y_{N3} - Y_{N1})} \tag{2.13}$$

$$Q = \frac{(Y_{N4} - Y_{N1})(Z_{N3} - Z_{N1}) - (Y_{N3} - Y_{N1})(Z_{N4} - Z_{N1})}{(X_{N4} - X_{N1})(Y_{N3} - Y_{N1}) - (X_{N3} - X_{N1})(Y_{N4} - Y_{N1})} \tag{2.14}$$

Then, the direction cosines of the normal $(\ell_{\text{normal}}, m_{\text{normal}}, n_{\text{normal}})$ to the contact surface are as follows:

$$n_{\text{normal}} = \sqrt{\frac{1}{(P^2 + Q^2 + 1)}} \tag{2.15}$$

$$m_{\text{normal}} = P \times n_{\text{normal}} \tag{2.16a}$$

$$\ell_{\text{normal}} = Q \times n_{\text{normal}} \tag{2.16b}$$

$$\left|\text{Absolute}(F_{\text{contact}})\right|_{t+\Delta t} = (\text{Panalty Factor}) \frac{AE}{\text{Element thickness}} (\delta_{\text{Penetration}}) \tag{2.17}$$

Thus, the vectorial components of penalty contact forces at the node at time step $(t + \Delta t)$ are determined as

$$\left([F_X]_{\text{contact}}\right)_{t+\Delta t} = -\left|\text{Absolute}(F_{\text{contact}})\right|_{t+\Delta t} \times \ell_{\text{normal}} \tag{2.18}$$

$$\left([F_Y]_{\text{contact}}\right)_{t+\Delta t} = -\left|\text{Absolute}(F_{\text{contact}})\right|_{t+\Delta t} \times m_{\text{normal}} \tag{2.19}$$

$$\left([F_Z]_{\text{contact}}\right)_{t+\Delta t} = -\left|\text{Absolute}(F_{\text{contact}})\right|_{t+\Delta t} \times n_{\text{normal}} \tag{2.20}$$

The above-computed terms are substituted in Eq. (2.27) before solving it for a given node.

2.3 CHARACTERISTICS OF EIGHT-NODED BRICK ELEMENTS WITH CONSTANT STRESS FIELD

In an explicit time marching forward integration scheme, the eight-noded bricks are most primitive elements, which use a single-point integration at the element centroid only. In effect, they carry only membrane stresses and do not offer any resistance to bending. The correct node numbering for the brick elements (N1, N2, N3, N4, N5, N6, N7, and N8) should be such that the determinant of the Jacobian is positive (see Figure 2.1). If the determinant is negative for any particular element, the connectivity should be reversed from (N1, N2, N3, N4) to (N5, N6, N7, N8) and vice-versa.

At the new time step of $(t + \Delta t)$ the vector components of nodal velocity terms such as $(\dot{u})_{t+\Delta t}$ and displacement $(u)_{t+\Delta t}$ terms in the global (X, Y, Z) directions are computed by direct-time integration of the corresponding acceleration vector $(\ddot{\boldsymbol{u}})_{t+\Delta t}$. In the explicit formulation of the time marching forward integration scheme, the most commonly used method is the fourth-order Runge–Kutta routine. For higher accuracy, a more numerically expensive method called sixth-order Runge–Kutta can also be used.

2.3.1 Progressive Failure Modeling During Impact

As discussed earlier in Section 1.3, depending upon the intensity and magnitude of shock loading between two colliding bodies, both projectile and target can fail in various ways such as penetration, fragmentation, or a combination of the two. In the explicit finite-element modeling technique, it is relatively easy to introduce strain-dependent failure criteria such that when the strain reaches a particular limit, based upon the material coupon test data, that eight-noded brick element can be considered failed from that time point onwards. It should be noted that even when the element has failed, the nodal masses are still active. In the numerical scheme, the "Failure" simply signifies that its stress-carrying capability has gone to zero, which in the stress formulation is accomplished by setting all six components of stress as zero for that eight-noded brick element, which in the mathematical form is accomplished as follows:

$$\sigma_{xx} = 0, \quad \text{SIG1} = 0$$

$$\sigma_{yy} = 0, \quad \text{SIG2} = 0$$

$$\sigma_{zz} = 0, \quad \text{SIG3} = 0$$

$$\tau_{xy} = 0, \quad \text{SIG4} = 0$$

$$\tau_{yz} = 0, \quad \text{SIG5} = 0$$

$$\tau_{zx} = 0, \quad \text{SIG6} = 0.$$

Element failure key: IFAIL = 1.

In the explicit time marching forward solver routine, once the element has been tagged "Failed" by setting the above condition, in the next time step onwards that element will not carry load anymore, and the terms containing $\boldsymbol{F}_{\text{elemental-stiffness-forces}}$ will all be equal to zero. In the dynamic situation, the load will move to the next surrounding element, and in this way, the dynamic crack will propagate through the structure.

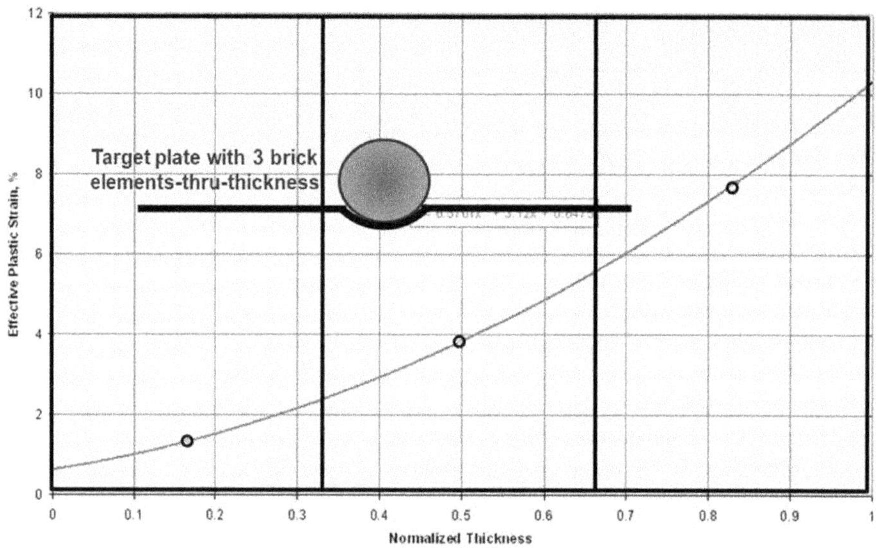

FIGURE 2.2 Example of plastic strain distribution in a target plate modeled with three brick elements through the thickness (1.2%, 3.8%, and 7.8% plastic strain, all tensile in nature).

It should be further noted that the majority of the mechanical or aerospace components are in the form of beams and plates, which during the impact event encounter the external dynamic loading mechanism as the flexural bending of the structural member. Ductile failure of such elastic–plastic materials takes place when the plastic strain has reached its limiting value in the outermost element. For this purpose, a study has been carried out for a plate structure being impacted by a pointed projectile attempting to penetrate the target (see Figure 1.7b). In this study, the target material is elastic–plastic and has been modeled by three elements through the thickness. During the deformation of the target, the typical plastic strain in the three elements through the thickness is shown in Figure 2.2. The analytically computed element centroidal tensile plastic strain varies from the element on the impact side with 1.2%–7.8% on the rear side element centroid. The corresponding numerically extrapolated value of the surface strain in the outermost fiber on the rear side of the plate obtained by curve fitting is 10.4%.

Based upon this analysis, the limiting failure strain in the FEA model with constant strain field element must be modified in order for the analytical model to predict the onset of failure or initiation of dynamic crack with precise accuracy. For example, if the material failure strain based upon the coupon test data is 10.4%, then the three-element through the thickness model should use only 7.8% as the failure limit in an FEA mesh with a simple single mid-point integration element model.

Using the above information about the strain distribution through the thickness as well as some empirical data, the recommended approach for the necessary correction in the failure strain to be used in the explicit finite-element model is as follows. If there are n number of solid brick elements through the thickness of a plate-type target or containment structure in an impact event, then the formula for the correction will be

$$\varepsilon_{\text{Failure strain in FEA model}} = \frac{(n-1)}{n} \varepsilon_{\text{Failure strain from material test data}} \tag{2.21}$$

2.4 HOURGLASSING TYPE OF NUMERICAL INSTABILITY

Hourglassing is a very common type of numerical instability in any eight-noded brick elements with a single-point integration, which is unable to capture bending the type of deformation or any other types of stress distribution within that individual element. Hourglass modes are nonphysical modes of deformation that occur in solid elements due to inherent assumption of constant stress field located at the element centroid and produce no stress distribution whatsoever, which is problematic in the thickness direction modeling of any such structural load-bearing component. The typical examples of such solid elements with a single-point integration at the centroid of that element are prone and very susceptible to hourglass instability even with a very small amount of bending deformation or a very rapid change in the stress field due to high-strain rates in that element. At the same time, the main advantage of using a typical eight-noded solid brick element, in an explicit time marching forward integration algorithm, is that they are extremely efficient in terms of computational cost of a large finite-element model. With the current state of digital technology, one can run a transient impact analysis simulation of very large million nodes and element finite-element models with explicit commercial codes like LSDYNA, without any difficulty even on a laptop computer.

Due to the above-mentioned reason, the recommended approach to tackle the problem of hourglass instability is not to use a higher order element with multiple points of integration, rather to still stay with the most commonly used eight-noded solid brick elements with a single-point integration scheme of constant stress distribution within the element. An eight-noded solid brick element with a single-point integration scheme can have up to 12 different hourglass deformation modes. One such mode in a 2-D view is shown in Figure 2.2.

However, to avoid the possibility of hourglass instability being encountered during the computation phase of the transient analysis, one should always adhere to the following recommended good meshing practices.

a. Always use more than one solid element in any direction including the thickness direction to model even a thin shell-type structural component.
b. The aspect ratio of any of the sides of any eight-noded solid brick elements should not be more than 1:4 anywhere in the entire finite-element mesh.
c. Always use extremely fine mesh, especially for the parts of the structure, where the dynamic loads due to contact or some other reasons may be highly concentrated. Another recommended approach is to avoid the point contact load situations as much as possible.

d. Any rapid loading situations causing high-strain rate in a localized part of the structure for any prolonged duration should also be avoided.
e. Hourglassing deformation can also be stopped by introducing a form of artificial numerical damping hourglass control parameter either by (i) viscous forms of hourglass control or (ii) stiffness form of hourglass control.

   In the viscous forms of hourglass control, the hourglass forces are generated proportionally to the components of nodal velocity contributing to hourglass modes.

   In the stiffness form of hourglass control, the hourglass forces are generated proportionally to the components of nodal deformation contributing to hourglass modes.

   However, any artificial damping in the numerical scheme will inevitably introduce errors, and the use of hourglass control parameter must be evaluated in advance for its effectiveness and accuracy, before one chooses to use it in his simulation.

2.5 METHOD OF CHARACTERISTICS

To demonstrate the use of the method of characteristics for numerically solving a typical hyperbolic wave equation, we will consider two coupled parameters (u, v) dynamical system in a one-dimensional spatial direction (x-axis). In an elastic wave problem, these two parameters can represent deformation in the structural member. The full complexity and capability of the methodology will be shown through a coupled set of two nonlinear partial differential equations of the following form:

$$a_{11}\frac{\partial^2 u}{\partial x^2}+a_{12}\frac{\partial^2 v}{\partial x^2}=\frac{\partial^2 u}{\partial t^2} \tag{2.22}$$

$$a_{21}\frac{\partial^2 u}{\partial x^2}+a_{22}\frac{\partial^2 v}{\partial x^2}=\frac{\partial^2 v}{\partial t^2} \tag{2.23}$$

On using the abbreviated notations such as $u_{,x}=\frac{\partial u}{\partial x}$, $v_{,x}=\frac{\partial v}{\partial x}$, $\dot{u}=\frac{\partial u}{\partial t}$, and $\dot{v}=\frac{\partial v}{\partial t}$, the above-coupled equations can be written as follows:

$$a_{11}(u_{,x}, v_{,x})u_{,xx}+a_{12}(u_{,x}, v_{,x})v_{,xx}=\ddot{u} \tag{2.24}$$

$$a_{21}(u_{,x}, v_{,x})u_{,xx}+a_{22}(u_{,x}, v_{,x})v_{,xx}=\ddot{v} \tag{2.25}$$

With the imposed velocity boundary conditions at ($x = 0$) and ($x = \ell$) as

$$\dot{u}(0,t)=\left[\frac{\partial u}{\partial t}\right]_{x=0}=\phi_1(t) \text{ and, } \dot{u}(\ell,t)=\left[\frac{\partial u}{\partial t}\right]_{x=\ell}=\phi_2(t) \tag{2.26}$$

$$\dot{v}(0,t)=\left[\frac{\partial v}{\partial t}\right]_{x=0}=\psi_1(t) \text{ and, } \dot{v}(\ell,t)=\left[\frac{\partial v}{\partial t}\right]_{x=\ell}=\psi_2(t) \tag{2.27}$$

and the initial conditions for the entire dynamical system under consideration to be in a prestressed state with the strain distribution of $f_1(x)$ and $f_2(x)$, for the two components of strains, respectively, and the entire dynamical system being at rest with no movement whatsoever,

$$u_{,x}(x,0)=\left[\frac{\partial u}{\partial x}\right]_{t=0}=f_1(x) \text{ and, } \dot{u}(x,0)=\left[\frac{\partial u}{\partial t}\right]_{t=0}=0 \tag{2.28}$$

$$v_{,x}(x,0)=\left[\frac{\partial v}{\partial x}\right]_{t=0}=f_2(x) \text{ and, } \dot{v}(x,0)=\left[\frac{\partial v}{\partial t}\right]_{t=0}=0. \tag{2.29}$$

Here, it should be noted that in a linear system, coefficients such as $(a_{11}, a_{12}, a_{21}, a_{22})$ are constants and usually depend only upon material parameters such as E, ν, ρ and structure's geometrical parameters such as cross-sectional area A, I, h . However, in a nonlinear stress wave problem, these coefficients will also depend upon deflection parameters such as (u, v) and strain parameters such as $(\varepsilon_{xx}.\varepsilon_{yy}.\gamma_{xy})$.

Equations (2.22) and (2.23) can be converted into a set of first-order partial differential equations. From the continuity condition, we can also express additional equations as follows:

$$\frac{\partial(u_{,x})}{\partial t}=\frac{\partial(u_{,t})}{\partial x}, \text{ i.e. } \frac{\partial(u_{,x})}{\partial t}-\frac{\partial(u_{,t})}{\partial x}=0 \tag{2.30}$$

$$\frac{\partial(v_{,x})}{\partial t}=\frac{\partial(v_{,t})}{\partial x}, \text{ i.e. } \frac{\partial(v_{,x})}{\partial t}-\frac{\partial(v_{,t})}{\partial x}=0 \tag{2.31}$$

$$\frac{\partial(u_{,x})}{\partial x}dx+\frac{\partial(u_{,x})}{\partial t}dt=d(u_{,x}) \tag{2.32}$$

$$\frac{\partial(v_{,x})}{\partial x}dx+\frac{\partial(v_{,x})}{\partial t}dt=d(v_{,x}) \tag{2.33}$$

$$\frac{\partial(u_{,t})}{\partial x}dx+\frac{\partial(u_{,t})}{\partial t}dt=d(u_{,t}) \tag{2.34}$$

$$\frac{\partial(v_{,t})}{\partial x}dx+\frac{\partial(v_{,t})}{\partial t}dt=d(v_{,t}) \tag{2.35}$$

The above set of eight equations (2.24), (2.25), and (2.30) to (2.31) in the matrix form can be written as

$$
\begin{bmatrix}
a_{11} & 0 & a_{12} & 0 & 0 & -1 & 0 & 0 \\
a_{21} & 0 & a_{22} & 0 & 0 & 0 & 0 & -1 \\
0 & 1 & 0 & 0 & -1 & 0 & 0 & 0 \\
0 & 0 & 0 & 1 & 0 & 0 & -1 & 0 \\
dx & dt & 0 & 0 & 0 & 0 & 0 & 0 \\
0 & 0 & dx & dt & 0 & 0 & 0 & 0 \\
0 & 0 & 0 & 0 & dx & dt & 0 & 0 \\
0 & 0 & 0 & 0 & 0 & 0 & dx & dt
\end{bmatrix}
\begin{Bmatrix}
\frac{\partial(u_{,x})}{\partial x} \\
\frac{\partial(u_{,x})}{\partial t} \\
\frac{\partial(v_{,x})}{\partial x} \\
\frac{\partial(v_{,x})}{\partial t} \\
\frac{\partial(u_{,t})}{\partial x} \\
\frac{\partial(u_{,t})}{\partial t} \\
\frac{\partial(v_{,t})}{\partial x} \\
\frac{\partial(v_{,t})}{\partial t}
\end{Bmatrix}
=
\begin{Bmatrix}
0 \\ 0 \\ 0 \\ 0 \\ d(u_{,x}) \\ d(v_{,x}) \\ d(u_{,t}) \\ d(v_{,t})
\end{Bmatrix}
\tag{2.36}
$$

The characteristic directions are determined by setting the determinant of the coefficient matrix of Eq. (2.36) equal to zero. Hence, the following equation results:

$$
(a_{11}a_{22} - a_{12}a_{21})\left(\frac{dt}{dx}\right)^4 - (a_{11} + a_{22})\left(\frac{dt}{dx}\right)^2 + 1 = 0 \tag{2.37}
$$

The above equation has four roots, which are

$$\left(\frac{dt}{dx}\right)_{1,2} = +\left[\frac{(a_{11}+a_{22})\mp\sqrt{\left[(a_{11}-a_{22})^2+4a_{12}a_{21}\right]}}{2(a_{11}a_{22}-a_{12}a_{21})}\right]^{1/2} \tag{2.38}$$

$$\left(\frac{dt}{dx}\right)_{3,4} = -\left[\frac{(a_{11}+a_{22})\pm\sqrt{\left[(a_{11}-a_{22})^2+4a_{12}a_{21}\right]}}{2(a_{11}a_{22}-a_{12}a_{21})}\right]^{1/2} \tag{2.39}$$

Since, in the system of equations (2.24) and (2.25), the coefficients a_{ij} are functions of the unknown derivatives, it will be assumed that the values of a_{ij} are such that equations (2.36) and (2.37) always yield four real roots. As the solution is marched forward, the values of the roots are monitored to see that this condition is satisfied. Hence, since the roots are real, the system is hyperbolic. The canonical form of a hyperbolic system along the characteristics (sometimes called either "compatibility equations" or "Riemann Invariant equations") can be determined by replacing any column of the coefficient matrix in equation (2.36) by the right-hand side column vector and setting the determinant equal to zero. In this case, the compatibility equations along the characteristics are determined by replacing the first column of the coefficient matrix in equation (2.36) by the column on the right-hand side of equation (2.36) and then setting the corresponding determinant equal to zero. The following equation results:

$$\left\{1-a_{22}\left(\frac{dt}{dx}\right)^2\right\}d(u_{,x})+\left\{a_{12}\left(\frac{dt}{dx}\right)^2\right\}d(v_{,x})-\left\{1-a_{22}\left(\frac{dt}{dx}\right)^2\right\}$$
$$\left(\frac{dt}{dx}\right)d(u_{,t})-\left\{a_{12}\left(\frac{dt}{dx}\right)^3\right\}d(v_{,t})=0 \tag{2.40}$$

In the difference form, equation (2.40) becomes

$$\left\{1-a_{22}\left(\frac{dt}{dx}\right)^2\right\}\Delta u_{,x}+\left\{a_{12}\left(\frac{dt}{dx}\right)^2\right\}\Delta v_{,x}-\left\{1-a_{22}\left(\frac{dt}{dx}\right)^2\right\}$$
$$\left(\frac{dt}{dx}\right)\Delta\dot{u}-\left\{a_{12}\left(\frac{dt}{dx}\right)^3\right\}\Delta\dot{v}=0 \tag{2.41}$$

Thus, the unknown values of $u_{,x}$, $v_{,x}$, $\dot{u}$ and $\dot{v}$ at any new point L, as shown in Figure 2.3, can be determined by knowing their numerical values at the points P, Q, R, S lying on the four characteristic lines passing through L, and then solving four simultaneous equations obtained from Eq. (2.41). Although the characteristics are curved due to the nonlinearity of Eqs. (2.22) and (2.23), it will be assumed that LP, LQ, LR, LS are straight lines.

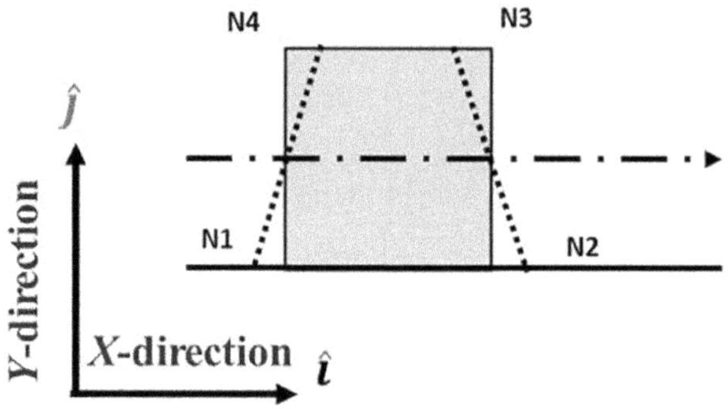

FIGURE 2.3 Hourglass type of zero energy mode deformation shown with a dotted line in a constant stress field schematically shown in a 2-D element.

Here, it must be noted that the characteristics have been assumed to be linear only between two consecutive grid points, but on an overall basis they are still nonlinear. The slope of each characteristic is computed for each known value of $u_{,x}$, $v_{,x}$. Hence, Eq. (2.41) yields,

$$\left[(u_{,x})_L - (u_{,x})_P\right]\left[1-(a_{22})_P\left(\frac{dt}{dx}\right)^2_{L,P}\right] + \left[(v_{,x})_L - (v_{,x})_P\right]$$
$$\left[(a_{12})_P\left(\frac{dt}{dx}\right)^2_{L,P}\right] - \left[(\dot{u})_L - (\dot{u})_P\right]\left[1-(a_{22})_P\left(\frac{dt}{dx}\right)^2_{L,P}\right]\left(\frac{dt}{dx}\right)_{L,P} \tag{2.42a}$$
$$-\left[(\dot{v})_L - (\dot{v})_P\right]\left[(a_{12})_P\left(\frac{dt}{dx}\right)^3_{L,P}\right] = 0$$

$$\left[(u_{,x})_L - (u_{,x})_Q\right]\left[1-(a_{22})_Q\left(\frac{dt}{dx}\right)^2_{L,Q}\right] + \left[(v_{,x})_L - (v_{,x})_Q\right]$$
$$\left[(a_{12})_Q\left(\frac{dt}{dx}\right)^2_{L,Q}\right] - \left[(\dot{u})_L - (\dot{u})_Q\right]\left[1-(a_{22})_Q\left(\frac{dt}{dx}\right)^2_{L,Q}\right] \tag{2.42b}$$
$$\left(\frac{dt}{dx}\right)_{L,Q} - \left[(\dot{v})_L - (\dot{v})_Q\right]\left[(a_{12})_Q\left(\frac{dt}{dx}\right)^3_{L,Q}\right] = 0$$

$$\left[(u_{,x})_L-(u_{,x})_R\right]\left[1-(a_{22})_R\left(\frac{dt}{dx}\right)^2_{L,R}\right]+\left[(v_{,x})_L-(v_{,x})_R\right]$$

$$\left[(a_{12})_R\left(\frac{dt}{dx}\right)^2_{L,R}\right]-[(\dot{u})_L-(\dot{u})_R]\left[1-(a_{22})_R\left(\frac{dt}{dx}\right)^2_{(L,R)}\right]\left(\frac{dt}{dx}\right)^2_{L,R} \tag{2.42c}$$

$$-\left[(\dot{v})_L-(\dot{v})_R][(a_{12})_R\left(\frac{dt}{dx}\right)^3_{(L,R)}\right]=0$$

$$\left[(u_{,x})_L-(u_{,x})_S\right]\left[1-(a_{22})_S\left(\frac{dt}{dx}\right)^2_{L,S}\right]+\left[(v_{,x})_L-(v_{,x})_S\right]$$

$$\left[(a_{12})_S\left(\frac{dt}{dx}\right)^2_{L,S}\right]-\left[(\dot{u})_L-(\dot{u})_S\right]\left[1-(a_{22})_S\left(\frac{dt}{dx}\right)^2_{L,S}\right]\left(\frac{dt}{dx}\right)_{L,S} \tag{2.42d}$$

$$-\left[(\dot{v})_L-(\dot{v})_S\right]\left[(a_{12})_S\left(\frac{dt}{dx}\right)^3_{L,S}\right]=0$$

Here, the four dynamic response parameters such as $(u_{,x})_L$, $(v_{,x})_L$, $(\dot{u})_L$, $(\dot{u})_L$ are unknowns to be computed at point L with the known values of these parameters at points P, Q, R, S shown by four equations (2.42a), (2.42b), (2.42c), and (2.42d), respectively. Here, we have four unknowns and four algebraic equations to be solved simultaneously, along with the respective slopes of the characteristics to be computed as follows:

$$\left(\frac{dt}{dx}\right)_{L,P}=\left[\frac{(a_{11}+a_{22})-\sqrt{\left[(a_{11}-a_{22})^2+4a_{12}a_{21}\right]}}{2(a_{11}a_{22}-a_{12}a_{21})}\right]_P^{1/2} \tag{2.43a}$$

$$\left(\frac{dt}{dx}\right)_{L,Q}=\left[\frac{(a_{11}+a_{22})+\sqrt{\left[(a_{11}-a_{22})^2+4a_{12}a_{21}\right]}}{2(a_{11}a_{22}-a_{12}a_{21})}\right]_Q^{1/2} \tag{2.43b}$$

$$\left(\frac{dt}{dx}\right)_{L,R}=-\left[\frac{(a_{11}+a_{22})+\sqrt{\left[(a_{11}-a_{22})^2+4a_{12}a_{21}\right]}}{2(a_{11}a_{22}-a_{12}a_{21})}\right]_R^{1/2} \tag{2.43c}$$

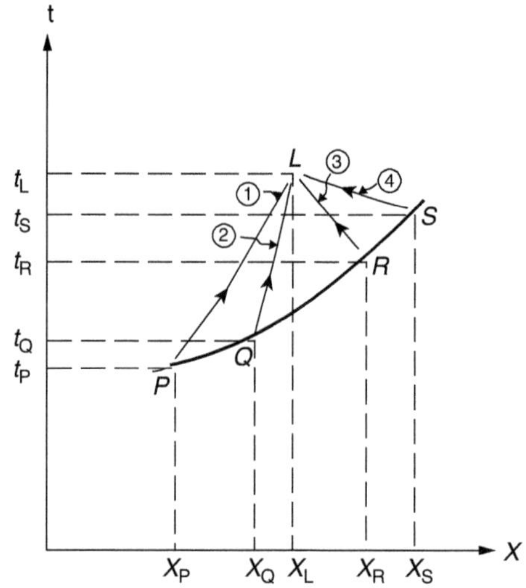

FIGURE 2.4 Numerical scheme for computing dependent parameters $u_{,x}$, $v_{,x}$, $\dot{u}$, and $\dot{v}$ at any new point L, by their known values at finite-difference grid points ***A, B, C***. (Reprinted with permission from Reference [2].)

$$\left(\frac{dt}{dx}\right)_{L,S} = -\left[\frac{(a_{11}+a_{22})-\sqrt{\left[(a_{11}-a_{22})^2+4a_{12}a_{21}\right]}}{2(a_{11}a_{22}-a_{12}a_{21})}\right]_S^{1/2} \tag{2.43d}$$

Here, Figure 2.4 shows typical known grid points A, B , C and also the unknown points L, P, Q, R, S. The numerical values of dynamic response parameters such as $u_{,x}$, $v_{,x}$, $\dot{u}$, and $\dot{v}$ are also assumed to be known at the existing grid points A, B, $C$and also $x_B = x_L$. The slope of the characteristics at A and C is known and as can be seen from Figure 2.5, the characteristics 1 and 2 passing through A and the characteristics 3 and 4 passing through C intersect the vertical line passing through B at four points. To remain inside the domain of dependence, the smallest value of t_L, obtained by the intersection of the characteristics with the vertical line through B, will be used. Hence,

$$t_L = (x_B - x_A)\left(\frac{dt}{dx}\right)_{1,A} + t_A \tag{2.44}$$

or

$$t_L = (x_B - x_C)\left(\frac{dt}{dx}\right)_{4,C} + t_C \tag{2.45}$$

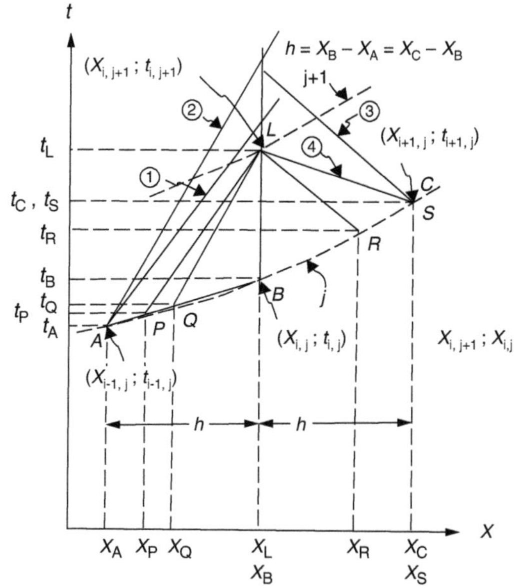

FIGURE 2.5 Determination of dependent parameters $u_{,x}$, $v_{,x}$, $\dot{u}$, and $\dot{v}$ at any new point L , by their known values at points P, Q, R, S. (Reprinted with permission from Reference [2].)

Figure 2.5 shows the case of where the characteristic number-4 passing through C intersects the vertical line at the smallest value of t_L. Hence, in this case point S will be coincident with C. Once the coordinates of point L are known as $(x_L\ ,\ t_L)$, the coordinates of points P, Q, and R can also be determined. The coordinates of point P, for example, are as follows:

$$x_P = \frac{(t_B - t_L)}{\left(\dfrac{dt}{dx}\right)_{1,P} - \dfrac{(t_B - t_A)}{(x_B - x_A)}} + x_B \tag{2.46}$$

$$t_P = \left(\frac{dt}{dx}\right)_{1,P} (x_P - x_L) + t_L \tag{2.47}$$

Similar expressions can be written down for points Q and R and in this case, since S is coincident with C, we have

$$x_S = x_C \ \text{ and}, t_S = t_C \tag{2.48}$$

It will now be assumed that

$$\left(\frac{dt}{dx}\right)_{1,P} = \left(\frac{dt}{dx}\right)_{1,A} \tag{2.49a}$$

$$\left(\frac{dt}{dx}\right)_{2,Q} = \left(\frac{dt}{dx}\right)_{2,A} \tag{2.49b}$$

$$\left(\frac{dt}{dx}\right)_{3,R} = \left(\frac{dt}{dx}\right)_{3,C} \tag{2.49c}$$

$$\left(\frac{dt}{dx}\right)_{4,S} = \left(\frac{dt}{dx}\right)_{4,C} \tag{2.49d}$$

The values of the coefficients $(a_{11}, a_{12}, a_{21}, a_{22})$ and values of $u_{,x}$, $v_{,x}$, $\dot{u}$ and $\dot{v}$ at the points P, Q, and R are then computed by linear interpolation. For example,

$$(a_{11})_P = \frac{\left((a_{11})_A - (a_{11})_B\right)}{(x_A - x_B)}(x_P - x_B) + (a_{11})_B \tag{2.50}$$

Similar expressions can be written down for other quantities, viz., $u_{,x}$, $v_{,x}$..., etc. Once the values of the coefficients a_{pq} $(p = 1,2; q = 1,2)$ and the dependent variables $u_{,x}$, $v_{,x}$, $\dot{u}$ and $\dot{v}$ at the points P, Q, R, and S are known, then the corresponding four equations (2.42a)–(2.42d) can be solved simultaneously to obtain their values at some new point L in the $(t - x)$ plane. The above procedure can then be continued to determine the values of variables $u_{,x}$, $v_{,x}$, $\dot{u}$ and $\dot{v}$ at new values of time $(t + \Delta t)$.

2.6 FINITE DIFFERENCE METHOD FOR HYPERBOLIC EQUATIONS

The first four equations outlined in the matrix form of Eq. (2.36) can be written in finite difference form for any grid point (i, j) as follows:

$$(a_{11})_{i,j}\left(\frac{\partial u_{,x}}{\partial x}\right)_{i,j} + (a_{12})_{i,j}\left(\frac{\partial v_{,x}}{\partial x}\right)_{i,j} - \left(\frac{\partial \dot{u}}{\partial t}\right)_{i,j} = 0 \tag{2.51}$$

$$(a_{21})_{i,j}\left(\frac{\partial u_{,x}}{\partial x}\right)_{i,j} + (a_{22})_{i,j}\left(\frac{\partial v_{,x}}{\partial x}\right)_{i,j} - \left(\frac{\partial \dot{v}}{\partial t}\right)_{i,j} = 0 \tag{2.52}$$

$$\left(\frac{\partial u_{,x}}{\partial t}\right)_{i,j+1} - \left(\frac{\partial \dot{u}}{\partial x}\right)_{i,j+1} = 0 \tag{2.53}$$

$$\left(\frac{\partial v_{,x}}{\partial t}\right)_{i,j+1} - \left(\frac{\partial \dot{v}}{\partial x}\right)_{i,j+1} = 0 \tag{2.54}$$

where $i = 1, 2, 3, \ldots\ldots\ldots n, n+1$ etc. refer to grid points in the x-direction and j and $j+1$ refer to old and new time steps in the t-direction, respectively. The initial conditions are

$$(u_{,x})_{i,1} = f_1(x) \text{ and } (v_{,x})_{i,1} = f_2(x) \tag{2.55}$$

$$(\dot{u})_{i,1} = 0 \text{ and } (\dot{v})_{i,1} = 0 \tag{2.56}$$

And the boundary conditions at ($x = 0$) and ($x = \ell$) as

$$\dot{u}(0,t) = (\dot{u})_{1,j} = \phi_1(t) \text{ and, } \dot{u}(\ell,t) = (\dot{u})_{n+1,j} = \phi_2(t) \tag{2.57}$$

$$\dot{v}(0,t) = (\dot{v})_{i,1} = \psi_1(t) \text{ and, } \dot{v}(\ell,t) = (\dot{v})_{n+1,j} = \psi_2(t) \tag{2.58}$$

Since the x-axis is divided into n parts, the distance between two consecutive grid points is

$$(x_i - x_{i-1}) = h = \frac{1}{n} \tag{2.59}$$

From the stability criterion of the finite-difference scheme, we should always remain inside the domain of dependence. That is, from Eq. (2.43), we have

$$\Delta t \leq h \left[\frac{(a_{11} + a_{22}) - \sqrt{\left[(a_{11} - a_{22})^2 + 4a_{12}a_{21}\right]}}{2(a_{11}a_{22} - a_{12}a_{21})} \right]_{i,j}^{1/2} \tag{2.60}$$

where the time-step size $\Delta t = (t_{j+1} - t_j)$. This is the Courant–Freidrichs–Lewy condition for finite difference approximations of a hyperbolic equation. This must be checked at each grid point before carrying out any further computations. Central difference approximations are used in the x-direction and forward differences are used in the t-direction at all intermediate grid points. Hence,

$$(\dot{u})_{i,j+1} = (\dot{u})_{i,j} + \frac{\Delta t}{2h}(a_{11})_{i,j}\left[(u_{,x})_{i+1,j} - (u_{,x})_{i-1,j}\right] + \frac{\Delta t}{2h}(a_{12})_{i,j}\left[(v_{,x})_{i+1,j} - (v_{,x})_{i-1,j}\right] \tag{2.61}$$

$$(\dot{v})_{i,j+1} = (\dot{v})_{i,j} + \frac{\Delta t}{2h}(a_{21})_{i,j}\left[(u_{,x})_{i+1,j} - (u_{,x})_{i-1,j}\right] + \frac{\Delta t}{2h}(a_{22})_{i,j}\left[(v_{,x})_{i+1,j} - (v_{,x})_{i-1,j}\right] \tag{2.62}$$

$$(u_{,x})_{i,j+1} = (u_{,x})_{i,j} + \frac{\Delta t}{2h}\left[(\dot{u})_{i+1,j+1} - (\dot{u})_{i-1,j+1}\right] \tag{2.63}$$

and

$$(v_{,x})_{i,j+1} = (v_{,x})_{i,j} + \frac{\Delta t}{2h}\left[(\dot{v})_{i+1,j+1} - (\dot{v})_{i-1,j+1}\right] \tag{2.64}$$

The values of $(\dot{u})_{i,j+1}$, $(\dot{v})_{i,j+1}$, …, etc. may also be obtained by another quite common finite-difference scheme, suggested by Lax and Wendroff, which yields

$$U_{i,j+1} = \left[1 - \left(\frac{\Delta t}{h}\right)^2\left(\frac{dx}{dt}\right)^2\right]U_{i,j} - \frac{1}{2}\frac{\Delta t}{h}\left(\frac{dx}{dt}\right)\left[1 - \frac{\Delta t}{h}\left(\frac{dx}{dt}\right)\right]$$

$$U_{i+1,j} + \frac{1}{2}\frac{\Delta t}{h}\left(\frac{dx}{dt}\right)\left[1 + \frac{\Delta t}{h}\left(\frac{dx}{dt}\right)\right]U_{i-1,j} \tag{2.65}$$

However, the Lax–Wendroff finite-difference scheme has been known to give poorer results for the impact type of problem than the simple finite-difference scheme given in equations (2.61)–(2.64). This difference of computational error may be attributed to the fact that in equations (2.63) and (2.64) the values of parameters such as $(\dot{u})_{i+1,j+1}$, $(\dot{u})_{i-1,j+1}$, $(\dot{v})_{i+1,j+1}$, and $(\dot{v})_{i-1,j+1}$ are used at the new time step of $(t + \Delta t)$, whereas in the Lax–Wendroff scheme, every time only old values of the respective parameters of the velocities are used.

Once the values of $(u_{,x})_{i,j+1}$, $(v_{,x})_{i,j+1}$, $(\dot{u})_{i,j+1}$, and $(\dot{v})_{i,j+1}$ have been computed at all the intermediate grid points for $(i = 2,3,4. \ldots\ldots N)$ by the above formulae outlined in Eqs. (2.61)–(2.64), their respective numerical values at the new time step $(t + \Delta t)$ for the end points denoted by subscripts $(i = 1)$ and $(i = N)$ are determined as follows:

a. at $(i = 1)$

$$(u_{,x})_{1,j+1} = (u_{,x})_{1,j} + \frac{\Delta t}{2h}\left[-3(\dot{u})_{1,j+1} + 4(\dot{u})_{2,j+1} - (\dot{u})_{3,j+1}\right] \tag{2.66}$$

$$(v_{,x})_{1,j+1} = (v_{,x})_{1,j} + \frac{\Delta t}{2h}\left[-3(\dot{v})_{1,j+1} + 4(\dot{v})_{2,j+1} - (\dot{v})_{3,j+1}\right] \text{and} \tag{2.67}$$

b. at $(i = n + 1)$

$$(u_{,x})_{n+1,j+1} = (u_{,x})_{n+1,j} + \frac{\Delta t}{2h}\left[(\dot{u})_{n-1,j+1} - 4(\dot{u})_{n,j+1} + 3(\dot{u})_{n+1,j+1}\right] \tag{2.68}$$

$$(v_{,x})_{n+1,j+1} = (v_{,x})_{n+1,j} + \frac{\Delta t}{2h}\left[(\dot{v})_{n-1,j+1} - 4(\dot{v})_{n,j+1} + 3(\dot{v})_{n+1,j+1}\right] \tag{2.69}$$

Thus, the new values of variables $u_{,x}$, $v_{,x}$, $\dot{u}$, and $\dot{v}$ can be computed at all the grid points for the new value of time step $(t + \Delta t)$ and by repeating a similar marching forward procedure, the computation can be carried out further for any new value time "t". However, it should be noted that any transient dynamic response solution using an explicit technique has its own limitations and as such cannot be carried out for an indefinite period of time. Depending upon the types of nonlinearities in the dynamic simulation of the majority of the impact problems, the transient numerical solution in the time domain will either become unstable or unacceptable cumulative errors may be introduced due to truncation or round-off approximations inherent in any algorithm. It is always advisable to verify the accuracy of any numerical solution at least by running some coupon tests under controlled conditions.

2.7 NUMERICALLY STABLE Δt TIME-STEP SIZE

In a nonlinear dynamic analysis, the transient response of the system is of great interest. We will consider a typical transient dynamics equation of a deformable structure in a matrix form as follows:

$$[\boldsymbol{M}]\{\ddot{\boldsymbol{u}}(t)\}+[\boldsymbol{C}]\{\dot{\boldsymbol{u}}(t)\}+\left[[\boldsymbol{K}]+[\boldsymbol{B}(t)]\right]\{\boldsymbol{u}(t)\}=\{\boldsymbol{F}(t)\} \tag{2.70}$$

To numerically solve the above transient dynamics equations in the time domain, we express them as follows:

$$\{\ddot{\boldsymbol{u}}(t)\}=\left[[\boldsymbol{M}]^{-1}\right]\left[\{\boldsymbol{F}(t)\}-[\boldsymbol{C}]\{\dot{\boldsymbol{u}}(t)\}-\left[[\boldsymbol{K}]+[\boldsymbol{B}(t)]\right]\{\boldsymbol{u}(t)\}\right] \tag{2.71}$$

The two most common techniques for integrating the equations of motion and computing the displacement and velocity responses of a dynamical system with prescribed initial conditions are as follows:

a. **Newmark's method**

   Newmark's method has already been discussed in Section 2.1.1. The respective equations for determining the displacement and velocity vectors are shown in Eqs. (2.4) and (2.5).

b. **Runge–Kutta method**

   There are different variations of the Runge–Kutta schemes for solving the nonlinear differential equations out of which the fourth order is most common for their numerical characteristics in terms of stability, computational cost, and accuracy. Here, we will discuss a highly accurate sixth-order Runge–Kutta–Fehlberg integration scheme [6]. The equation of motion outlined in Eq. (2.71) easily accounts for a time-varying external load vector $\{\boldsymbol{F}(t)\}$, because in the Runge–Kutta method, the contact load is updated continuously even at the mid-time-points. All contact forces are determined by the penalty method, where the longitudinal deformations are in the blade represented by a radial rotating Timoshenko beam, and the outer containment case acts like a rigid wall. Referring to Eqs. (1.88) and (1.89), the external force column vector $\{\boldsymbol{F}(t)\}$ contains the term pertaining to components of forces due to the radial contact load F_a from the outer circular containment surface acting in the local longitudinal direction of the Timoshenko beam. In the numerical integration algorithm of nonlinear equations of motion, the radial load on the beam-column F_a is computed according to the direction of the instantaneous velocity of the tip at the contact location. In this sixth-order scheme for better accuracy, the derivatives of the dependent variables on the right-hand side of the equation are computed at several different mid-points for a given time-step size Δt. With the introduction of a new column vector $\{\boldsymbol{f}(t)\}=\{\boldsymbol{u}(t)\}$, we will briefly discuss here in a generic sense to describe the implementation of the sixth-order integration scheme in the current problem. The equations of motion shown in Eq. (2.70) are converted into a set of first-order coupled differential equations by introducing a new set of variables called $\{\boldsymbol{g}(t)\}$ such that $\{\boldsymbol{g}(t)\}=\{\dot{\boldsymbol{f}}(t)\}$. This enables us to express Eq. (2.71) as the following two sets of first-order coupled differential equations:

$$\{\dot{\boldsymbol{g}}(t)\}=\left[[\boldsymbol{M}]^{-1}\right]\left[\{\boldsymbol{F}(t)\}-[\boldsymbol{C}]\{\boldsymbol{g}(t)\}-\left[[\boldsymbol{K}]+[\boldsymbol{B}(t)]\right]\{\boldsymbol{f}(t)\}\right] \tag{2.72}$$

$$\{\dot{\boldsymbol{f}}(t)\}=\{\boldsymbol{g}(t)\} \tag{2.73}$$

Here, the number of terms in the column vectors representing the displacement vector $\{\boldsymbol{f}(t)\}$ and velocity vector $\{\boldsymbol{g}(t)\}$ will be twice that of the number of generalized coordinates. For example, for "3N" number of deformation variables for a typical Timoshenko beam-column (ψ, w, u) described in Eqs. (1.90), (1.91), and (1.92), we can assume the Rayleigh–Ritz displacement functions in a series form as follows:

$$\text{Cross-section rotation: } \psi(x,t)=\sum_{n=1}^{N} U_n(x)X_n(t) \tag{2.74}$$

$$\text{Lateral displacement } \quad w(x,t)=\sum_{n=1}^{N} V_n(x)Y_n(t) \tag{2.75}$$

$$\text{Axial extension } \quad u(x,t)=\sum_{n=1}^{N} W_n(x)Z_n(t) \tag{2.76}$$

The corresponding time-derivative terms to represent the respective velocities of distortion, transverse movement, and axial stretching would be as follows:

$$\text{Velocity of cross-section rotation: } \dot{\psi}(x,t)=\sum_{n=1}^{N} U_n(x)\dot{X}_n(t) \tag{2.77}$$

$$\text{Velocity of lateral movement } \quad \dot{w}(x,t)=\sum_{n=1}^{N} V_n(x)\dot{Y}_n(t) \tag{2.78}$$

$$\text{Velocity of longitudinal stretching} \quad \dot{u}(x,t) = \sum_{n=1}^{N} W_n(x)\dot{Z}_n(t) \tag{2.79}$$

Thus, we would have total "6N" number of first-order differential equations as shown in Eqs. (2.71) and (2.72). It should be noted that each dependent variable is described by a sum of "N" number of generalized coordinate terms. As such, after the above transformations are introduced in the equation of motion, the total number of first-order equations becomes twice the original equations of motion. At this point, we introduce another new set of variables called $\{y(t)\}$, which in the subscripted index notations can be written as

$$y_j(t) = \{y(t)\} \equiv \begin{Bmatrix} X_1, \dot{X}_1, X_2, \dot{X}_2, X_3, \dot{X}_3, X_4, \dot{X}_4, \ldots\ldots\ldots X_k, \dot{X}_k, \ldots\ldots X_{N-1}, \dot{X}_{N-1}, X_N, \dot{X}_N, \\ Y_1, \dot{Y}_1, Y_2, \dot{Y}_2, Y_3, \dot{Y}_3, Y_4, \dot{Y}_4, \ldots Y_k, \dot{Y}_k, \ldots\ldots Y_{N-1}, \dot{Y}_{N-1}, Y_N, \dot{Y}_N, \\ Z_1, \dot{Z}_1, Z_2, \dot{Z}_2, Z_3, \dot{Z}_3, Z_4, \dot{Z}_4, \ldots\ldots\ldots Z_k, \dot{Z}_k, \ldots\ldots Z_{N-1}, \dot{Z}_{N-1}, Z_N, \dot{Z}_N, t \end{Bmatrix} \tag{2.80}$$

The corresponding entire set of first-order differential equations in a generic form is written as

$$\frac{dy_j}{dt} = \Phi_j\left(y_1, y_2, y_3, y_4, y_5, \ldots\ldots\ldots y_{6N}, t\right) \tag{2.81}$$

The initial conditions at time $t=0$ to be used for computing the transient dynamic response described by the above set of equations are obtained as the displacements and velocities from the steady-state solution of the rotating beam [7]. These displacement and velocity terms associated with all the generalized coordinates are expressed as follows:

$y_1(0), y_2(0), y_3(0), y_4(0), y_5(0), y_6(0), y_7(0), \ldots\ldots\ldots y_{6N}(0)$, etc. In the sixth-order Runge–Kutta–Fehlberg integration scheme, the actual numerical values of the coefficients relating the mid-point values of j's are quite inconvenient and lengthy. For the purpose of the present discussion, the computed value of a typical variable $\left(y_j\right)_{t+\Delta t}$ at the time step $(t+\Delta t)$ in a finite-difference form symbolically may be written as

$$\left(y_j\right)_{t+\Delta t} = \left(y_j\right)_t + (\Delta t) \times \begin{bmatrix} \left(\Phi_j\right)_t, \left(\Phi_j\right)_{t+\left(\frac{1}{4}\right)\Delta t}, \left(\Phi_j\right)_{t+\left(\frac{3}{8}\right)\Delta t}, \left(\Phi_j\right)_{t+\left(\frac{1}{2}\right)\Delta t}, \\ \left(\Phi_j\right)_{t+\left(\frac{12}{13}\right)\Delta t}, \left(\Phi_j\right)_{t+\Delta t} \end{bmatrix} \tag{2.82}$$

The numerically stable value of Δt for the entire set is established by computing the highest frequency of the following homogeneous equation for undamped vibration of the beam after dropping the velocity-dependent coefficient matrix **C** and the force and the time-dependent part of the stiffness matrix $[\boldsymbol{B}]$ from Eq. (2.61), which yields

$$[\boldsymbol{M}]\{\ddot{f}\}+[\boldsymbol{K}]\{f\}=\{0\}. \tag{2.83}$$

As opposed to the homogeneous form of the original equations of motion given in Eq. (2.70), the above eigenvalue problem is very simple to solve with standard techniques such as the Householder method, Jacobi method, and Matrix-iteration scheme. It yields 3N numbers of imaginary roots $\pm i\lambda_i$ giving the undamped frequencies of the simple dynamical system. Using the highest eigenvalue of the set called $|\lambda_i|_{\text{highest}}$ and taking the 1/8 of the time period of the highest mode as

$$\text{Time-step size}: \quad \Delta t \le \frac{1}{8}\frac{2\pi}{|\lambda_i|_{\text{highest}}} \tag{2.84}$$

In this particular numerical procedure, it is fully confirmed that the numerically computed dynamic response will be accurate and will not run into any numerical instability.

2.8 SENSITIVITY COMPARISON OF TRANSIENT RESPONSE FOR DIFFERENT NUMERICAL SCHEMES

In this section, we will study and compare as to how the characteristics of analytically computed numerical values of various dynamic parameters get affected even with the same stable value of time-step size Δt. For this discussion, we will simulate a problem of a helical spring being held at one end and being impacted at the other end. For this study, we will compare the two methods presented earlier in Section 2.5 as (i) the method of characteristics and in Section 2.6 as (ii) the finite difference method for hyperbolic equations. The results of this section are based upon the author's previously published results in the *International Journal for Numerical Methods in Engineering* [2], and its key summarized findings are being reproduced here with permission from the journal's publisher. The reader is referred to the original paper for a more detailed discussion of this topic. The governing second-order coupled hyperbolic partial differential equations used in this study are shown earlier in Eqs. (2.22) and (2.23) and are as follows:

$$a_{11}\frac{\partial^2 u}{\partial x^2}+a_{12}\frac{\partial^2 v}{\partial x^2}=\frac{\partial^2 u}{\partial t^2} \tag{2.85}$$

$$a_{21}\frac{\partial^2 u}{\partial x^2}+a_{22}\frac{\partial^2 v}{\partial x^2}=\frac{\partial^2 v}{\partial t^2} \tag{2.86}$$

In a helical spring formulation, the coefficients a_{11}, a_{12}, a_{21}, and a_{22} are expressed in terms of external axial force F being applied on the spring along its axis and T is the external torque. Then, with helical spring of mean radius r and helix angle α and with (EI) as the flexural rigidity of the helical coil cross section, we determine the coefficients of the hyperbolic equations such as a_{11}, a_{12}, a_{21}, and a_{22} as follows:

$$a_{11} = \frac{r^2}{EI}\frac{\partial F}{\partial \varepsilon} = \left(1 - \frac{\nu}{1+\nu}\cos^2\alpha\right)\sin\alpha \tag{2.87}$$

$$a_{12} = a_{21} = \frac{r^2}{EI}\frac{\partial F}{\partial \beta} = \frac{r}{EI}\frac{\partial T}{\partial \varepsilon} = -\frac{\nu}{1+\nu}\sin^2\alpha\,\cos\alpha \tag{2.88}$$

$$a_{22} = \frac{r}{EI}\frac{\partial T}{\partial \beta} = \left(1 - \frac{\nu}{1+\nu}\sin^2\alpha\right)\sin\alpha \tag{2.89}$$

Two separate FORTRAN computer programs were run on an IBM 360/75 computer. The same problem is solved using the following methods:

a. 120 grid points for the method of characteristics and
b. 200 grid points for the finite-difference method.

The numerical results are obtained based on the following spring parameters:

$h = 19$ in (original length of spring) (48.26 cm)
$\alpha = 0.141815$ radians (helix angle)
$r = 7.06$ in (17.932 cm)
Number of coils = 3
$\nu = 0.29$
wire radius = 0.594 in (1.509 cm)
$E = 30 \times 10^6$ psi (20.685×10^6 Newtons/cm^2)
$\Delta = 6.5$ in (initial compression (16–51 cm)
Mass of the helical spring = 0.10935 lb-sec^2/in (19.146 kg)

The initial conditions in the helical spring for strain terms are as follows:

$$\frac{\partial u}{\partial x}(x,0) = -\frac{\Delta}{h} \tag{2.90}$$

$$\frac{\partial v}{\partial x}(x,0) = 0 \tag{2.91}$$

Similarly, the initial conditions in the helical spring for velocity terms are as follows:

$$\frac{\partial u}{\partial t}(x,0) = 0 \tag{2.92}$$

$$\frac{\partial v}{\partial t}(x,0) = 0 \tag{2.93}$$

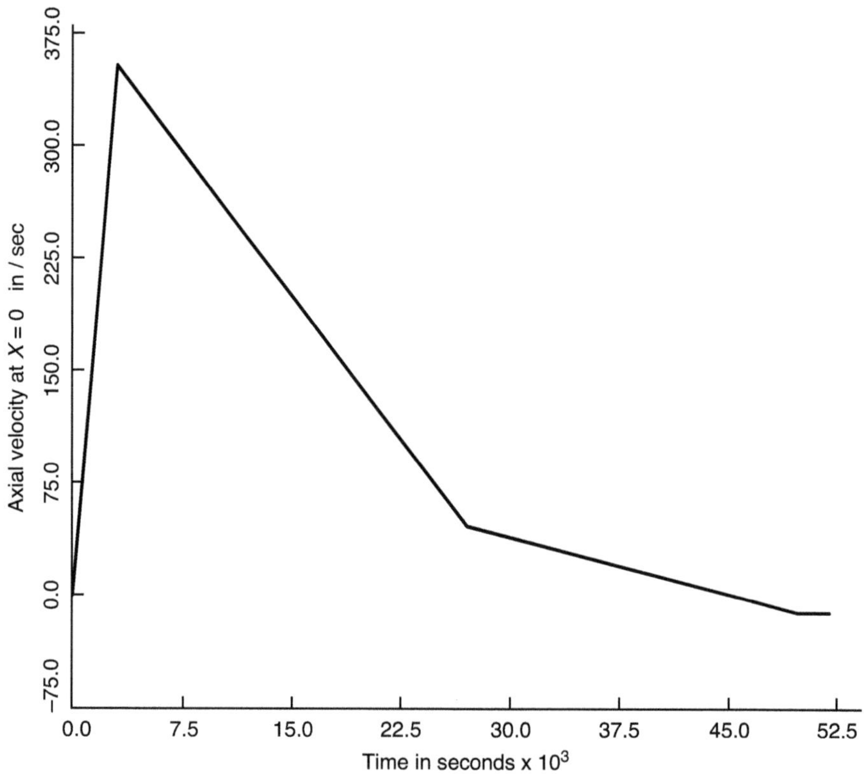

FIGURE 2.6 Impact condition of velocity time history $\phi(t)$ at the impacted end of the helical spring located at $(x = 0)$. (Reprinted with permission from Reference [2].).

The boundary conditions at the impacted end of $(x = 0)$ for the helical spring are as follows:

$$\frac{\partial u}{\partial t}(0,t) = \phi(t) \text{ (see the imposed velocity condition in Figure 2.6)} \tag{2.94}$$

$$\frac{\partial v}{\partial t}(0,t) = 0 \tag{2.95}$$

Similarly, the boundary conditions at the clamped end of $(x = h)$ for the helical spring are as follows:

$$\frac{\partial v}{\partial t}(h,t) = 0 \tag{2.96}$$

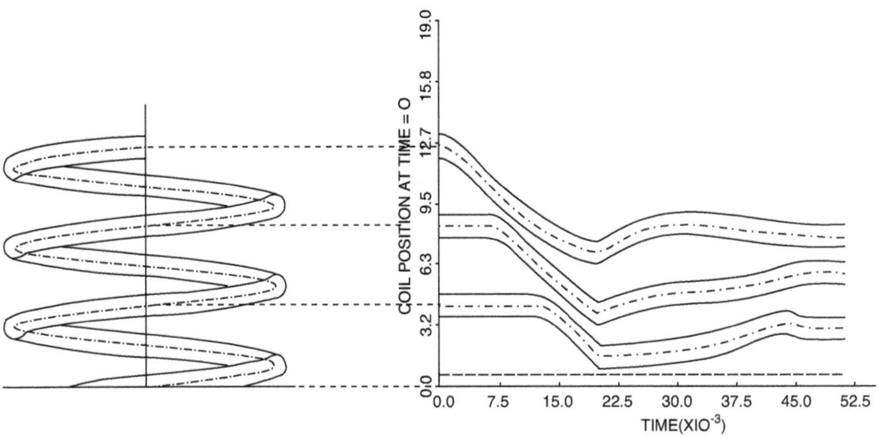

FIGURE 2.7 Helical spring at time $t=0$, and its deformation time history due to axial impact. (Reprinted with permission from Reference [2].)

$$\frac{\partial v}{\partial t}(h,t)=0 \tag{2.97}$$

The model of the helical spring considered in this numerical simulation is shown in Figure 2.7. This illustration also shows the time-history plot of the axial deformation of all three coils considered in this analysis.

The results of the method of characteristics and finite differences along with the linear solution are shown in Figures 2.8 and 2.9. The axial strain and the rotational strain are at the impacted end ($x=0$). Of the two methods employed, the method of characteristics seems to be the superior one. There is less computer cost involved with this method. Moreover, as can be seen from the plots, even with a comparatively large number of grid points, the finite difference method results in too many oscillations and does not give smooth results.

The results of the plots also show the necessity of solving the nonlinear equations of motion for the spring under this type of loading. The linear theory is reasonably accurate as far as the axial strain is concerned but is in considerable error for investigating the rotational strain. This means that the linear theory is adequate for predicting the axial force in the spring but can lead to erroneous results in predicting the axial twisting moment and radial expansion of the spring. It should be pointed out that once the axial and rotational strains are known, the stresses can be computed from the elementary strength of material formula.

In summary, currently, almost all time marching forward integration schemes use the fourth-order Runge–Kutta method. However, for better accuracy, the sixth-order Runge–Kutta [6] is preferred and recommended. Furthermore, for any two-point boundary value problem in all nonlinear dynamics applications, the method of Miller [7] is highly recommended.

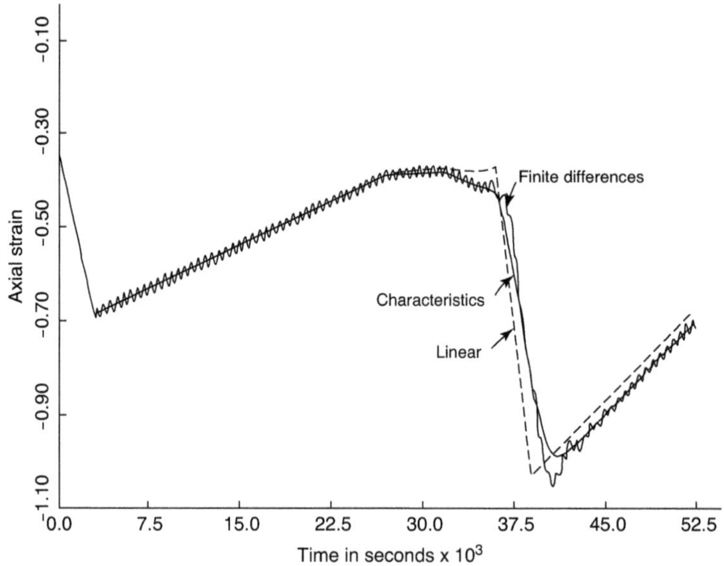

FIGURE 2.8 Time history of axial strain at the impacted end of the helical spring. (Reprinted with permission from Reference [2].)

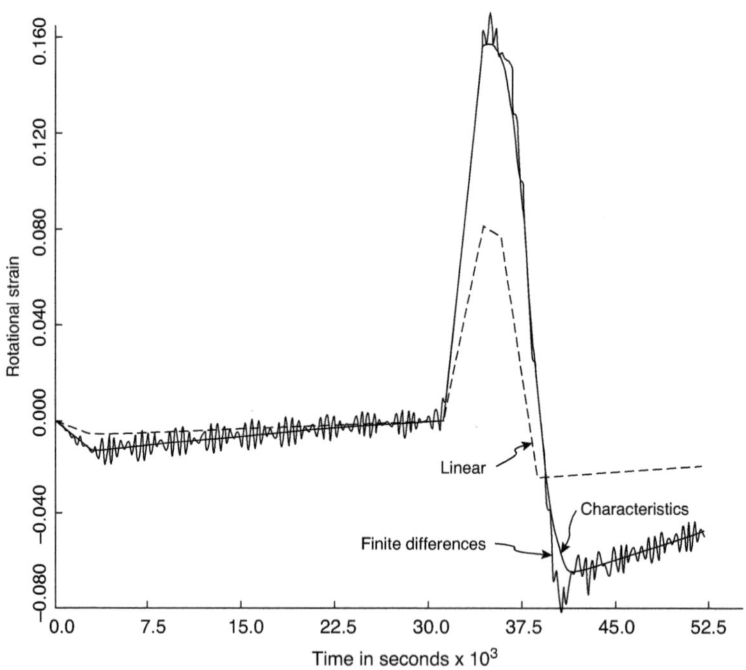

FIGURE 2.9 Time history of rotational strain at the impacted end of the helical spring. (Reprinted with permission from Reference [2].)

REFERENCES

1. S. K. Sinha (1977) "Non-linear Wave Propagation in Helical Springs by the Method of Characteristics and Finite Element Method", Ph.D. Thesis, University of Illinois at Urbana-Champaign, IL, July 6, 1977. https://hdl.handle.net/2142/68492
2. S. K. Sinha and G. A. Costello (1978) "The Numerical Solution of the Dynamic Response of Helical Springs", *International Journal for Numerical Methods in Engineering*, Vol. 12, pp. 949–961. https://dx.doi.org/10.1002/nme.1620120607
3. S. K. Sinha (2005) "Non-linear Dynamic Response of a Rotating Radial Timoshenko Beam with Periodic Pulse Loading at the Free-End", *International Journal of Non-linear Mechanics*, Vol. 40, No. 1, pp. 113–149. https://dx.doi.org/10.1016/j.ijnonlinmec.2004.05.019
4. S. K. Sinha (1988) "On Free Vibrations of a Thin Spinning disk stiffened with an Outer Reinforcing Ring", *ASME Journal of Vibration, Acoustics, Stress and Reliability in Design, Trans. ASME,* Vol. 110, pp. 507–514. http://dx.doi.org/10.1115/1.3269558
5. S. K. Sinha (1988) "Free Vibrations of a Thick Spinning Annular Disk with distributed Masses at the Outer edge", *Journal of Sound and Vibration*, Vol. 122, No. 2, pp. 217–231. http://dx.doi.org/10.1016/S0022-460X(88)80350-X
6. E. Fehlberg (1964) "New High Order Runge-Kutta Formulas with Stepsize Control for Systems of First- and Second-Order Differential Equations", *Zeitschrift für Angewandte Mathematik und Mechanik* (*ZAMM*), Vol. 44, pp. 17–29. https://doi.org/10.1002/zamm.19640441310
7. R. E. Miller (1967) "Use of the Field Method for Numerical Integration of Two-Point Boundary-Value Problems", *AIAA Journal*, Vol. 5, No. 4, pp. 811–813. https://doi.org/10.2514/3.4075

3 Soft-Body Impact and Bird Strike on Turbofan Engine

3.1 INTRODUCTION AND APPLICATIONS OF SOFT BODY IMPACT ANALYSES

In the soft-body impact analysis of jet engine components, there are two important impact scenarios that are the main focus of attention: bird strike and ice hailstone impact. Soft-body impacts, as opposed to hard-body impacts, create additional challenges in any numerical technique due to the large distortion of the Lagrangian mesh used to simulate the soft-body impactor.

Bird strike is the most common form of flight hazard condition, which commercial aviation encounters almost on a daily basis around the world at various airports. Due to inherent issues of flight safety and damage to critical engine components caused by the bird strike, the importance of predictive analytical modeling and design methodology in aerospace applications cannot be overemphasized. Because of increased demand of number of flights and also due to noise concerns, the current trend is to build newer airports away from the heavily populated urban areas. As such, the likelihood of birds being ingested in a running engine, especially during landing and takeoff, and resulting in some severe damage to engine components, vastly increases. On the other hand, due to significant advancements in numerical simulation of high-velocity impact events and enhanced capability of analytical predictions, it is possible to design the critical engine components, which are better bird-strike tolerant than previous generation engines. The peak magnitude of dynamic impact forces being generated during a bird strike on an engine blade is extremely large, which is typically in the range of 100–200 kN values. For loads of these magnitudes, the effective bearing stiffness of the rolling element bearing is highly nonlinear, which makes the rotordynamic response of the fan rotor nonlinear as well. Until now, no attempt has been made to determine these loads on the rotor shaft and the structural components of the engine mounts and support system in any physics-based analytical model. Nearly all the analytical studies done to date have been carried out using finite-element techniques based upon time marching forward integration schemes with explicit formulation solver such as LS-DYNA® and PAM-CRASH®.

In a typical bird-strike scenario on a turbofan rotor, it is invariably an "oblique impact" event with the bird acting as the projectile or impactor, and the fan blade being the target, which also acts as a slicer. In this process, an incoming bird gets cut into multiple slices by the rotating blades, which dynamically load the fan blades as each individual bird slice travels from the leading edge (LE) of the fan blade toward

DOI: 10.1201/9781003749752-3

its trailing edge (TE). Theoretically, the "oblique impact" generates both normal and tangential components of the dynamic contact loads; however, because of the very short duration of the bird-strike event, the normal forces due to inertia of the target are an order of magnitude higher than the corresponding tangential forces, which are resisted by the dynamic coefficient of friction at the blade pressure surface. In the high-velocity bird-strike simulations, it is a common practice to neglect the effect of friction at the contact surface, which has been found to be adequate through many numerical simulations and verified with the test measurements of predicted peak strains for both near and far-field responses of the blade. The time duration of a typical contact-impact bird-loading on a fan blade may last anywhere from a fraction of milliseconds to a few milliseconds. The peak magnitude of these dynamic loads can be of the order of 10^5 Newtons. However, the precise time history of the transient dynamic loads has never been determined accurately due to its very short duration and highly nonlinear nature of the bird-strike event.

In the following section, it is shown how to determine the dynamic load history on a fan blade in a semiclosed form by numerically solving the highly nonlinear partial differential equations. The solution of the partial differential equation helps in understanding the complex dynamics of a bird strike on a twisted curved plate and also to carry out parametric studies, which can be difficult to complete using a finite element-based approach. In the present discussion, the bird has been modeled as a fluid with a mixture of water and air, which comes in contact with the airfoil blade that is modeled as an elastic material rotating in a centrifugal force field. The transient dynamic contact load being applied by the bird material on the pressure (concave) surface of the fan blade is treated as a traveling load vector acting in the direction normal to the surface. Here, it should be noted that due to a relatively very short duration of bird-loading time being in the range of 1–2 ms, the effect of the dynamic coefficient of sliding friction is very minimal. As such, in any typical bird-strike type impact simulation, the tangential component of the relative velocity is usually neglected. The instantaneous location of this external load vector with respect to the blade coordinates is determined by the nonlinear trajectory analysis of a moving mass on a curved and twisted surface rotating in space about the engine axis. Here, we also discuss as to how these impulsive loads generated during the bird strike can be used to develop a forcing function for determining the dynamic response of any generic pretwisted fan blade in an aeroengine.

As mentioned earlier, in all fan blade-related numerical simulations, the incoming bird is modeled in the shape of a circular cylinder made up of 10% air and 90% water. The current bird modeling approach has been verified by simulating, in LS-DYNA®, the impact event of a controlled panel impact test. It is demonstrated that the substitute bird model used in generating the dynamic loads on the fan blade, presented in this paper, one can replicate the test data very well.

3.1.1 Hailstone and Ice Impact Analyses

Hailstone and ice impacts are other examples of foreign object soft-body impacts in a turbo engine application. Ice impacts can be of many types, which usually happen near the ground due to cold weather icy conditions. However, ice formation and accumulation have been observed even at the cruising altitude of 30,000–40,000 feet.

3.1.2 Ice Shedding from Inlet

Ice shedding from the inlet happens near the ground and can become a serious problem at the time of takeoff. Since ice becomes water under the effect of high pressure near the contact area, ice can be numerically modeled as a slug of water impacting the pressure surface of the fan blade. Thus, the constitutive equation of water under pressure is an adequate hydrodynamic material model to analyze the effect and dynamic response of ice shedding and impacting the fan blade.

3.2 BIRD STRIKE MODELING AND ANALYSIS

The incoming bird is considered a cylindrical impactor made up of hydrodynamic material. Similar modeling schemes with some other geometrical variations of impactor shapes have become almost an aerospace industry standard and are routinely used by all engine manufacturers in their design analysis activity [see Reference 1]. The cylindrical shape gelatin bird has two main advantages.

a.1. The standard geometrical shape is relatively simple to analyze and also to produce it in the form of an artificial bird.
a.2. There are ample historical data with the engine manufacturers from previous test done on legacy engines. It should be noted that most of the actual engine test data are of a proprietary nature and are not available in the public domain.

The axially moving bird makes contact with the fan blade at the LE of the airfoil. The fan blades of the running engine cut the ingested bird into multiple slices, and in that process, generate a very large dynamic slicing-impact contact load on the blade. Typically, a metal blade made up of titanium alloy deforms plastically and forms a permanent bulge in the shape of "cusp" at the thin LE. However, a composite fan blade, commonly used on large-size high-bypass ratio turbo-engines, deforms under the dynamic loads, but also recovers very quickly. To determine the dynamic loads on the rotor shaft due to bird impact, the magnitude of deformation of such large blades with respect to their geometrical size can be considered almost negligible. Thus, in the current formulation, the blades are not rigid, rather they deform elastically under impact loading called $F_{\text{Impact}}(t)$. This is exactly what happens and is true in composite fan blades toward which this paper is applicable to.

However, to compute the time history of $F_{\text{Impact}}(t)$, it is assumed that the elastic deformation of the blade under the hydrodynamic impactor used for bird modeling is relatively so small that any local deformation of blade can be neglected. During the short duration of 1–2 ms, typical for bird-strike events, the blade deformation is negligibly small with respect to other dimensions of the fan rotor. This assumption is consistent with the classic groundbreaking work of bird impact modeling and testing presented by Wilbeck on a flat rigid plate. Since the analytically computed impact load imparted by a moving bird projectile on a plate with clamped edges will always produce a conservative numerical value, this assumption is a perfectly valid approach from blade design considerations.

3.2.1 Bird as Hydrodynamic Material (Bird Shapes, L/D = 2)

Bird has traditionally been modeled as a cylindrical impactor made up of 90% water and 10% air mixture. This modeling approach is based upon the classic work of Barber, Wilbeck, Rand, and their coworkers [see References 2, 3] done during late 1970s and early 1980s. In any experimental test setup, the artificial bird is made up of gelatin, which has the same mass density as the real bird, but has no hygiene-related issues with handling dead bird debris. The main advantage of the cylindrical gelatin impactor is the simplicity of its geometrical shape and due to the identical mass density parameter, which captures the dynamical characteristics of bird-strike event very well. Over the 40 years or so, this bird testing and modeling approach in both shape and size of the impactor has become practically an aerospace industry standard and is commonly used to validate the bird-strike design parameters during the development phase of fan blades by all engine manufacturers. The actual test result related to bird-strike testing is of a highly proprietary nature of major engine manufacturers, and no test results of any significance are available in any published literature. Meanwhile, in academic institutions, many other parameters of bird-strike modeling and analysis methodology are being developed. For instance, other nicely shaped impactors of different sizes and geometrical shapes such as ellipsoids or cylindrical projectiles with rounded ends have also been tried (see Figure 3.1).

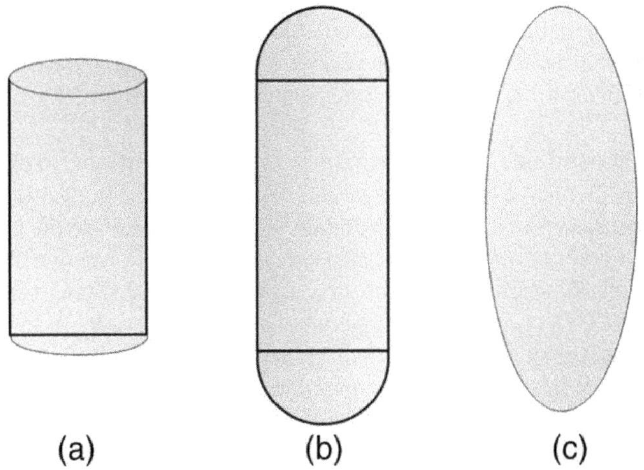

FIGURE 3.1 Different geometrical shape impactors used in bird-strike modeling and analysis: (a) cylindrical impactor ($B_L/B_D = 2$); (b) cylindrical impactor with hemispherical ends; and (c) ellipsoid. (Reprinted with permission from Reference [8].)

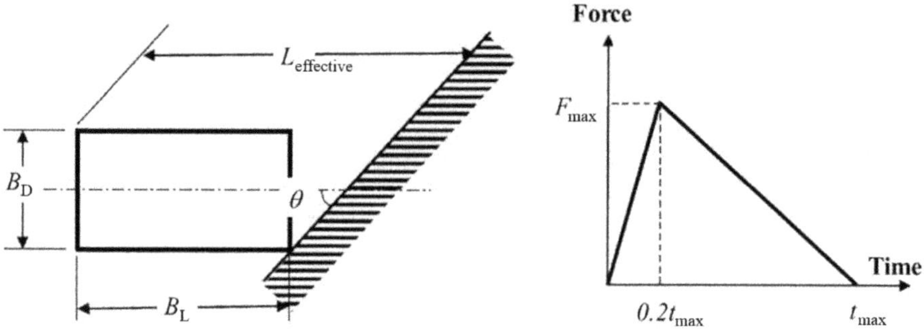

FIGURE 3.2 Typical experimental test setup for oblique impact at an incidence angle θ used to validate the accuracy of bird-strike results. (Reprinted with permission from Reference [8].)

In terms of any experimental tests, different researchers have published results based upon normal impact scenarios against a rigid wall as well as metal plates as the target with clamped edges. Since an actual bird-strike event on a running engine involves much more complex physics than these simple tests would represent, the results obtained from such tests have no practical use at all. As such, now academic researchers are focusing their effort on more complex, but realistic test parameters such as oblique impact test conditions (see Figure 3.2).

The key focus of the current section is to study the transient dynamic response of the fan rotor due to bird strike. For such applications, slicing action is very critical (see Figure 3.3), which any numerical scheme must capture very well. It should be noted that for this particular illustration we have used the symbol φ to represent the blade pretwist angle. The typical peak magnitude of dynamic loads on the fan blade during the slicing action is in the range of 100–200 kN. The numerical accuracy of the bird slice size calculation and the relative velocity approach for determining the peak impact load on the blade have independently been verified by Larsson in his thesis.

As the right-hand side forcing function $\{F(t)\}$ in the matrix form of the dynamical equation is directly dependent upon the magnitude of the impact load on the fan blades, it is necessary to discuss the validation and verification of the bird modeling methodology versus known test data. However, sufficient information about the validated bird models, specifically about their shape, size, and slicing refinements, by running impact tests on fixed plate type targets, is readily available and can be used to verify the accuracy of the numerical techniques. Such information can also be used to further improve the modeling procedure and numerical simulation in terms of material model, element size, and mesh density as well. Depending upon the complexity of the fan blade design requirements, some of the bird modeling approaches, such as smooth particle hydrodynamics (SPH) and arbitrary Lagrangian Eulerian (ALE), are still evolving.

FIGURE 3.3 A typical plastically deformed bulge of "cusp" in the leading edge of a metal turbofan blade after slicing action during the bird strike. (Reprinted with permission from Reference [1].)

3.2.2 Numerical Errors Due to Large Deformation of Bird Mesh

Since the bird behaves like a fluid, one needs a good knowledge of fluid mechanics. Most of the modern fan blades are made up of multilayered laminated composites, which behave completely in a different way from ductile metals. Under impact loading, they exhibit complex modes of damage such as very large deformation of the bird cylinder Lagrangian finite-element mesh issues and the resulting numerical errors cannot be overemphasized. The challenges of numerical simulation and accuracy of finite-element results due to extremely fine mesh size should also be considered. As the mesh gets finer, the time-step size also becomes smaller. The following is a step-by-step process to tackle this complex multiphysics problem.

a. As the bird comes into contact at the LE of the airfoil with relative velocity V_{Relative}, the spinning blade starts slicing it. During the slicing action, the bird behaves like a hydrodynamic fluid and applies a surface normal vector component of the stagnation pressure of magnitude $0.5\rho_B V_{\text{Relative}}^2$ on the pressure surface of the airfoil over a wetted area in the vicinity of the bird foot-print. The time-dependent impact force $F_{\text{Impact}}(t)$ on the fan blade can be computed by integrating the pressure distribution over the active foot-print area of the airfoil pressure surface. Based upon the simple impulse-momentum balance equation, it can result in a very large impulsive shock type of torque loading on the fan rotor shaft. It should be further noted that the bird strike-induced impulsive torque $T(t)$ is invariably a load torque on the fan shaft, as it is always acting opposite to the direction of rotation, and lowers the instantaneous RPM of the rotor.

b. The impact force $F_{\text{Impact}}(t)$ deforms the blade in the vicinity of impact at the span height s, creating a local cusp, especially in metal blades made up of titanium alloy. The local blade out-of-plane deformation w, in the direction normal to the (ξ,η) airfoil mid-surface, and the combination of bird-slice mass B_m attached to the blade, produces significant unbalance force F_U on the fan rotor shaft. In this way, the sum of vector components (F_X, F_Y, F_Z) of all the external transient dynamic forces in the fixed frame of reference acting on the fan-disk location of the rotor shaft at $(Z = \ell)$ can be computed.
c. Here, the computed time-history values of out-of-plane deformation $w(\xi,\eta,t)$ enable us to determine the local spanwise and chordwise strains in the blade airfoil during bird impact.
d. Then using the time-marching-forward solution of Runge–Kutta integration scheme, we solve the matrix form representation of the rotor-shaft nonlinear dynamical equation. The recommended approach is actually a sixth-order Runge–Kutta algorithm, which is much more accurate than the commonly used traditional fourth-order Runge–Kutta scheme.
e. The solution method outlined in the above step (d) and using the initial conditions of time $t = 0$ yields the numerical values of time-dependent deformation every new time step of $(t + \Delta t)$.

3.2.3 Bird Modeling in Arbitrary Lagrangian Eulerian (ALE) Mesh

In this approach of simulating the bird-strike impact event on a turbofan airfoil blade, the bird is modeled as an Eulerian fluid flowing freely from one cell to another. In this case, the bird Eulerian mesh is connected to the rotating blade Lagrangian mesh. Hence, the bird material moves from cell to cell.

3.2.4 Smooth Particle Hydrodynamic (SPH) Modeling of Bird

This is a relatively newer approach to model the bird. Here again, the bird is modeled like a hydrodynamic fluid material property, but it is made up of sand-like smooth spherical particles of individual masses called SPH modeling of bird. The main advantage of this modeling approach is that there is no concern about mesh distortion, usually prevalent in Lagrangian mesh with the eight-noded brick element modeling method.

3.3 EFFECT OF BIRD STRIKE ON TURBOFAN AIRFOIL BLADE

As the bird gets ingested in a running engine, it gets sliced into several pieces by the LE of the fan blade airfoils. The slicing load is extremely large, and in a typical bird-strike event, even a medium-size bird can generate up to 200–350 kilo-Newtons of dynamic forces for a few milliseconds.

3.3.1 Slicing Action of a Bird

The physics and the current methodology of accurate modeling of bird-slicing action and the resulting damage to fan blades have been built upon the groundbreaking work of Barber carried out at the University of Dayton Research Institute in the early 1970s [see References 2 and 3]. Since the original 1977 publication of their fundamental model of impact of soft bodies on jet engine fan blades, the slicing model has been verified and confirmed by abundant field experiences as well as multiple rig tests conducted by various engine manufacturers and the regulatory agencies like FAA. Barber and his coworkers were the first ones to study the dynamic response of jet engine fan blades during a bird strike, where they emphasized the slicing action by the lead edge of the rotating airfoil-shaped blades (see Refs. [2] and [3]). Through a series of actual tests, they had demonstrated that such a gelatin impactor will generate a dynamic load time history on fan blades that is representative of a real bird. As a result, in the current analysis, we will consider the incoming bird as made up of a mixture of water and air in the form of a circular cylinder with its length-to-diameter ratio of $\left(\frac{B_L}{B_D}\right) = 2$. In the turbofan design analysis, it is a standard practice to position the incoming bird in such a way that the bird-cylindrical axis is oriented along the engine axis.

In their schematic model, they considered the bird as a moving projectile, which is sliced by a blade oriented at an oblique angle of 25° with respect to the bird's trajectory. Referring to Figure 3.3 as the projectile is sliced, the impacting side or the blunt edge of the blade is continuously loaded from underneath, thus bending it upwards. This insight was the very first attempt to describe as to how a slicing action can load the blade and will deform it locally by forming a "cusp" at the impacting edge of the blade (see Figure 3.3).

To determine the magnitude of the transient dynamic loads generated by the impacting bird mass, it is important to distinguish the different phases of high-velocity impulsive loading due to soft-body impact and establish certain mathematical parameters that would capture the mechanics of the bird-slicing action in full detail. As soon as the bird cylinder comes into contact with the blade, it generates a high-intensity shock front called "Hugoniot pressure wave" that travels through the bird material. However, in terms of the dynamic response of the airfoil blade, its effect is very minimal during the initial phase of the loading. Although magnitudewise the pressure in the shock-front of the Hugoniot wave could be very high, the effective dynamic load felt by the blade is not significant because the initial bird contact loading area on the blade pressure surface is almost like a point loading, which is negligibly small and thus its contribution in the overall scheme of bird-strike

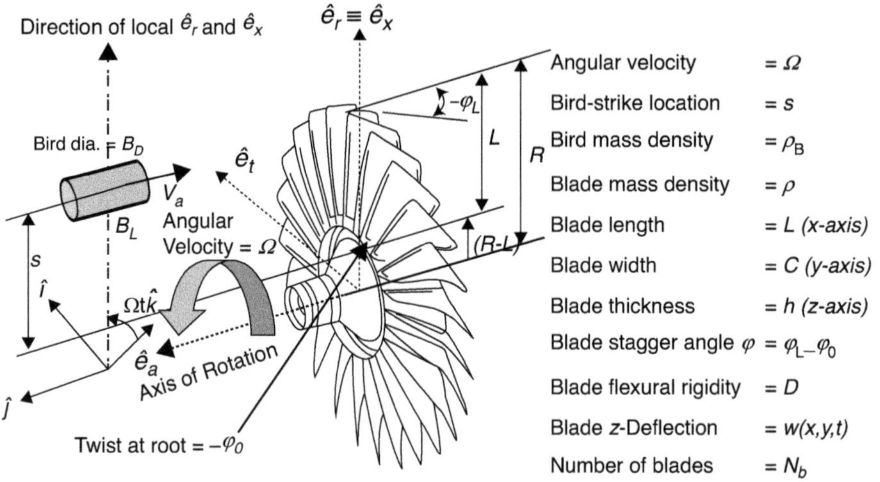

FIGURE 3.4 Bird cylinder coming in contact with rotating turbofan blades and local coordinate systems. (Reprinted with permission from Reference [1].)

dynamic loading can easily be ignored. For mathematical formulation, we start with the slice size, which is a function of various parameters such as the number of blades on the fan rotor, rotational speed, and aircraft speed. For the purpose of analytical formulation (see Figure 3.4), we consider a case where the bird of diameter B_D comes into contact with the airfoil blade at a span height "s" on a rotating bladed-rotor with N_b number of blades on it. The bird cylinder of length B_L moving with an axial velocity of V_a gets sliced into several pieces with the axial length B_S of each slice given by

$$B_S = \frac{60V_a}{(RPM)N_b} \tag{3.1}$$

where (RPM) is the rotational speed of the fan rotor $= 60\Omega/(2\pi)$.

From the geometrical profile of the blade, the "stagger angle $\varphi(s)$" at the impact location due to pretwist in the airfoil is expressed as

$$\varphi(s) = \varphi_0 + \varphi' s \tag{3.2}$$

Referring to Figure 3.4, we start with a rectangular Cartesian coordinate system $(\hat{i},\hat{j},\hat{k})$ fixed in space with $\hat{j}$-direction coinciding with the engine axis. We also introduce two additional rotating curvilinear coordinate systems, viz., $(\hat{e}_t,\hat{e}_a,\hat{e}_r)$ attached to the center of the fan shaft and the other $(\hat{e}_x,\hat{e}_y,\hat{e}_z)$ to the clamped end of the cantilevered fan blade. Using the velocity triangle formed by the tangential velocity of the rotating blade V_t and the axial velocity of the aircraft V_a, the angle of the slice $\beta = \tan^{-1}(V_a/V_t)$ can easily be calculated from Figure 3.2. Here, the magnitude of tangential velocity V_t at the impact location of the blade's LE is given by

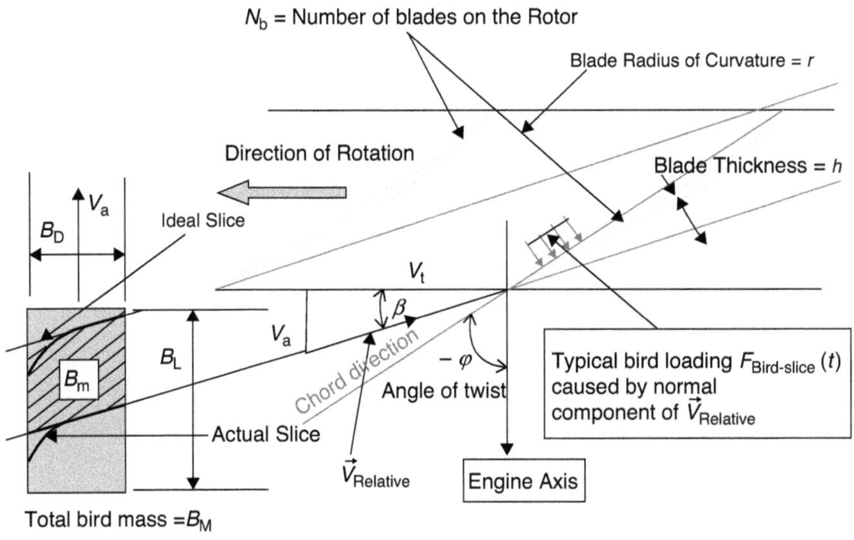

FIGURE 3.5 Schematic representation of bird slicing action due to the relative velocity of rotating blades and the axially moving bird. (Reprinted with permission from Reference [1].)

$$V_t = \Omega(R - L - s) = \frac{2\pi(RPM)}{60}(R - L - s) \tag{3.3}$$

For the majority of bird-strike scenarios, the typical magnitude of V_t is usually much larger than V_a, which is usually in the range of only 100 m/s.

In Figure 3.5, two different cut-slice shapes called "Ideal slice" and "Actual slice" are shown, which are caused by the difference in the "rigid blade" versus "deformable blade" formulations. The "ideal slice" in the shape of a parallelogram can be formed, if the blade were been "perfectly rigid". However, as a result of deformable blade behavior, a transient dynamic bulge is formed in the vicinity of the initial contact with the blade LE as the slicing process continues for a finite time-duration called "$t_{slicing}$".

Let us assume that for a typical bird-strike analysis we are considering a bird of mass B_M with its diameter being B_D and the length of the cylinder as B_L, then using ρ_B as the density of the bird material, we can write

$$\text{Bird mass } B_M = \rho_B\left[\left(\frac{\pi}{4}\right)B_D{}^2\ B_L\right] \tag{3.4}$$

Then using $B_L/B_D =$, we can describe the bird's cylindrical shape parameters as diameter and length as

$$B_D = \left[\frac{2B_M}{\pi\rho_B}\right]^{1/3} \text{ and } B_L = 2\left[\frac{2B_M}{\pi\rho_B}\right]^{1/3} \tag{3.5}$$

It should be noted that a large-size bird will have a longer cylindrical length B_L, which may result in multiple blades being fully or partially hit in one sector of the fan-rotor by a single bird. In that situation, one needs to determine the number of fan blades coming in contact with the bird as

$$\text{Number of blades coming in contact with the bird} = \frac{B_L + C\cos(\varphi_0 + \varphi' s)}{V_a} \left[\frac{(RPM)N_b}{60}\right] \tag{3.6}$$

However, in a typical bird-strike analysis for turbofan design, the interest is usually only in the blades which take the brunt of full-size slice as they might experience maximum damage. For this reason, the number of blades being damaged or impacted is approximated as N_B, which is given by

$$N_B = \frac{B_L}{B_S} = \frac{B_L}{V_a}\left[\frac{(RPM)N_b}{60}\right] \tag{3.7}$$

Also, from Eq. (3.1), we have

$$\text{Bird slice mass } B_m = \rho_B \left[\left(\frac{\pi}{4}\right) B_D{}^2 \frac{60V_a}{(RPM)N_b}\right] \tag{3.8}$$

The bird slice or the point mass "B_m" experiences a complex combination of forces such as Coriolis forces, centripetal forces, and turning forces. To keep this analysis amenable to solution, we have made several assumptions, where the results may slightly deviate from a real-life bird-strike problem:

a. Real blade deforms and a bulge is created near the impact area; thus, changing the curvature of blade locally in the vicinity of the bird; however, in the present analysis, the curved surface is assumed to be rigid and as such the radius of curvature "r" for a given span location "s" is constant.
b. Real bird has a finite size and spreads over a large area after the impact called bird's foot-print, but in the current analysis it is assumed to be concentrated at a point.
c. Real bird starts coming into contact with the blade gradually until the entire bird has been squashed up against the pressure (concave) surface; however, in the current analysis, it is assumed that the entire mass of the bird slice loads the concave surface instantaneously.

3.3.2 Transient Dynamic Loads Applied by the Bird Slice

The transient dynamic loads on the blade during the bird strike can be divided into two distinct phases: (i) slicing impact load at the LE of the blade and (ii) bird-slice turning load during sliding from the LE to the TE of the blade. The dynamic characteristics of these two loads are discussed in detail in the following two subsections.

3.3.2.1 Slicing-Impact Load at the Leading Edge of the Blade

During the slicing event as the first phase of impact loading, the dynamic forces are due to the stagnation pressure generated by a stream of moving fluid when it is stopped suddenly by a rigid wall. For determining the slicing stagnation pressure $p_{\text{stagnation}}$ being generated at the contact location of the concave surface of the blade, we compute the impact velocity of the bird slice as the normal component called " V_{impact}" with respect to the blade, which yields

$$p_{\text{stagnation}} = \frac{1}{2}\rho_B V_{\text{impact}}{}^2 \tag{3.9}$$

where the usual value of the fluid density term ρ_B for the bird material is 0.9 g/cm^3. The transient magnitude of the slicing load, henceforth called "F_{slicing}," is determined by computing the contact area of the blade pressure surface, referred to as the "bird-foot-print-area" represented by B_a, which in the numerical scheme is updated as the slicing action progresses, i.e.,

$$F_{\text{slicing}}(t) = B_a(t)\, p_{\text{stagnation}} \tag{3.10}$$

The total duration of slicing is computed by the bird cylinder diameter B_D and the tangential velocity V_t of the blade LE at the contact location is given as follows:

$$t_{\text{slicing}} = \frac{B_D}{V_t} \tag{3.11}$$

It should be noted that due to the contribution from the bird's axial velocity V_a, the relative velocity vector of the bird with respect to the blade changes direction, which results in the shape of the slice being in the form of a parallelogram with an angle of inclination being β. Thus, the largest bird-foot-print area on the blade pressure surface is generated when the bird cylinder is fully sliced from one side of the bird cylinder to the other side by the relative velocity vector and is computed as follows:

$$\text{Max. } B_a = \frac{B_A}{\cos\beta} = \frac{(\pi/4)B_D{}^2}{\cos\beta} \tag{3.12}$$

It should be noted that the cross-sectional area of the bird cylinder B_A is constant for a given bird size; however, B_a representing the "bird-foot-print-area" is a function of time as the bird cylinder gets sliced. Although for mathematical simplification purposes, the shape of the ideal slice is determined along a straight line cut in the direction of the relative velocity vector, the actual slice takes place on a slightly curved path due to the local deformation in the shape of the bulge of the blade LE during the slicing action itself. It can be seen from Figure 3.3 that due to this additional time-dependent nonlinearity of the slicing action, the effective slice-mass in a test environment is somewhat bigger than the "ideal" slice-size calculated in this manner. Depending upon the span location of the bird impact, the LE velocity in the circumferential direction of the blade V_t at the initial contact location would typically be in the range of 200–400 m/s.

The magnitudes of other relevant velocities in a typical bird-strike event on a turbofan rotor blade are in the following ranges:

a. Normal component V_{impact} of relative velocity V_{Relative} is usually 200 m/s.
b. Axial velocity of incoming bird V_a with respect to the fan blade is usually 70–120 m/s.
c. Peak out-of-plane velocity of the blade pressure surface due to local elastic deformation is $[\dot{w}]_{\max} = 20 - 70$ m/s.

3.3.2.2 Bird-Slice Turning Load During Sliding from the Leading Edge to the Trailing Edge of the Blade

The bird-slice coordinate location "*c*" changes as a function of time as it travels from the blade LE to the blade TE. Thus, the loading time for the blade is computed by the time it takes the bird slice to travel or sweep through the chord of the blade, which is approximated by

$$t_{\text{travel}} = \frac{C\cos\varphi}{V_a} \tag{3.13}$$

As the bird mass slides on the concave surface of the rotating blade after being sliced and moves from the LE toward the TE of the blade, it generates large Coriolis forces.

These Coriolis forces, represented by the term F_{travel}, are acting in the direction normal to the pressure surface of the blade; the other complimentary component of which propels the bird-mass toward the higher radius trajectory. Thus, the total blade loading time t_{loading} is the sum of two distinct actions on the bird, which are the slicing and the sweep time, namely,

$$t_{\text{loading}} = t_{\text{slicing}} + t_{\text{travel}} = \frac{B_D}{V_t} + \frac{C\cos\varphi}{V_a} \tag{3.14}$$

Thus, the total transient loading on the pressure surface of the impacted blade, henceforth called $F_{\text{Bird-slice}}$, during the bird-strike event is the vector sum of these two loads, which is written as

$$F_{\text{Bird-slice}} = F_{\text{slicing}} + F_{\text{travel}} \tag{3.15}$$

In the following section of this paper, we will use a rigid-body nonlinear trajectory analysis to determine the time history of the load magnitude $F_{\text{Bird-slice}}$ and the corresponding instantaneous bird-slice location (s, c) to solve for transient dynamic response of the rotating fan blade. In the twisted-plate equation of motion, described in Section 3.5 of this paper, the normal load vector $F_{\text{Bird-slice}}$ is applied as an external forcing function on the rotating blade in the form of a moving load.

3.4 BIRD-SLICE TRAJECTORY AND CONTACT-IMPACT DYNAMIC LOAD ON THE BLADE

In this section, we will develop some kinematical relationships to determine the bird-slice trajectory as it slides and travels from the front to the rear of the curved and pretwisted airfoil. Here, the dynamics of a point mass representing a bird slice, traveling on a concave curved and twisted surface is analyzed. We will also determine the relative velocity vector that applies the contact-impact load F_{Travel} on the concave surface of the blade in that process. With that objective in mind, suppose the aircraft is moving with velocity V_a, then the components of the bird velocity vector $\vec{V}_a$ in the blade coordinate system $\left(\hat{\boldsymbol{e}}_x, \hat{\boldsymbol{e}}_y, \hat{\boldsymbol{e}}_z\right)$ are written as (see Figure 3.5)

$$\vec{V}_a = -\ V_a \cos\varphi\ \hat{\boldsymbol{e}}_x + V_a \sin\varphi\ \hat{\boldsymbol{e}}_y \tag{3.16}$$

Thus, the velocity of bird-impact with respect to the rotating blade coordinate system is

$$V_a \cos\varphi\ \hat{\boldsymbol{e}}_y\ -\ V_a \sin\varphi\ \hat{\boldsymbol{e}}_z = V_a\left[\cos\left(\varphi_0+\varphi' s\right)\hat{\boldsymbol{e}}_y\ -\sin\left(\varphi_0+\varphi' s\right)\ \hat{\boldsymbol{e}}_z\right] \tag{3.17}$$

Let us assume that the bird makes initial contact with the LE of the rotating fan blade at a coordinate location (s,c) such that the spanwise location $x = s$ and zA $y = c = -r\sin\theta$. Thus, the instantaneous position vector of the bird slice $\vec{\boldsymbol{r}}(s,\theta)$ in the blade curvilinear coordinate system (see Figure 3.4) can be written as

$$\vec{\boldsymbol{r}}(s,\theta) = s\ \hat{\boldsymbol{e}}_x\ -r\ \sin\theta\ \hat{\boldsymbol{e}}_y + \left[-r(1-\ \cos\theta) + w\right]\ \hat{\boldsymbol{e}}_z \tag{3.18}$$

where $\theta = \sin^{-1}\left(\dfrac{c}{r}\right)$

In the above equation, "w" is the out-of-plane elastic deformation of the blade due to the contact-impact traveling load vector defined in Eq. (3.15). In general, the magnitude of the elastic deformation of the blade, "w," will be very small with respect to the other parameters "s" and "c" and therefore will be disregarded for trajectory determination. Under this assumption, the instantaneous position vector $\vec{\boldsymbol{r}}(s,\theta)$ of the bird slice in terms of blade coordinates (see Figure 3.6) can be simplified as

$$\vec{\boldsymbol{r}}(s,\theta) = s\ \hat{\boldsymbol{e}}_x - r\ \sin\theta\ \hat{\boldsymbol{e}}_y + \left[-r(1-\ \cos\theta)\right]\ \hat{\boldsymbol{e}}_z \tag{3.19}$$

Also, the chord direction limits on the blade between the TE and the LE vary between the limits of $-\dfrac{C}{2} \le c \le +\dfrac{C}{2}$. In the current system, the LE is at $c = +\dfrac{C}{2}$ and the TE of the blade is at $c = -\dfrac{C}{2}$. The corresponding limits in the local shell coordinates of θ-direction will be $-\dfrac{\vartheta}{2} \le \theta \le +\dfrac{\vartheta}{2}$, such that

$$\vartheta = -2\sin^{-1}\left[\frac{(C_0 + C' s)}{2(r_0 + r' s)}\right] \tag{3.20}$$

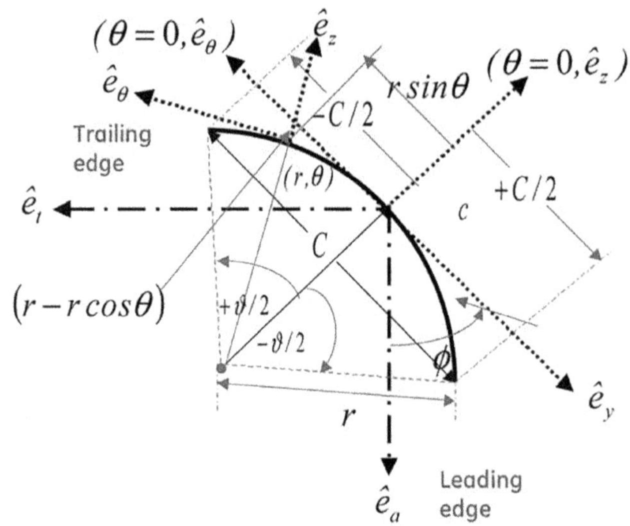

FIGURE 3.6 A pretwisted and curved airfoil surface as viewed from the blade tip side. (Reprinted with permission from Reference [1].)

The curved surface, the cross-section of which is assumed to be in the form of an arc of a circle with radius "r" is considered spinning at a constant angular velocity "Ω" about the global Y axis in the stationary frame of reference. The spinning surface representing the blade pressure surface is both curved and twisted in space such that both the radius of curvature "r" and the twist angle "φ" are functions of span height "s," which for $x = s$ can be written as

$$r = r(x) = r_0 + r's = r_0 + \frac{r_L - r_0}{L} s \tag{3.21}$$

$$\varphi = \varphi(x) = \varphi_0 + \varphi's = \varphi_0 + \frac{\varphi_L - \varphi_0}{L} s \tag{3.22}$$

also,

$$x = s(t) \tag{3.23}$$

For the sake of convenience in mathematical derivations, we will use previously defined coordinate systems (see Figure 3.6) attached to the running engine, henceforth called the "tangential-axial-radial" system $\left(\hat{e}_t, \hat{e}_a, \hat{e}_r\right)$ and the plate coordinates called the "length-width-thickness" system $\left(\hat{e}_x, \hat{e}_y, \hat{e}_z\right)$, respectively. Here, unit vectors $\left(\hat{e}_t, \hat{e}_a, \hat{e}_r\right)$ are in the cylindrical coordinate system and as such are only functions of angular velocity Ω, whereas $\hat{e}_z$ is normal to the mid-surface pointing away from the center of curvature, $\hat{e}_y$ is along the chord to the arc forming the curved surface, and $\hat{e}_x$ is radially outward from the engine centerline. The transformational relationships between the two local coordinate systems are as follows:

$$\hat{e}_t = \cos\Omega t\hat{i} - \sin\Omega t\ \hat{k} \tag{3.24}$$

$$\hat{e}_a = \hat{j} \tag{3.25}$$

$$\hat{e}_r = \sin\Omega t\ \hat{i} + \cos\Omega t\ \hat{k} \tag{3.26}$$

Thus,

$$\dot{\hat{e}}_t = -\Omega\hat{e}_r\ \ \dot{\hat{e}}_a = 0 \text{ and, } \dot{\hat{e}}_r = +\Omega\hat{e}_r \tag{3.27}$$

The corresponding transformational equations in the matrix form between the two coordinate systems $(\hat{e}_t,\ \hat{e}_a,\ \hat{e}_r)$ and $(\hat{e}_x,\ \hat{e}_y,\ \hat{e}_z)$ are written as follows:

$$\begin{Bmatrix} \hat{e}_t \\ \hat{e}_a \\ \hat{e}_r \end{Bmatrix} = \begin{bmatrix} 0 & -\sin(\varphi_0+\varphi's+\theta) & -\cos(\varphi_0+\varphi's+\theta) \\ 0 & \cos(\varphi_0+\varphi's+\theta) & -\sin(\varphi_0+\varphi's+\theta) \\ 1 & 0 & 0 \end{bmatrix} \begin{Bmatrix} \hat{e}_x \\ \hat{e}_y \\ \hat{e}_z \end{Bmatrix} \tag{3.28}$$

The time derivatives of the respective unit vectors $(\hat{e}_x,\ \hat{e}_y,\ \hat{e}_z)$ can be expressed as

$$\dot{\hat{e}}_x = -\Omega\left[\sin(\varphi_0+\varphi's+\theta)\hat{e}_y + \cos(\varphi_0+\varphi's+\theta)\hat{e}_z\ \right] \tag{3.29}$$

$$\dot{\hat{e}}_y = \left[\varphi'\dot{s} + \dot{\theta}\right]\hat{e}_z + \Omega\sin(\varphi_0+\varphi's+\theta)\hat{e}_x \tag{3.30}$$

$$\dot{\hat{e}}_z = -\left(\varphi'\dot{s} + \dot{\theta}\right)\hat{e}_y + \Omega\cos(\varphi_0+\varphi's+\theta)\hat{e}_x \tag{3.31}$$

The corresponding absolute velocity vector $\dot{\vec{r}}(s,\theta)$ of a point (in the global frame of reference) on the blade with (s,c) representing the time-dependent coordinate location of the bird-slice mass is derived as

$$\begin{aligned}\dot{\vec{r}}(s,\theta) = &\left[\dot{s} + \Omega r\cos\theta\ \cos(\varphi+\theta) - \Omega r\ \cos(\varphi+\theta) - \Omega r\sin\theta\ \sin(\varphi+\theta)\right]\hat{e}_x \\ &-\left[\ r'\dot{s}\sin\theta + r\cos\theta\dot{\theta} + \left(\varphi'\dot{s}+\dot{\theta}\right)r(\cos\theta-1) + \Omega(R-L+s)\sin(\varphi+\theta)\right]\hat{e}_y \\ &+\left[r'\dot{s}(\cos\theta-1) - r\sin\theta\dot{\theta} - \left(\varphi'\dot{s}+\dot{\theta}\right)r\sin\theta\ -\ \Omega(R-L+s)\cos(\varphi+\theta)\right]\hat{e}_z\end{aligned} \tag{3.32}$$

Using the components of the absolute velocity vector of the bird slice $\vec{V}_a$ shown in Eq. (3.17), its relative velocity vector with respect to an observer sitting at the rotating blade is given by

$$\vec{V}_{\text{Relative}} = V_x\ \hat{\boldsymbol{e}}_x + V_y\ \hat{\boldsymbol{e}}_y + V_z\ \hat{\boldsymbol{e}}_z = \begin{bmatrix} -\dot{s} - \Omega r\cos\theta\ \cos(\varphi+\theta) + \Omega r\ \cos(\varphi+\theta) \\ +\Omega r\sin\theta\ \sin(\varphi+\theta) \end{bmatrix} \hat{\boldsymbol{e}}_x$$

$$+ \begin{bmatrix} -V_a\cos\varphi + \Omega(R-L+s)\sin(\varphi+\theta) + \\ (r'\sin\theta - \varphi' r(1-\cos\theta))\ \dot{s} + (2\cos\theta - 1)r\dot{\theta} \end{bmatrix} \hat{\boldsymbol{e}}$$

$$+ \begin{bmatrix} V_a\sin\varphi + \Omega(R-L+s)\cos(\varphi+\theta) + \\ (\varphi' r\sin\theta + r'(1-\cos\theta))\dot{s} + 2r\sin\theta\dot{\theta} \end{bmatrix} \hat{\boldsymbol{e}}_z \tag{3.33}$$

In the local blade curvilinear coordinates "normal-tangential-radial" $(\hat{\boldsymbol{e}}_t, \hat{\boldsymbol{e}}_a, \hat{\boldsymbol{e}}_r)$ system, Eq. (3.33) enables us to express the components of the relative velocity of the bird-slice mass in the directions normal, tangent, and radial to the blade curved surface as

$$V_n = -V_y\sin\theta + V_z\cos\theta \tag{3.34}$$

$$V_\theta = -V_y\cos\theta - V_z\sin\theta \tag{3.35}$$

$$V_r = V_x \tag{3.36}$$

Here, the normal component of the relative velocity vector V_n in Eq. (34) becomes the V_{impact} per Eq. (3.9). Thus, according to Eq. (3.10), the dynamic contact-impact load $F_{\text{slicing}}(t)$ during the slicing action, with $B_a(t)$ being the instantaneous bird-foot-print-area, is computed as

$$F_{\text{slicing}}(t) = B_a(t)\left[\frac{1}{2}\rho_B(V_n)^2\right] \tag{3.37}$$

Also, the tangential component of the relative velocity vector V_θ in Eq. (3.35) causes the bird-slice mass to slide across the chord and is active for a time duration called t_{travel} according to Eq. (3.13). After some lengthy algebraic manipulation, the corresponding relative acceleration vector $\ddot{\vec{r}}(s,\theta)$ of the bird-slice mass on the blade curvilinear surface is obtained as

$$\ddot{\vec{r}}(s,\theta) = A_x\ \hat{\boldsymbol{e}}_x + A_y\ \hat{\boldsymbol{e}}_y + A_z\ \hat{\boldsymbol{e}}_z \tag{3.38}$$

where

$$\begin{aligned} A_x &= \ddot{s} - \Omega^2(R-L+s) - 2\Omega\dot{\theta}\,(r_0+r's)\sin(\varphi_0+\varphi's+2\theta) + 2\Omega(\varphi'\dot{s}+\dot{\theta}) \\ &(r_0+r's)\sin(\varphi_0+\varphi's+\theta) - 2\Omega(\varphi'\dot{s}+\dot{\theta})(r_0+r's) \\ &\sin(\varphi_0+\varphi's+2\theta) - 2\Omega r'\dot{s}\cos(\varphi_0+\varphi's+\theta) + 2\Omega r'\dot{s}\cos(\varphi_0+\varphi's+2\theta) \end{aligned} \tag{3.39}$$

$$A_y = \ddot{s}\left[r'\sin\theta+(r_0+r's)(\cos\theta-1)\varphi'\right]+\ddot{\theta}(2\cos\theta-1)(r_0+r's)+2\varphi' r'(\dot{s})^2$$
$$(\cos\theta-1)-(\varphi'\dot{s})^2(r_0+r's)\sin\theta+4r'\dot{s}\dot{\theta}\cos\theta-4(\dot{\theta})^2(r_0+r's)\sin\theta$$
$$-4\varphi'\dot{s}\dot{\theta}(r_0+r's)\sin\theta+2\Omega\,\dot{s}\sin(\varphi_0+\varphi's+\theta)-2r'\dot{s}\dot{\theta}-\Omega^2(r_0+r's)$$
$$\sin(\varphi_0+\varphi's+\theta)\times\left[\cos(\varphi_0+\varphi's+\theta)-\cos(\varphi_0+\varphi's+2\theta)\right] \tag{3.40}$$

$$A_z = \ddot{s}\left[r'(\cos\theta-1)-\varphi'(r_0+r's)\sin\theta\right]-2(r_0+r's)\sin\theta\,\ddot{\theta}-2\varphi' r'(\dot{s})^2\sin\theta$$
$$-4(r_0+r's)\cos\theta(\dot{\theta})^2-4r'\dot{s}\dot{\theta}\sin\theta-2\Omega\,\dot{s}\cos(\varphi_0+\varphi's+\theta)-4\varphi'\dot{s}\dot{\theta}(r_0+r's)$$
$$\cos\theta+(\varphi'\dot{s}+\dot{\theta})^2(r_0+r's)-(r_0+r's)(\varphi'\dot{s})^2\cos\theta-\Omega^2(r_0+r's)$$
$$\cos(\varphi_0+\varphi's+\theta)\times\left[\cos(\varphi_0+\varphi's+2\theta)-\cos(\varphi_0+\varphi's+\theta)\right] \tag{3.41}$$

$$\text{Thus,}\quad A_n = -A_y\sin\theta+A_z\cos\theta \tag{3.42}$$

$$A_\theta = -A_y\cos\theta-A_z\sin\theta \tag{3.43}$$

$$A_r = A_x \tag{3.44}$$

Since the bird slice is free to move in the span and the chord directions on the curvilinear surface of the pretwisted blade without any resistance, we set $A_r = 0$ and $A_\theta = 0$. Thus, we obtain a set of coupled second-order nonlinear differential equations with two unknowns (s,θ) as a function of time "t," which determine the time history of the bird-slice mass trajectory. Once the coupled equations have been solved for the time-dependent parameters (s,θ), they are used to determine the acceleration "A_n" of the bird-slice mass B_m in the direction normal to the concave pressure surface of the blade, which yields the time history of Coriolis forces as

$$F_{\text{travel}}(t) = B_m A_n \tag{3.45}$$

The coupled set of nonlinear differential equations of motion with $A_r = 0$ and $A_\theta = 0$ are solved numerically by a sixth-order Runge–Kutta method. The initial conditions for the time-marching-forward solution are obtained from the impact conditions described in the bird-strike problem. As shown in Figure 3.6, suppose the bird cylinder makes initial contact with the turbofan blade at the span location of s_i, then the initial conditions for the numerical solution of the contact-impact forces would be described as

$$s(0) = s_i \tag{3.46}$$

$$\theta(0) = -\frac{\vartheta}{2} \tag{3.47}$$

$$\dot{\theta}(0) = \frac{V_\theta\left(s_i, -\vartheta/2\right)}{r} \tag{3,48}$$

$$\dot{s}(0) = \frac{\left[V_n - V_a \sin(\varphi+\theta) - \Omega(R-L+s_i)\cos(\varphi+2\theta) - r\sin\theta\dot{\theta}\right]}{\left[\varphi' r \sin\theta - r'(1-\cos\theta)\right]} \tag{3.49}$$

In the numerical scheme, it is assumed that the radius of curvature, "r", of the pressure side of the blade changes linearly from the root of the airfoil toward the tip. Similarly, blade pretwist angle "φ" changes linearly as the blade twist angle changes from φ_0 being at the root to twist angle of φ_L being at the tip. The time histories of the bird loading $F_{\text{Bird-slice}}$ from the current analytical model are being produced here in a nondimensional form for a typical turbofan airfoil blade with the following parameters:

Fan outer radius $R = 150\,\text{cm}$
Span length of the blade $L = 100\,\text{cm}$
Blade average aspect ratio $(L/C)\ \dfrac{L}{C} = 2.0$
Blade average camber radius $r = 150\,\text{cm}$
Total twist in the blade $(\varphi_L - \varphi_0) = -60^\circ$
Number of fan blades on the disk $N_b = 20$
Rotational speed $RPM = 3000$
Angular velocity $\Omega = 314.159$ radians/s
Blade tip tangential velocity $V_t = 471.885\,\text{m/s}$
Blade material density $\rho = 4.443\ \text{g/cm}^3$
Bird material density $\rho_B = 0.9\ \text{g/cm}^3$
Young's modulus of the blade material $E = 117\,\text{GPa}$
Aircraft velocity $V_a = 100\,\text{m/s}$

The typical trajectory of the bird slice on the concave surface of the above hypothetical blade for a 75% span-shot is shown in Figure 3.7. The graphical result shows that for this blade the bird slice can move up by as high as 85.7% in the spanwise direction. The slice is propelled upward for about 2/3 of the chord travel and then starts moving downward until it exits through the TE at 84% span height, which is 9% higher than the initial entry contact location. In the plot, the bird slice is exiting at 104% of the initial chord, because the data has been normalized with respect to chord size at the initial span height. In the current formulation of the blade, the chord size is increasing at the rate of C' as one moves from the root of the airfoil toward the blade-tip. In other words, if it is a constant chord blade, then the rate of chord change parameter $C' = 0$.

A set of dynamic loads on the fan blade has been generated for different size birds such as 1-, 2-, 3-, and 4-kg bird-strike scenarios (see Figure 3.8). Here, Figure 3.8a–c

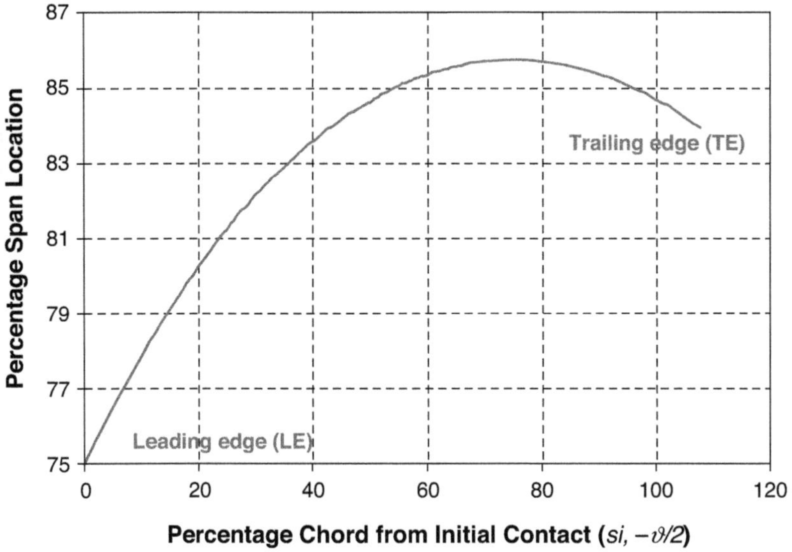

FIGURE 3.7 Analytically computed bird-slice trajectory on the concave surface of an airfoil blade for a typical 75% span shot. (Reprinted with permission from Reference [1].)

shows the family of curves for different bird sizes, for the initial impact location of 25%, 50%, and 75% span, respectively. In these plots, for presenting the analytically predicted dynamic load time history for a given size bird slice, we define the nondimensional time term called $\bar{t}$ as

$$\bar{t} = t\left[\frac{\Omega}{2\pi}\right] \tag{3.50}$$

The bird-loading time history shown here for varying impact locations on the blade LE enables us to identify the critical span location, which would result in the maximum damage or the largest strain to the fan blade. The current analytical method of solving nonlinear coupled partial differential equations provides a tremendous advantage in narrowing down the search for the critical location on the blade both for important design decisions about the LE thickness and for any test planning. As an example, for this hypothetical fan configuration with $N_b = 20$ and $\Omega = 0.058$ radians/s, the critical span location has been determined as 43% span (see Figure 3.8), in which case the peak dynamic contact-impact-slicing load for a typical 2-kg bird strike can reach as high as 232 kN.

These dynamic load curves can easily be scaled for evaluating some other fan blade designs using dimensional scaling relationships. The peak dynamic load for a similar blade tip-velocity and a given bird-size will be

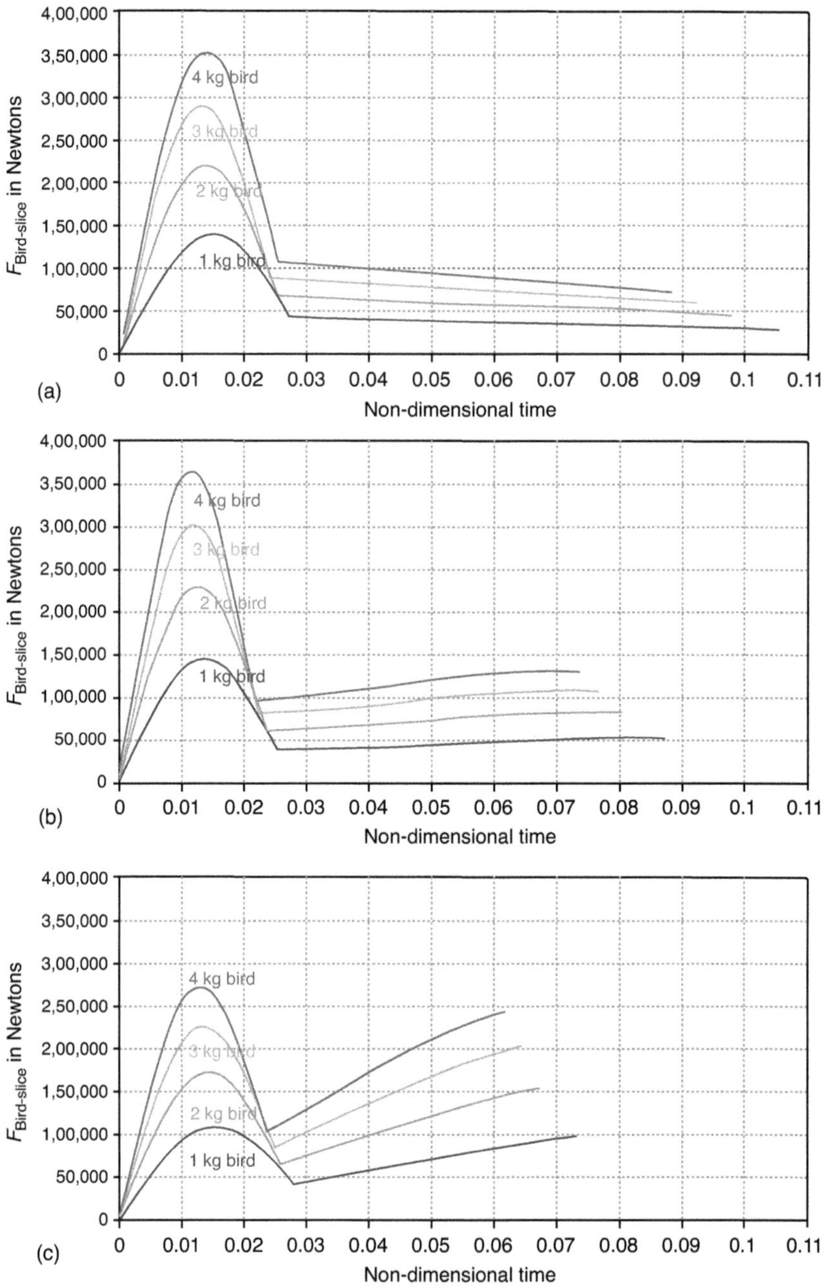

FIGURE 3.8 (a) 25% span impact, (b) 50% span impact, and (c) 75% span impact. Time history of contact-impact loads ($F(t)_{Bird-slice}$) on a typical fan blade $\frac{L}{C} = 2.0$, by different size birds at the leading edge span locations (a) 25%, (b) 50%, and (c) 75%. (Reprinted with permission from Reference [1].)

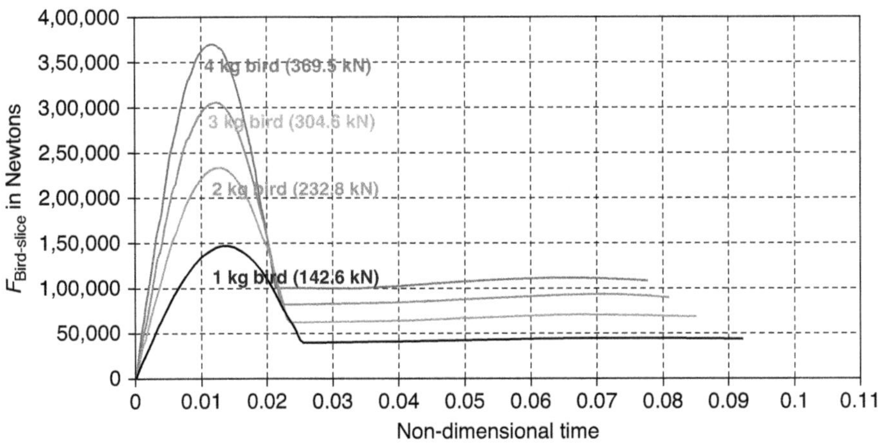

FIGURE 3.9 Time history of contact-impact loads $F(t)_{Bird-slice}$ on a typical fan blade ($L/C=2$), by different size birds at the critical impact location of 43% of span. (Reprinted with permission from Reference [1].)

a. directly proportional to the square of the rotational angular velocity ratio and
b. inversely proportional to the number of fan blades on the rotor.

For example, the dynamic load for a different rotational velocity of the fan, suppose Ω_0, can be scaled as a square of the speed ratio given by

$$F(t) = F(t)_{\text{Bird-slice}}\left(\frac{\Omega_0}{314.159}\right)^2 \tag{3.51}$$

The peak near-field spanwise strain at the blade LE develops at the end of the bird-slicing event. For example, in the critical bird-strike event shown in Figure 3.9 for a 2-kg bird-strike scenario, the peak strain at the lead edge would develop at the instant $\bar{t} = 0.023$, whereas the bird loading due to sliding on the pressure surface of the blade continues until $\bar{t} = 0.085$. After the travel time t_{travel}, bird slice has exited beyond the blade-surface and the blade goes into free elastic oscillations in the centrifugal force field as the postimpact dynamic response. Depending upon the blade design, sometimes the blade may exhibit higher strains at far-field locations as its postimpact dynamic response of free vibrations even after the primary bird-strike impact event is over.

3.5 BIRD-STRIKE LOADS FROM THE ANALYTICAL MODEL VERSUS FEA RESULTS

For example, one can also establish the accuracy of the current analytical model in predicting the peak magnitude of transient dynamic loads by running a highly detailed LS-DYNA® simulation of the hypothetical blade described earlier. The LS-DYNA® model developed for this purpose is shown in Figure 3.10.

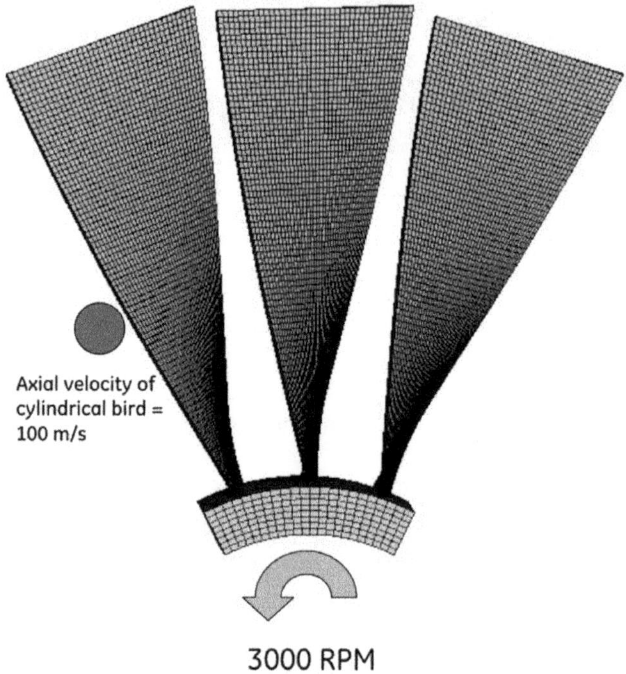

FIGURE 3.10 LS-DYNA® simulation of a 2-kg bird strike on a three-bladed sector of a typical turbofan rotor. (Reprinted with permission from Reference [1].)

In this simulation, a 2-kg bird hits a three-bladed sector of 20-bladed turbofan rotor at 43% span. In this analysis, the bird is modeled as a cylinder with $B_L/B_D = 2$ and approaching the fan rotor blades at 100 m/s. The bird cylinder gets sliced as it is described earlier in Section 3.2 of this chapter. In a multiple-blade slicing event, the first two blades get full-size slice and the last blade gets impacted only by the remaining partial slice. The other relevant bird and slicing parameters for this impact condition are summarized as follows:

Bird cylinder diameter $B_D = 11.23$ cm
Bird cylinder length $B_L = 22.46$ cm
Axial length of bird "Ideal slice" $B_s = 10.00$ cm
Mass of the bird cylinder $B_M = 2.0$ kg
Bird slice mass $B_m = 891.4$ g
Bird material density (10% air +90% water) $\rho_B = 0.9$ g/cm^3
Tangential velocity $V_t = 292.17$ m / s
Impact velocity $V_{\text{impact}} = 198.48$ m / s
Aircraft velocity $V_a = 100$ m/s
Leading edge loading time $B_s / V_{\text{impact}} = 0.504 \times 10^{-3}$ seconds

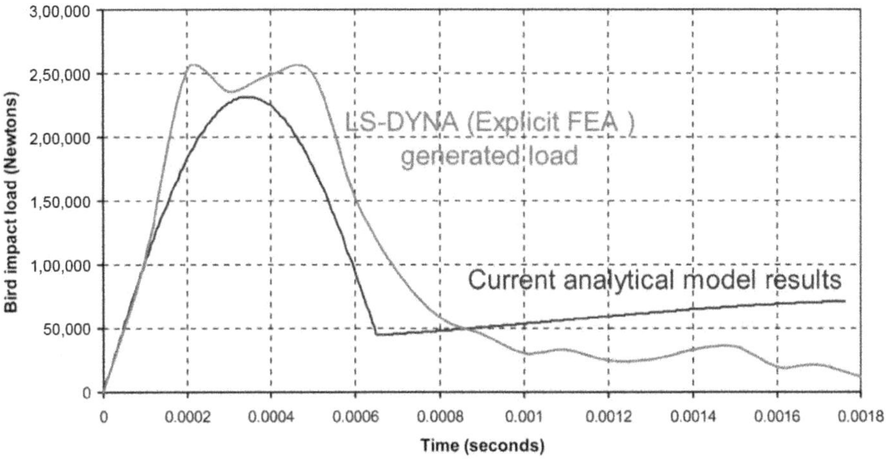

FIGURE 3.11 Time history of the bird-slice load in the tangential direction of the fan rotor on a 20-bladed rotor due to a 2-kg bird at 100 m/s. (Reprinted with permission from Reference [1].)

Bird slicing time $t_{\text{slicing}} = 0.384 \times 10^{-3}$ seconds
Number of blades with full-size slice $N_B = 2$
Bird spanwise impact location on blade (75% span) $s = 75$ cm

Figure 3.11 shows the comparison of the load time history from the current physics-based partial differential equation versus LS-DYNA®-generated results. The force-history plot clearly shows that for this hypothetical fan blade the loading time $t_{\text{loading}} = 1.8$ ms out of which first 0.7 ms shows impulsive characteristics, which is acting at the blade LE in the vicinity of the initial contact location. The results indicate that for a typical fan blade in a running condition, the present physics-based solution captures the qualitative time-domain characteristics of the bird-strike impact force history reasonably well. This is especially true during the slicing phase, where it has been shown to capture the individual peak dynamic load to within 9% as compared to the LS-DYNA®-simulated peak load. The impact load shown in Figure 3.11 being the tangential component of the total bird-load produces an impulsive torque on the fan shaft, which in a multiple-blade bird-strike scenario may be computed by superimposing these individual load-histories data with the initial offset on the time scale to take into account the blade separation in the circumferential direction of the rotor. Thus, the time history of the total integrated load on the blade can be produced using the following relationship:

$$\left[F(t)_{\text{Bird-slice}}\right]_{\text{total}} = \frac{\left[F(t)_{\text{Bird-slice}}\right]_{\text{tangential}}}{\cos\varphi} \tag{3.52}$$

3.6 TRANSIENT RESPONSE OF ROTATING AIRFOIL BLADES DUE TO BIRD-STRIKE LOADS

The transient dynamic response of the structure due to bird-strike loads is the main interest during the mechanical design of the airfoil blades. The out-of-plane deflection and the resulting strains in the blade can be calculated analytically from a physics-based model, which needs to be validated by experimental measurements as well. In this section, we will briefly describe as to how the contact-impact loads called $F_{\text{Bird-slice}}$ can be used as an external forcing function for determining the time-domain dynamic response of the deformable turbofan blade during the bird strike. For this discussion, we will use standard Timoshenko's plate theory notations. Based upon the classical Kirchhoff–von Karman's plate bending formulation and boundary conditions for thin shallow shells, the general equation for the flexural motion $w(x,y,t)$ in the transverse out-of-plane direction (z-axis) along with the rotary and warping inertias of a pre-twisted rectangular plate with the in-plane distributed membrane forces N_x and N_y, respectively, is written as follows:

$$D\left[\frac{\partial^4 w}{\partial x^4}+2\frac{\partial^4 w}{\partial x^2\partial y^2}+\frac{\partial^4 w}{\partial y^4}\right]+D(\varphi')^2\left[3\frac{\partial^2 w}{\partial x^2}+\frac{\partial^2 w}{\partial y^2}\right]-D(\varphi')^4\begin{bmatrix} x^2\dfrac{\partial^2 w}{\partial x^2}+\dfrac{(1-\nu)}{2} \\ x^2\dfrac{\partial^2 w}{\partial y^2}+2x\dfrac{\partial w}{\partial x}\end{bmatrix}$$

$$-\frac{\partial}{\partial x}\left[N_x\frac{\partial w}{\partial x}\right]-\frac{\partial}{\partial y}\left[N_y\frac{\partial w}{\partial y}\right]-\rho h\Omega^2\cos^2\varphi w+2\rho h\Omega x\frac{\partial \dot{w}}{\partial x}z=-\rho h\frac{\partial^2 w}{\partial t^2}$$

$$+\rho\frac{h^2}{12}\frac{\partial^2}{\partial t^2}\left[\frac{\partial^2 w}{\partial x^2}+\frac{\partial^2 w}{\partial y^2}\right]-\rho h(\varphi' x)^2\left[\frac{\partial^2 w}{\partial t^2}-\nu\left(\frac{Cy}{L}\right)^2\frac{\partial^4 w}{\partial y^2\partial t^2}\right]+q(x,y,t) \qquad (3.\ 53)$$

Here, the in-plane distributed membrane forces N_x and N_y are caused by the centrifugal forces. The Coriolis force $2\rho h\Omega x\frac{\partial \dot{w}}{\partial x}$ in the blade is generated due to the velocity component caused by its movement along the engine axis. Here, it should be noted that in the rotating condition of a pretwisted airfoil, while the longitudinal component (spanwise) of the centrifugal forces $(\sigma_{cf})_x$ result in stiffening of the blade, its chordwise component $(\sigma_{cf})_y$ tries to soften the blade by untwisting the airfoil cross-section. We obtain the relationships for $(\sigma_{cf})_x$ and $(\sigma_{cf})_y$ due to the angular velocity Ω in a rectangular plate with a pre-twist of $\varphi(x)$ at any typical point located at (x,y) as follows:

$$\left.\begin{aligned} \frac{N_x}{h}&=(\sigma_{cf})_x=\frac{\rho\,\Omega^2}{2}\left[L(2R-L)-2(R-L)x+x^2\right] \\ \frac{N_y}{h}&=(\sigma_{cf})_y=-\frac{\rho\,C^2\,\Omega^2\sin^2(\varphi_0+\varphi' x)}{8}\end{aligned}\right\} \qquad (3.54)$$

In Eq. (3.53), the right-hand-side external load vector term $q(x, y, t)$ being expressed as pressure applied per unit area is obtained by converting the dynamic contact-impact load $F_{\text{Bird-slice}}$ due to the bird strike at a location (s, c) of the plate pressure surface in shell coordinates using Dirac's δ-impulse functions $\delta(x - s)$ and $\delta(y - c)$ as follows:

$$q(x,y,t) = \left[-B_m \ddot{w} + F_{\text{Bird-slice}}\right] \delta(x-s) \quad \delta(y-c) \tag{3.55}$$

In the above equation, the time-dependent coordinate location (s, c) of the bird slice is determined by solving the trajectory equation as shown in Figure 3.7. Here, it should be recalled that by definition the dimensions of the Dirac's δ function are always L^{-1}. In the present discussion, our focus is on determining the time-dependent forcing function $F_{\text{Bird-slice}}$ for different size birds come in contact with a rotating engine.

3.7 SAMPLE NUMERICAL RESULTS AND SUMMARY

It is a common practice to simulate the airfoil blades used in rotating machinery by a pretwisted cantilever plate to its exact shape and profile using finite-element representation. In normal operation, these blades may be subjected to various types of impact loading scenarios known as foreign object damage (FOD). To evaluate the dynamic response of these blades under FOD-related impact conditions, one needs a precise description of the time-dependent history of the forcing function. These contact-impact forcing functions are highly nonlinear in nature and have been difficult to determine by any analytical means, especially in soft-body impact situations like a bird strike on a fan blade [see References 4–10].

To demonstrate the applicability and accuracy of the analytical model shown here, a sample problem has been created and numerically solved for a typical bird-strike scenario on a turbofan rotor. The impact conditions used in this hypothetical numerical simulation are listed earlier. The corresponding span and chordwise strain values in the blade airfoil in the vicinity of bird-impact are shown in Figures 3.12 and 3.13, respectively. The mathematical formulation presented here enables us to determine the transient deformation and cusp type of bulge formation at the LE of the airfoil under the contact-impact load time history of $F_{\text{Bird-slice}}$ in a physics-based analytical model (Figure 3.11).

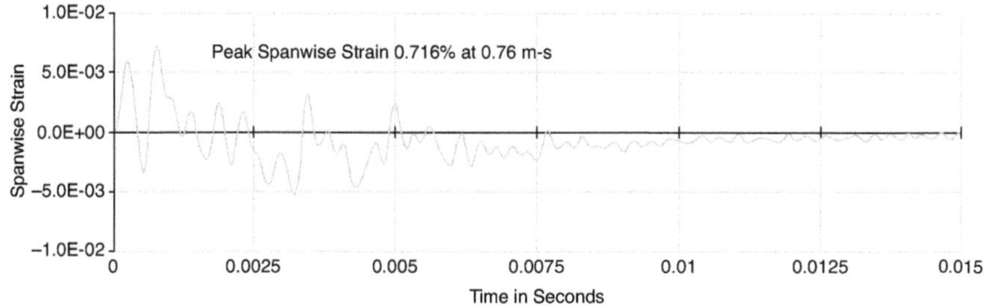

FIGURE 3.12 Transient spanwise strain in the airfoil at the leading edge of the airfoil (peak strain = 0.716% at 0.76 ms). (Reprinted with permission from Reference [8].)

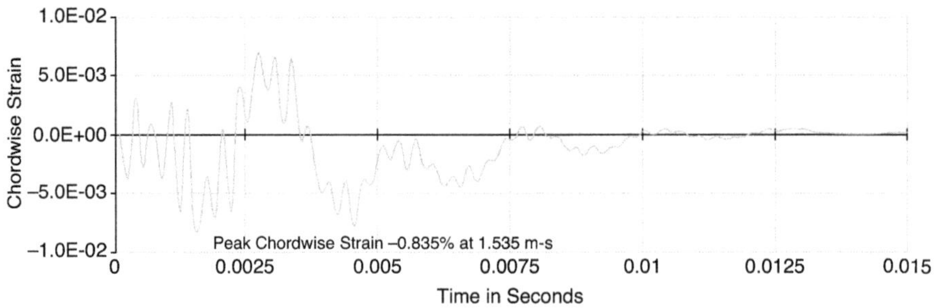

FIGURE 3.13 Transient chordwise strain in the airfoil at the leading edge of the airfoil blade (peak value = −0.835% at 1.535 ms). (Reprinted with permission from Reference [8].)

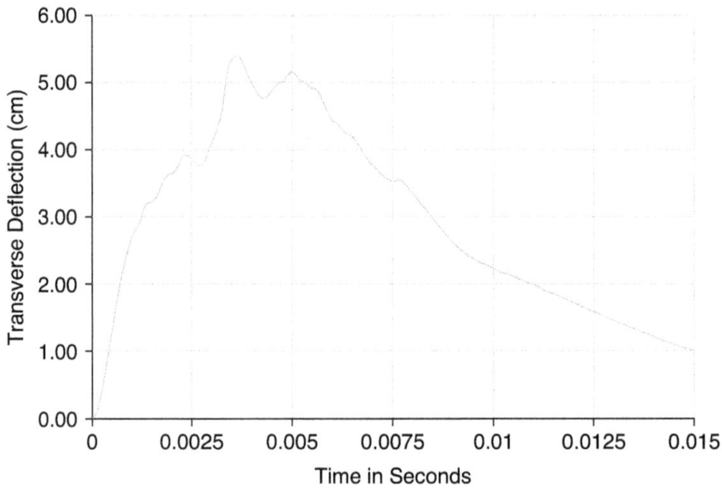

FIGURE 3.14 Transient elastic bulge cusp formation at the leading edge of the airfoil peak transverse deflection = 5.34 cm at 3.75 ms after initial contact). (Reprinted with permission from Reference [8].)

The time-history plots of blade airfoil strains, essentially at the LE in the vicinity of initial contact, enable us to evaluate the LE thickness required to withstand the impact load of the bird strike by a typical 2-kg bird being ingested in the running engine. The peak value of strains analytically obtained by the numerical simulation was consistent with typical dynamic characteristics observed in this size airfoil. Because the peak magnitudes of strains on the order of 0.716% and 0.835% were well within the material capability of typical blade material, the 2-kg bird was not predicted to cause any significant damage to the airfoil.

This modeling also provides the critical impact location, which is an important requirement as established for any ingestion test acceptable to regulatory agencies like the FAA (Federal Aviation Administration). The critical location for the bird-strike test is different for each blade design and the FAA requires that it be identified accurately by the engine manufacturer before the bird-ingestion test plan gets approved. This chapter also briefly discusses the flexural stress-wave differential equation approach to determine the transient response of the airfoil blade in terms of dynamic stresses being developed during the bird strike. Furthermore, it is conceivable that the bird-strike loading determined in this fashion can also be used to determine the dynamic response of turbofan blades and even full engines [see Ref. 10]. We will discuss it later in Chapter 5 of this book.

REFERENCES

1. S. K. Sinha, K. E. Turner and N. Jain (2011) "Dynamic Loading on Turbofan Blades due to Bird-Strike", *ASME Journal of Engineering for Gas Turbines and Power*, Vol. 133, No. 12, pp. 122504-1–122504-13. https://dx.doi.org/10.1115/1.4004126
2. J. P. Barber, P. F. Fry, J. M. Klyce and H. R. Taylor (1977) "Impact of Soft Bodies on Jet Engine Fan Blades", Rep. No. AFML-TR-77-29. Wright-Patterson AFB, OH, Air Force Materials Lab. https://apps.dtic.mil/sti/html/tr/ADA058194/
3. J. S. Wilbeck and J. L. Rand (1981) "The Development of a Substitute Bird Model", *ASME Journal of Engineering Power*, Vol. 103, pp. 725–730. https://doi.org/10.1115/1.323
4. H. Larsson (2015) "Bird Strike Analysis: An Analytical Approach", Master's Thesis in Aerospace Engineering, KTH Royal Institute of Technology, Trollhattan, Stockholm, Sweden. https://www.diva-portal.org/smash/get/diva2:1057190/FULLTEXT01.pdf
5. Y. Song (2016) "Development of Comprehensive Dynamic Damage Assessment Methodology for High-Bypass Air Breathing Propulsion Subject to Foreign Object Ingestion", Doctoral dissertation, Virginia Polytechnic Institute and State University, Blacksburg, VA. https://vtechworks.lib.vt.edu/server/api/core/bitstreams/93f2260e-bc91-485f-aa96-5c415d170196/content
6. J. Li, Y. Lou, X. Chai, Z. Ma and X. Jin (2022) "Numerical Simulation of Bird Strike on Jet Engine Considering Bird Ingestion Requirements", *Journal of Aircraft*, Vol. 59, No. 3, pp. 761–773. https://doi.org/10.2514/1.C036076
7. T. Lyons (2022) "Stress State Characterization of Ti-6Al-4V Fan Blades Under Impact Loading", Doctoral dissertation, Department of Mechanical and Aerospace Engineering, The Ohio State University, Columbus, OH. https://rave.ohiolink.edu/etdc/view?acc_num=osu1650218564307963
8. S. K. Sinha (2014) "Transient Response of a Multi-Layered Composite Rotating Airfoil Under Slicing-impact Loading", *AIAA Journal*, Vol. 52, No. 12, pp. 2701–2711. https://dx.doi.org/10.2514/1.J052888
9. S. K. Sinha (2018) "Transient Dynamics of Slicing-Impact Loading on Jet Engine Fan Blades During a Bird-Strike Event", *15th International LS-DYNA Users Conference*, Detroit, June 10–12, 2018, Paper No. 3. https://pdfs.semanticscholar.org/d024/ead306fa5716d5f0979806b46b0f38be4c02.pdf
10. S. K. Sinha (2021) "Transient Vibratory Response of Turbofan Engine Rotor Impacted by Bird-Strike", *ASCE Journal of Aerospace Engineering*, Vol. 34, No. 4, pp. 04021027. https://dx.doi.org/10.1061/(ASCE)AS.1943-5525.00012500795

4 Hard Body Impact and Fan Blade Containment System

4.1 INTRODUCTION TO THE FAN BLADE CONTAINMENT SYSTEM

The fan blade-out (FBO) test is a very complex nonlinear transient dynamics event with large deformation of the release blade, large deflection of the trailing blade, and progressive failure and fragmentation of blades, vanes, and casing components. The failure of rotor and stator components is typically classified under two categories called primary and secondary damage. To correctly simulate the entire FBO event in a finite element model (see Figure 4.1), one must have proper insight in the full physics of the blade-out event and have a finite-element model with adequate accuracy to capture all relevant dynamics starting from the instant the blade is released until the rotor has slowed down to a sufficiently low speed such that the dynamic loads would not cause any additional significant damage. In a typical large-size commercial engine, the initial blade release and primary containment damage, involving

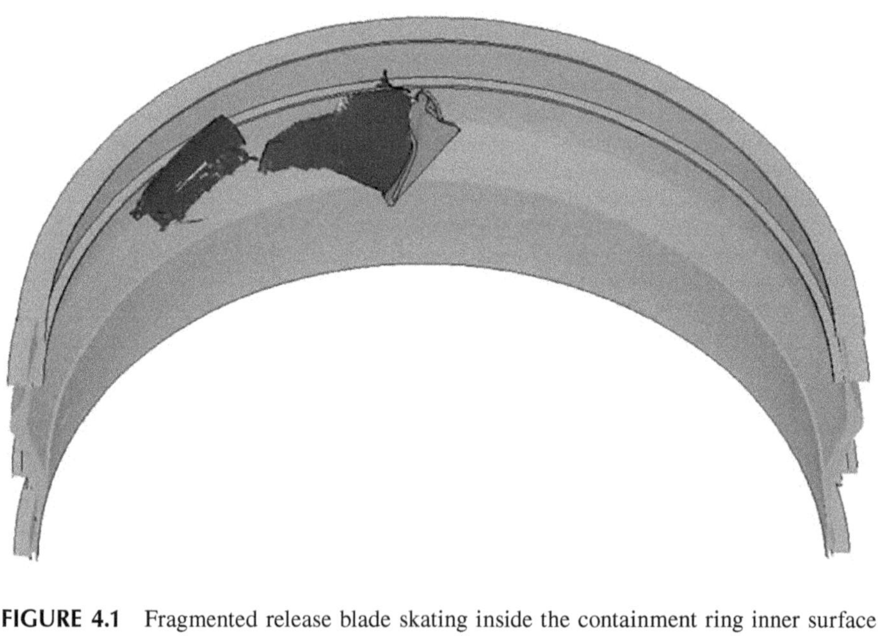

FIGURE 4.1 Fragmented release blade skating inside the containment ring inner surface simulating a fan blade-out event in LS-DYNA® analysis. (Reprinted with permission from Reference [2].)

DOI: 10.1201/9781003749752-4

large deformation and failure, may last anywhere from 20 to 100 ms. The time duration of the full event from the initial blade release through engine rundown may usually last from 2 to 10 seconds. The technical contents of this chapter are completely based on the author's own independent research work as well as his groundbreaking contributions originally presented in the ASCE Conference [1] and later on published as a detailed full-length paper in the *ASCE Journal of Aerospace Engineering* in July 2009 [2]. The author's original *ASCE journal* publication has been of such an archival nature that it has been the basis for many doctoral theses all across the world [see References 3–7]. Due to immense continuing interest in this topic, the body of knowledge has also been extended by recent scholarly journal publications of other researchers [see References 8–10].

When the blade is released, its initial velocity vector is tangential to the case, with its CG (Center of Gravity) moving freely along a direction perpendicular to its instantaneous radial direction, while rotating about the released blade CG to conserve angular momentum. It should be noted that in the analytical simulation the blade is physically rotated about the engine axis inside the fan case, so that when it is released it has the correct translational and rotational components of motion. Once the released blade's tip makes contact with the case, its velocity vector starts changing direction. There will be certain special effects due to specific blade and case design details, but in general the FBO event can be broken into five distinct phases, which will be discussed in detail later in Section 4.2. The differences in the actual physics of the blade-out event are due to different philosophies used in designing the containment capability, which are categorized as follows:

a. *Hard-wall containment*: Hardwall containment system is usually a multi-layer structure, which includes a layer of soft abradable material next to the fan blade tips with the outermost ring being the containment case. Actual cross-section of a hardwall containment system may vary considerably (see Figure 4.2a) due to various designs of mid-layer filler material among different engine manufacturers, which would depend upon their respective design philosophies, the details of which are highly proprietary information.

   In this type of fan case structure, no puncturing of a hole in the containment ring is permitted during the primary impact phase. At most, a plastic bulge in the main impact area without any through-hole will be acceptable. Any hole punched during this phase will most likely result in uncontained pieces or fragments of the release blade exiting the engine structure.

   Depending on the running tip clearance, the tip of the release blade hits the case about 12°–15° circumferentially from the release location and skates for approximately 30° circumferentially on the case inner surface. During this time, the top 1/3 of the airfoil is bent and broken. The tip fragment of the release blade has the tendency to move forward toward the inlet, while the root part of the airfoil typically moves aft.

   After the release blade upper panel is broken, the remaining 2/3-span of the release blade airfoil hits the inner surface of the trench-filler system at an oblique angle and depending upon its impact strength the blade may plow out the inner layer of the abradable or the filler layers (Figure 4.2a).

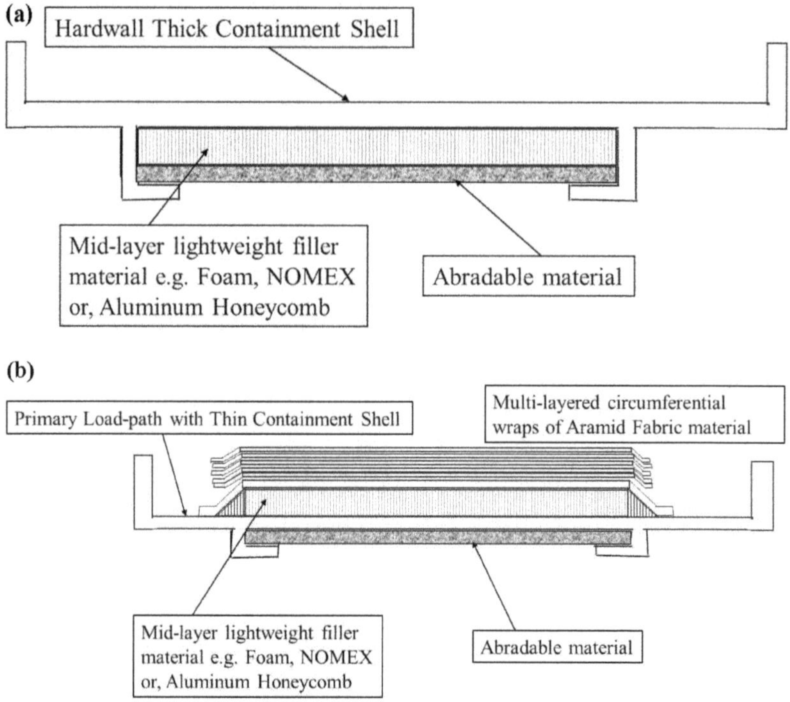

FIGURE 4.2 (a) A typical cross-section of a generic hardwall fan containment system. (b) A typical cross-section of a generic softwall fan containment system.

This damage can extend in the circumferential direction between 45° and 90° from the original release location. During this phase, the release blade is also propelled radially outward by the trailing blade pressure surface, which feeds additional kinetic energy to the release blade root. The contact forces applied on the blade in that process are responsible for added fragmentation of the blade and turning the overall velocity vector of the release blade CG from full tangential to mostly normal to the case. During the tumbling, the translational energy of the release blade fragments gets converted into rotational energy. The release blade fragments are pushed by the trailing blade tip. As it tumbles inside the case, its translational velocity is reduced considerably due to friction and eventually it continues to spiral and slide with low rotational velocity until it may come to rest by hitting some other barrier in its path. The containment case will bulge in the region of impact damage but will not be punctured.

b. *Soft-wall containment*: In this type of fan casing structure, the inner shell is relatively thin such that the release blade tip can easily punch a hole during the primary impact phase and therefore relies on the large movement of the multilayer Aramid fabric wrap to provide the main containment capability (see Figure 4.2b). In this type of design, the release blade may remain practically intact as it punctures the fan case and nests inside the Aramid belt, which behaves like a large catcher's hand gloves. In this process, the blade tip may cut through several inner layers of Aramid. However, as long as the outermost layers of the belt are undamaged, the release blade is considered fully contained. Obviously, since the release blade does not fragment and remains significantly intact, the local movement and resulting bulge in the Aramid belt generates very large strains, which the analytical model must be able to capture correctly.

The tip of the release blade hits the case about 12°–15° circumferentially from the release location and skates for approximately 30° circumferentially on the case inner surface. As the blade skates, the fan case bulges significantly and the blade tip typically curls without breaking. When the casing material's ultimate strength is reached, the blade tip penetrates through the casing and impacts the Aramid belt. Typically, several inner layers of the belt are cut as the tip nests into the belt. The blade continues to rotate until the root impacts expanding the hole in the case. In arresting the release blade, the Aramid wrap bulges as it conforms to the blade and is stretched tight around the fan case.

In this type of containment, the blade fragmentation is very minimal, with the majority of the blade remaining almost intact. The release blade tip starts curling in the direction of rotation as it skates on the case. In metal blades, most of the deformation of the release blade is limited to bending and twisting of the airfoil without any significant size metal debris being liberated. The release blade is impacted by the trail blade and is forced radially outward, punching the release blade through the casing shell. The tip of the release blade is caught in the Aramid belt and may cut through the inner layers of the Aramid wrap (see Figure 4.2b). This damage forms a local sack-line bulge in the wrap and in a contained event eventually brings the release blade to a complete stop.

4.2 DYNAMICS OF BLADE-OUT EVENT AND ITS SEQUENTIAL PHASES

The following sequence provides a general overview of the generic FBO event. In this discussion, the first three phases of the blade-out physics deals with the release of blade-containment aspects of the event and the last two phases focus on the rotor-dynamic response of the fan rotor and the resulting dynamic loads on various parts of the structure.

4.2.1 Fan Blade Loss and Containment Penetration

To pass the certification, for any safe and successful containment design regardless of either hardwall or softwall configurations, the containment penetration of the final outmost barrier of the containment system is not permitted.

- *First phase*: The first phase involves the primary blade containment without exiting and breaching the final outmost barrier of the containment system design made up of multiple barriers as shown in Figure 4.2a and b, respectively.
- *Second phase*: Depending on the containment as well as blade airfoil design, during this phase of a typical FBO event, the release blade may fragment into multiple pieces or may even remain intact in one piece while encountering large plastic deformation. The regulations require that all individual blade fragments including the small debris must be contained over the prime impact zone in the radial direction of the containment system ring. An acceptable LS-DYNA® model must be able to predict the release blade fragments reasonably well. Figure 4.3 reproduces the illustration of LS-DYNA®.
- *Third phase*: Rotordynamic response during the safe shutdown period: Following blade release, the unbalance force drives the heavy side of the fan rotor off-center. This causes the tips of the undamaged heavy side blades to start rubbing against the fan case. The rub digs out rub strip material and causes the fan rotor to slow down and start orbiting. During this phase, the containment structure (fan case) starts responding dynamically to the rotating unbalance load vector. The resulting response is essentially the superposition of the decaying response from the first three phases of impact loading with the imbalance-driven forced response during rotor rundown. This phase of the FBO event continues until the engine has been

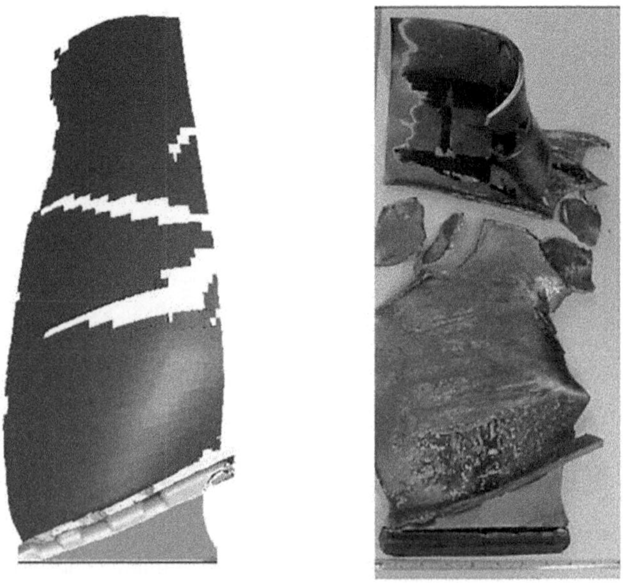

FIGURE 4.3 LS-DYNA® predicted and actual fragmentation of the release blade in the rig test (analysis versus test). (Reprinted with permission from Reference [2].)

shutdown and the fan rotor has slowed to a speed that does not create significant unbalance response.

- Final fourth and fifth phases of the post-FBO of safe landing: Wind-milling or rotordynamic response due to a sustained engine imbalance: This phase of the FBO event occurs after the safe shutdown of the engine (15 seconds for the pilot to initiate shutdown plus the engine rundown time) and relates to steady-state low-speed windmilling as the aircraft is flown home. Although the engine has sustained significant damage, the aerodynamic forces acting on the fan are typically more than adequate to prevent rotor seizure. The asymmetric rotor damage resulting from an FBO event equates to a very large imbalance, which even at low rotor speeds (i.e., 10%–20% of the maximum permissible speed) can generate high unbalance loads. During this phase, the fan-blade tips may intermittently rub against the containment case as the fan and full LP-rotor orbit with large amplitude.

4.2.2 Disk Burst and Containment

In the hardwall containment system of an aerospace application, another type of impact event must also be considered. This is called disk burst condition due to overspeed of the rotor. Most of the fan, compressor, and turbine disks are designed to withstand 10% of the maximum rated speed of the rotor. If the overspeed occurs and exceeds 110% of the rated maximum speed, the plastic deformation of the disk initiates at the inner radius and usually bursts in three segments. Some containment designs require that each of these pi-shaped segments must be contained.

4.3 DETAILS OF FULL ENGINE MODEL FOR FAN BLADE LOSS SIMULATION

As mentioned earlier, any high-velocity impact event must be analyzed using an explicit scheme time marching forward finite-element model to capture the full dynamic characteristics of the highly nonlinear physics involved in a FBO situation in a running jet engine. Currently, there is only one commercially available solver code called LS-DYNA®, which is fully capable of numerical simulations of analyzing the FBO-related dynamic response either in a whirligig rig or in a full running engine. In this section, we will focus on the meshing details and modeling and aspects of a full-engine 3-D simulation.

4.3.1 Finite Element Meshing and Modeling Considerations

In the LS-DYNA® simulation of a blade-out event, the recent practice by Shmotin and his coworkers [see Reference 11] has been to model only a 180° sector of the fan case. This simplified model produces reasonably good results about the impact resistance capability of the containment structure as well as the release blade fragmentation under primary impact. In this approach, one usually considers only the three blades representation of the fan rotor and a 180° sector of the fan casing. However, it is not an adequate model for analyzing the rotor-dynamic characteristics and the resulting dynamic response of the entire system. For determining the unbalance response of the fan rotor, it is necessary that the transient analytical simulation be carried out for several revolutions of the LP shaft. The biggest challenge in any long-running LS-DYNA® simulations in the 40–100 ms range is to obtain a time-domain solution without encountering numerical instability. The basic workhorse of the LS-DYNA® finite-element meshing scheme is a constant-strain eight-noded brick element. However, due to the reduced number of integration points, these brick elements are prone to hourglass instability, a type of numerical instability seen when mesh resolution is not high enough to accommodate the applied loading. This propensity to have hourglass issues is a result of the explicit integration scheme and causes additional complexities when running LS-DYNA® simulations.

In this section, we are introducing some of the acronyms, commonly used in the turbomachinery industry, such as low-pressure turbine (LPT), high-pressure turbine (HPT), high-pressure compressor (HPC), and low-pressure compressor (LPC) or booster, respectively. One needs the fully bladed rotor including the LPT shaft and disk for simulating the FBO event in LS-DYNA® and one needs to run it for a sufficient length of time with a normal termination so that the rotor response due to unbalance forces can be fully investigated. These models are fairly large in terms of their number of nodes and elements, but are capable of virtually simulating the actual FBO event as close as possible to the real test. For such analyses, a typical LS-DYNA® model will comprise about 1 million nodes (see Figure 4.1) and approximately 1/2 million elements.

The blades as well as the fan containment structure are modeled with sufficient detail using eight-noded brick element called *ELEMENT_SOLID. This being a constant strain element, one must use a minimum of two elements through the thickness, although the recommended number is three elements through thickness for a more accurate representation of strain distribution in major structural components of the engine. The blades are rotated about the engine axis using *BOUNDARY_PRESCRIBED_MOTION_NODE for the nodes, where the shaft or the disk is being driven by some external means and the *INITIAL_VELOCITY_NODE card is used for all other nodes, which will simply rotate as a result of prescribed angular velocity at the boundary nodes. It is recommended that ahead of the LS-DYNA® transient analysis, the stress initialization for the rotating blades be performed using an implicit code such as ANSYS® or through the implicit version of LS-DYNA® itself. All the inner surfaces of the containment structure as well as any other filler material surfaces should be specified in separate sets using the *SET_SEGMENT card. The contact interface between the blades and inner surfaces of the fan case must be defined by using the *CONTACT_ERODING_NODES_TO_SURFACE command. Furthermore, it is imperative that due to the very high strain rates involved in these simulations one uses strain rate-dependent material properties whenever they are available. The material model for both the blades and the fan casing should be specified using one of the two following options:

1. *MAT_STRAIN_RATE_DEPENDENT_PLASTICITY; and
2. *MAT_PIECEWISE_LINEAR_PLASTICITY.

All the material models in LS-DYNA® are used with FAILURE– option "ON." In our simulation, we did not use any specific hourglass control parameter or a penalty factor other than the default values built in to the code already. If the LS-DYNA® results indicate the presence of the hourglassing instability, it is best handled by making the mesh finer in critical areas, rather than using different types of hourglass control parameters. Whenever needed, the remeshing should be performed in such a way that

1. The point load situation during the entire duration of the transient analysis is eliminated by refining the mesh locally, such that a concentrated load is always shared by multiple surrounding nodes;
2. Slave side segments are finer than the master side segments; and
3. The mesh is fine enough so that it can capture true strain distribution accurately even in the areas of very high strain gradients.

For this FBO model, it is not necessary to consider the gas loads or thermal gradients in the analysis. The reason is that the dynamic loads generated during the FBO contact-impact loading are orders of magnitude higher than the pressure loading due to airflow or due to temperature variations in different parts of the engine. Since the release blade tip fragments have a tendency to go forward, it is also necessary to include the flight inlet in this simulation. The same argument applies to the rear end of the containment structure; it needs to be included in the analytical modeling. Depending on the requirements, one may model the bolts and the bolt holes in the flanges explicitly or one can use the *CONTACT_SURFACE_TO_SURFACE interface definition. Bolt pretension will be handled later as to how to prestress the bolt shank, which also introduces compressive stress in the flanges.

To handle the rigid body rotations of the spinning rotor, the major challenge is how to deal with the interface involving the bearing surfaces between the rotating shaft and the outer-race bearing surface connected to the stator structure. In a typical LS-DYNA® model, these interfaces allow free spinning of the shaft inside the respective bearing surfaces without locking the rotor due to mesh discretization of a curved cylindrical surface. In typical commercial jet engines, there are two types of bearings as follows:

1. Ball bearing: it takes both radial and axial load; and
2. Roller bearing: it can take only radial load.

In the LS-DYNA® model, both these bearings are simulated using the *CONTACT_NODE_TO_SURFACE type of interface. The stator side of the ball bearing is represented by a U-shape groove for the rotor side elements to sit inside two axial walls preventing its axial movement [see Reference 11]. On the other hand, the stator side of the roller bearing is represented by a cylindrical surface, such that the rotor side elements are free to move axially.

In this modeling approach, one represents the core of the engine with two concentric rotors involving the fan-LPT rotor as the inner rotor and HPC-HPT rotor as the outer rotor. It is necessary that this type of model be capable of capturing the

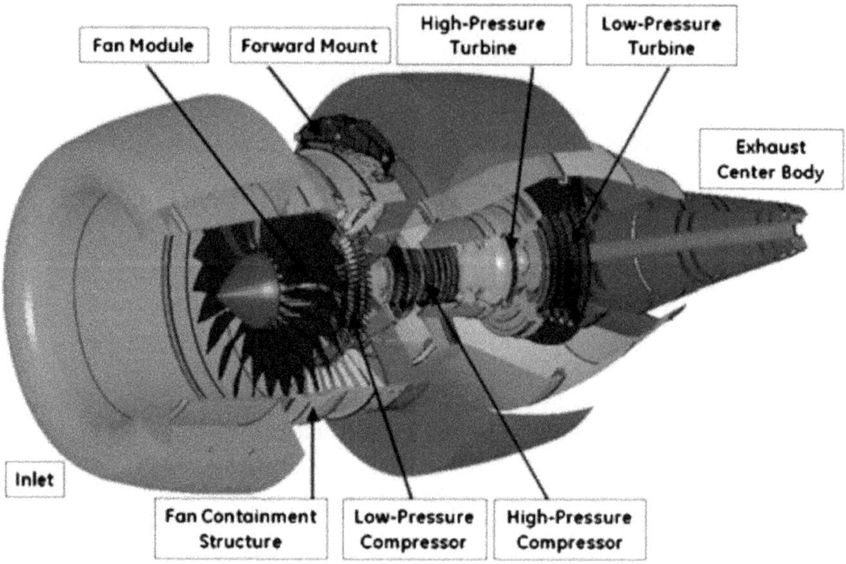

FIGURE 4.4 LS-DYNA® model of a typical high by-pass ratio turbofan engine (shown with cut cross section of stator to highlight the details of rotor components). (Reprinted with permission from Reference [2].)

deformation in the shaft due to the relative rubbing of the two concentric rotating shafts. In this modeling approach, all blades are modeled using shell elements with the ELEMENT_SHELL option, except for the fan blades. It is very important for dynamic accuracy that the individual blades on both rotors be modeled in such a way that their deformation due to rub or their clashing with any stator components be properly captured in the LS-DYNA® model. In a typical model, one needs to take special precautions for combining the shell nodes on one side with six degrees of freedom and the brick nodes with three degrees of freedom on the other side. The LS-DYNA® analytical model of the full engine shown in Figure 4.4 also contains the following components of the engine.

- Inlet;
- Outlet guide vanes and fan frame;
- Fan containment structure;
- Fan aft case;
- Bearing housings;
- LPC or booster casing;
- HPC casing;
- Combustor structure;
- HPT casing;
- LPT casing;
- Thrust links;
- Exhaust nozzle;

- Exhaust center body;
- Thrust reverser;
- All stator vanes;
- Cowling and Nacelle; and
- Forward and aft-mount systems.

Other accessory items, such as oil tank and auxiliary gearbox (AGB), may be represented by separate rigid masses or explicitly if the dynamic load on the mounting system is of interest. These different components in an engine are connected through several bolted joints, which form a convoluted load path during the blade-out event. In simulating a complicated system like a large commercial jet engine, one needs to make several simplifying assumptions in the model so that its size remains manageable while capturing the full physics of a very complex dynamic event. For example, we have neglected the effect of the airflow as well as such aerodynamic events as engine stall after the FBO. Other items, such as oil tank, AGB, full authority digital electronic control (FADEC), and fuel metering unit (FMU), may be represented by separate rigid masses or as individual modular components connected to the fan containment structure through some linkages and rubber isolators. If the possibility of fire prevention during the FBO event exists and needs to be predicted, then all the tubes carrying the inflammable liquids should also be included in the LS-DYNA® analytical model representation. These tubes can be modeled as beam elements in some limited locations of the full-engine FEA mesh.

A model of this size in the LS-DYNA® explicit solution invariably requires a massively parallel-processing MPP computing environment. Precise prediction of dynamic loads, in terms of both magnitude and their time history after an FBO event, is very critical. The typical LS-DYNA® model of the full engine, shown in Figure 4.4, captures the entire dynamic response of the engine for four full revolutions without any numerical instability. This investigation also analyzed the dynamic characteristics of various accessories attached to the fan-blade containment structure. The rig simulation is done for validating and calibrating the LS-DYNA® modeling technique to test data, as well as a complete engine model for analytically predicting its dynamic response prior to the actual FBO certification engine test. In the real engine simulation, they include components such as AGB, oil tank, FADEC, and FMU. After the FBO event, all these additional accessories connected to the fan casing structure go through rather large movements caused by the impact forces of the release blade. These components may also experience additional forced excitation caused by the rotating load vector as a result of the rotor imbalance due to the missing blade. The harmonic excitation caused by rotating fan imbalance can result in damage to one of the links or may produce excessive deformation of the brackets holding these accessories as well. The results of the analytical model help in evaluating and fixing any potential design problems ahead of the actual FBO test.

In this representation of the engine, one can see that the transient dynamic loads reach the forward mount system very early in a blade-out event. Since the mount system is usually at the 12:00 O'clock position for wing-mounted engines, the impulsive loads generated during the release blade impact on the fan casing structure are no longer symmetric. For this purpose, it is necessary to simulate a medium-size

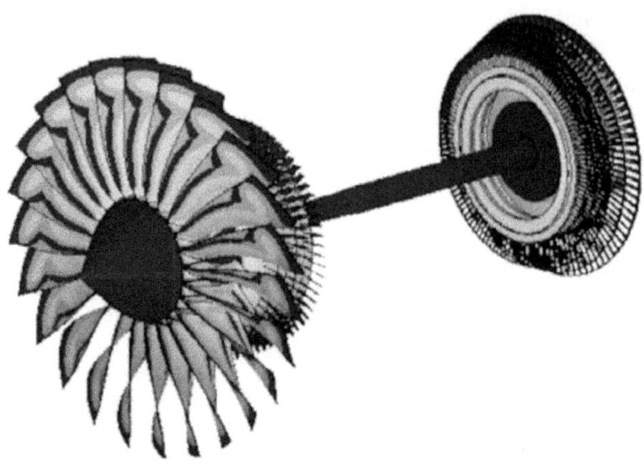

FIGURE 4.5 Steady-state stress distribution caused by centrifugal forces in the inner rotor of two concentric rotors joining fan module and low-pressure turbine (LPT) rotor. (Reprinted with permission from Reference [2].)

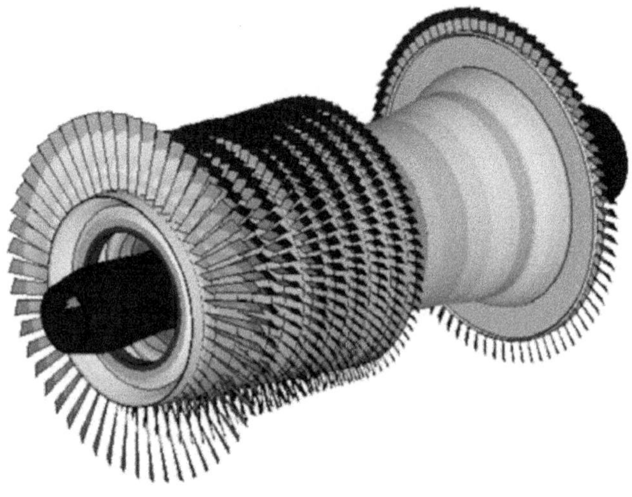

FIGURE 4.6 Steady-state stress distribution caused by centrifugal forces in the outer rotor joining high-pressure compressor and high-pressure turbine (HPT) rotor. (Reprinted with permission from Reference [2].)

high-bypass engine. It has two concentric rotors: (i) fan and the LPT rotor running at a lower rpm (see Figure 4.5) and (ii) the core of the engine (HPC spool and the HPT rotating at about three times (see Figure 4.6) the rotating speed of the fan/LPT rotor.

It is very important to obtain the centrifugal force-induced steady-state deformed state of the two rotors before an explicit time-dependent analysis in LS-DYNA® can be started. The significance of this is discussed in detail in Section 4.3.3 of this chapter.

4.3.2 Contact Definition for Different Interface Surfaces

Generally, in LS-DYNA® modeling, the contact between two impacting bodies is defined by an interface set made up of slave and master sides. In any explicit-integration scheme, it is imperative that for proper load transfer between the two bodies in contact the slave side mesh must be finer than the master side mesh. From the load transfer considerations, the correct definition of interface requirements such as absolutely not having initial interference or penetration between the slave and the master sets is extremely important. Other critical interfaces such as between the fan blade tips and the inner surface of the casing, root of the blade airfoil versus disk post, and between the shaft nodes and bearing surfaces must also be verified so that they do transfer the dynamic loads correctly. The containment trench-filler system and the fan blade tips are defined in such a way that their respective interfaces must simulate the casing material removal and account for resulting plowing loads acting at the blade tip as well important. Other critical interfaces such as between the fan blade tips and the inner surface of the casing, root of the blade airfoil versus disk post, and between the shaft nodes and bearing surfaces must also be verified so that they do transfer the dynamic loads correctly. The containment trench-filler system and the fan blade tips are defined in such a way that their respective interfaces must simulate the casing material removal and account for resulting plowing loads acting at the blade tip as well.

4.3.3 Stress Initialization due to Spinning

Stress initialization as a stress field at time $t=0$ in LS-DYNA® due to centrifugal forces in a rotating component is the first phase of any transient analysis of a dynamic event on the engine. In this analysis. we account for the centrifugal forces; however, due to relatively small magnitudes, we disregard the effect of the gas loads as well as thermal effects. The importance of stress initialization in various rotating components of the LS-DYNA® model can never be overemphasized. One must ensure that due to stagger and twist in the airfoil and the stress-stiffening effect, the steady-state deformed shape of the blade airfoil has been determined correctly using a proper convergence technique. It must be established the steady-state stress distribution in the fan blade using a commercially available implicit code such as ANSYS® and is verified in LS-DYNA® time $t=0$ that the overall fringe pattern is identical between the two codes (see Figure 4.7) Some minor variations between ANSYS® and LS-DYNA® codes should be expected as the eight-noded brick element formulations are different in the two solvers. For instance, in a recent paper published by Shmotin and his co-workers [see Reference 12], the differences in stress fringe patterns are too large to be used correctly in running LS-DYNA®. Such differences in the initialized state introduce large elastic oscillations during the transient dynamics run of LS-DYNA®, which when dynamically magnified can become a significant source of numerical errors in the analytical results.

Other stress initialization is needed in the static part of the containment structure at the bolted joints. Since every bolt used in the containment structure has a pretension in the bolt shank and a compressive stress in the flanges in contact, they also must be properly represented in the analytical model.

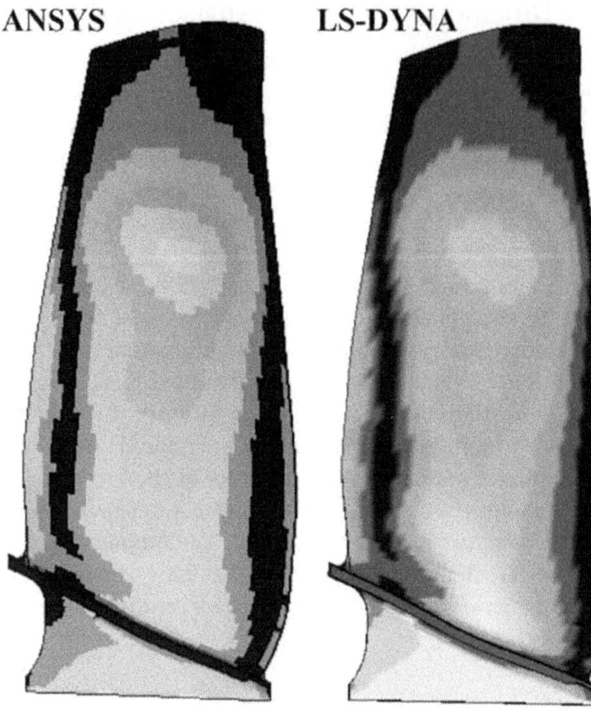

FIGURE 4.7 Comparison of von Mises stresses fringe pattern between ANSYS® and LS-DYNA® at time = 0 second. (Reprinted with permission from Reference [2].)

4.4 IMPACT ANALYSIS OF CONTAINMENT STRUCTURE DUE TO BLADE LOSS

Under the primary damage analysis, one usually looks into the dynamic loads which the release blade tip applies to the fan case and a few milliseconds later the release blade root impacts the pressure surface of the adjacent trailing blade. As the magnitude of the dynamic load increases at the blade tip, it starts deforming, and eventually breaks such that the tip-side fragment goes forward toward the inlet, and the root side starts moving aft. As the trailing blade first the following blade behind the release blade pushes the release blade root, its remaining fragments are propelled upward such that the heavy pieces slam against the aft-end of the fan case in a place, which is more susceptible to the release blade root impact. The maximum plastic strain in the casing material develops during this phase of the blade impact against the case. Figure 4.8 shows one such snapshot of transient deformation in the case for the sample problem being considered in this book. In the following sections, we discuss the results of the analytical model and compare them with the rig test data. The validated rig model helped us in developing the modeling guidelines to be used in generating the full-engine model, and the details of which are described in this section.

FIGURE 4.8 Fringe pattern showing the plastic bulge formed in casing due to the impact of the release blade. (Reprinted with permission from Reference [2].)

4.5 LOADING AND DAMAGE TO CONTAINMENT STRUCTURE

The impact force of the release blade on the case deforms the containment structure making a visible plastic bulge in the metal fan case. In the near vicinity of the initial impact, the tip of the blade gouges the metal in the casing. However, the entire blade and its liberated fragments are fully contained without punching any through hole in the case. The impact force of the release blade is also large enough to remove the inner surface abradable coating material from the trench filler system locally in the vicinity of initial impact. The magnitude of the dynamic load on the containment structure is verified by comparing the strain gauge data collected during the test and plotting against the results from the LS-DYNA® engine model. Figure 4.9 illustrates one such typical comparison of the axial strain time history between the rig test and the analytical model for the strain in the axial direction of the casing.

4.6 RELEASE BLADE TRAJECTORY AND FRAGMENTATION

To fully contain all the debris and fragments of the release blade as well as any trailing blade pieces, it is necessary to track every liberated piece of the blade. The LS-DYNA® model enables the analyst to follow the release blade plastic deformation and fragmentation within the casing and allows us to track the trajectory of debris

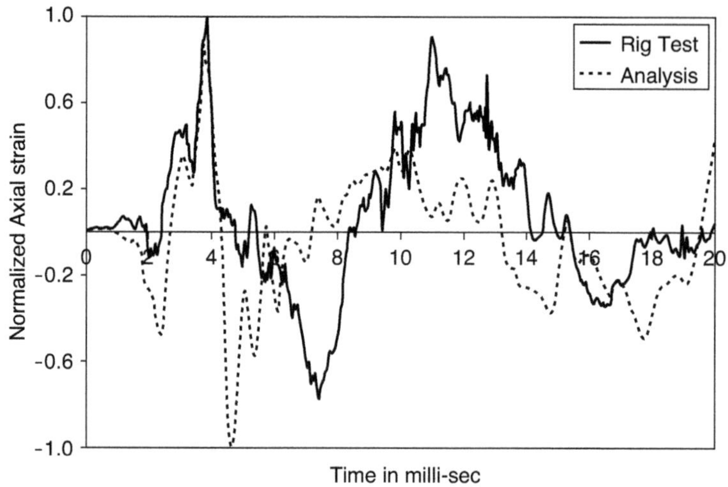

FIGURE 4.9 Time history of typical containment axial strain after impact analysis versus test. (Reprinted with permission from Reference [2].)

and fragments generated during that process. Figure 4.1 illustrates the motion of the major fragments of the release blade as it rotates inside the casing.

The analysis shows that the release blade breaks into three major fragments with two large pieces from the top and the root part of the blade. As far as the trajectories of the fragments are concerned, the LS-DYNA® analysis highlights that the bottom part containing the platform gets nested in the containment structure, which is consistent with the previous rig test observations. At the end of the rig test, we also collect all the debris and the release blade fragments and we compare them with the LS-DYNA® predicted shapes of the fragments. The LS-DYNA® model predicts the shape and size of the major fragments extremely well (see Figure 4.3).

4.6.1 Release Blade Energy Dissipation

Using the LS-DYNA® model, we also track the total energy of the release blade, which enables us to analytically determine the amount of the release blade kinetic energy consumed in various parts of the engine component after the blade loss as well as the rate at which the energy is dissipated through the system. Figure 4.10 shows the time-history plots of the release blade and in the containment structure and the following blade first trail blade on the fan rotor spool. The plot (see Figure 4.10) clearly shows that the release blade gains energy in the beginning as it is propelled radially outward by the trailing blade. The plot shown in Figure 4.11 highlights an interesting observation that the major part of the release blade kinetic energy is consumed in deforming the first trail blade and not by the containment structure. These analytical results allow us to improve the design of the airfoil, as any potential separation of the following blade would result in a significant increase in the unbalance load on the rotor.

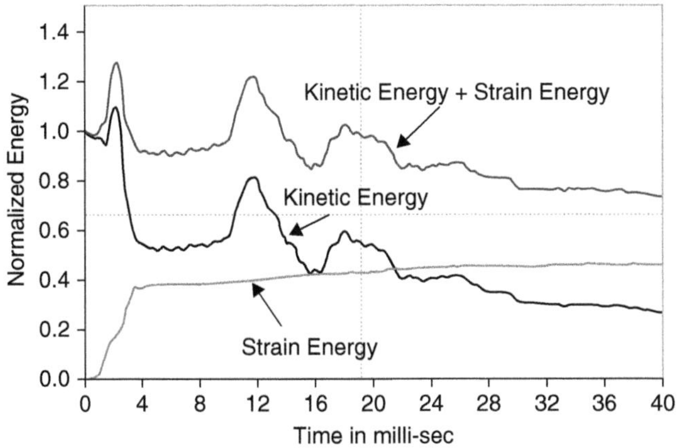

FIGURE 4.10 Time history of first following fan blade kinetic energy and strain energy. Energy values normalized with respect to initial KE of the release blade. (Reprinted with permission from Reference [2].)

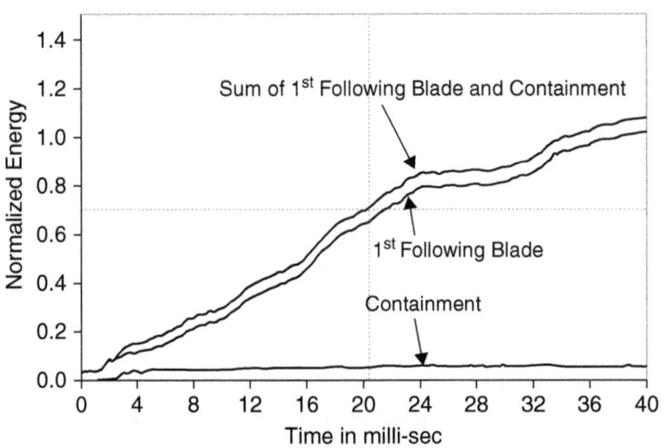

FIGURE 4.11 Time history of first following fan blade and containment strain energy. Energy values normalized with respect to initial KE of the release blade. (Reprinted with permission from Reference [2].)

4.7 TRANSIENT DYNAMIC RESPONSE OF ROTOR AFTER BLADE LOSS

As soon as a fan blade is released in a running engine, the LP rotor is subjected to an impulsive force at the fan disk acting radially outward on the heavy side of the rotor due to the introduction of a sudden imbalance in the dynamical system.

4.7.1 Bearing Loads

After the blade is released, as the fan disk rotor tends to move off-center from the engine axis, the load starts building up on the bearing (called Bearing No. 1 in

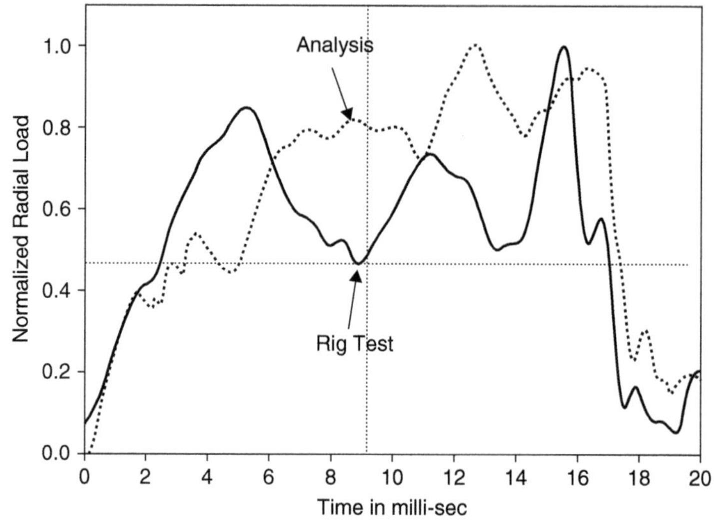

FIGURE 4.12 Analysis vs. test comparison of time history of load in Bearing No. 1 support cone. (Reprinted with permission from Reference [2].)

Figure 4.12) closest to the overhung part of the fan rotor. The Bearing No. 1 support cone is designed in such a way that at some critical load, the bearing support cone collapses and the load path changes such that it is transferred to the second nearby bearing. Figure 4.12 shows the time-history plot of the typical radial load up to 20 m-s on the bearing # 1 support cone as it as compared to the measured test versus predicted by the LS-DYNA® simulation.

4.7.2 Fan Rotor Orbit

As the rotor goes off balance and its center of rotation starts moving the fan disk center starts orbiting in a helical trajectory. In the rotor-dynamic analysis, the focus is on the dynamic loads developing at various bearing locations as well as the rub loads being generated between the rotating blade tips and the inner surface of the casing [see Reference 13].

Figure 4.13 shows the analytically computed orbit plot of the fan rotor shaft for 40 m-s from the time of the blade release at two different axial locations, viz., the center of the fan disk and the Bearing No. 1 housing. The respective movement has been compared with the high-speed movie taken at the front of the engine and the analytical results are in very good agreement with the test observation.

4.7.3 Dynamic Loads due to Tips of Fan Blades Rubbing Against the Case

After the fan disk orbit has become so large that the available running blade tip clearances with respect to the containment casing have been consumed, the blade tips start rubbing hard against the outer casing. Once it happens during the fast-moving dynamic event after the blade loss, they generate a new load path for the eccentric shaft and the load sharing starts. This results in a very complex nature of dynamic

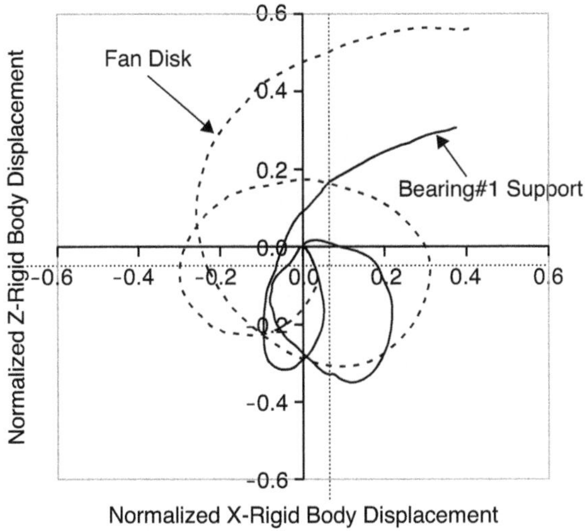

FIGURE 4.13 Fan-LPT rotor orbit at two different axial locations of rotor: (a) fan-disk center; (b) Bearing No. 1 support. (Reprinted with permission from Reference [2].)

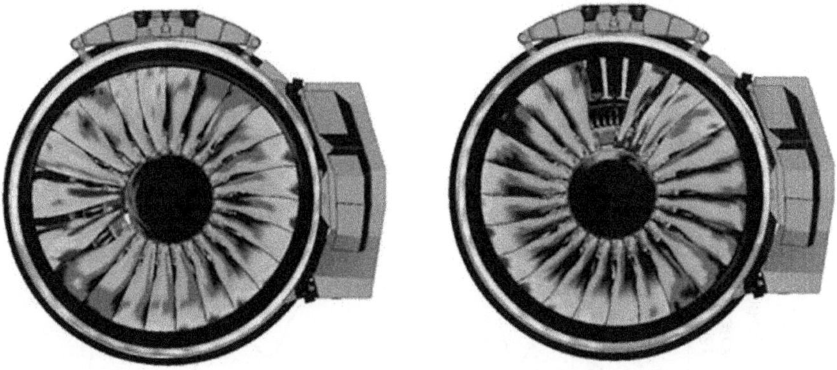

FIGURE 4.14 Fan blades rubbing differently at two different times after FBO showing the nonlinear effect of tip rub. (Reprinted with permission from Reference [2].)

loading of the blade tip. Figure 4.14 highlights the highly complex dynamic event and changing load path by showing the changing fringe pattern of von Mises equivalent stress distribution at and near the tip of the blades.

4.7.4 Dynamic Response of Engine Accessories

As mentioned earlier, the actual engine has various accessories hanging from the containment structure. The LS-DYNA® model of the engine presented here includes accessories such as the AGB, FADEC, FMU, and oil tank, to name a few. In the majority of the cases where the dynamic loads are concerned, the interest is in determining

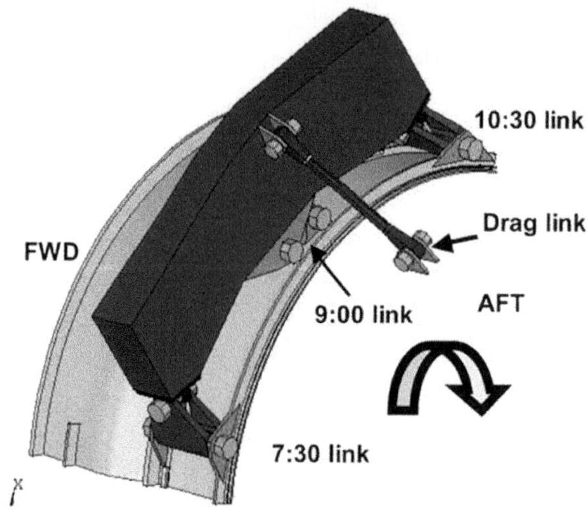

FIGURE 4.15 LS-DYNA® model of auxiliary gearbox and its linkages. (Reprinted with permission from Reference [2].)

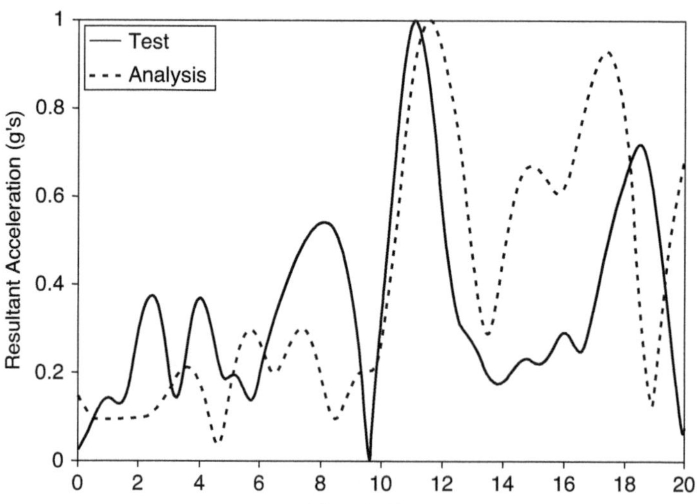

FIGURE 4.16 Normalized acceleration time history of auxiliary gearbox after an FBO event on engine. (Reprinted with permission from Reference [2].)

the acceleration in terms of g/s of the component or in the bracket holding it might be subjected to. Sometimes, these components also have rubber isolators to absorb the shock. As a typical example, we have included the AGB (see Figure 4.15) and its analytically predicted acceleration, which has also been compared with the rig test data. Figure 4.16 illustrates the LS-DYNA® predicted values of the normalized acceleration with respect to the corresponding measured test data.

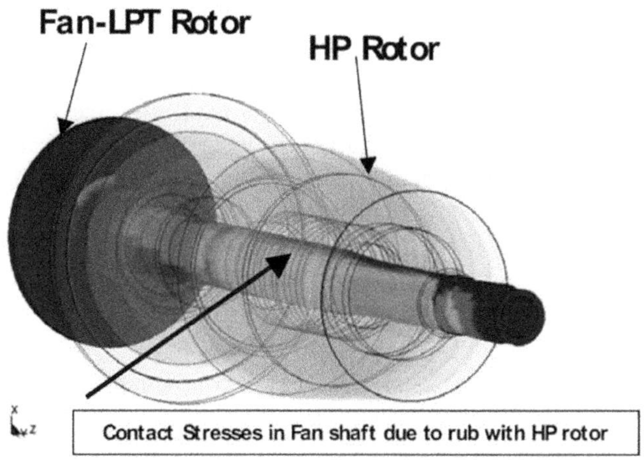

FIGURE 4.17 Effect of rubbing action highlighted by contact stresses being generated in shaft in rub location. (Reprinted with permission from Reference [2].)

4.8 RUBBING ACTION OF TWO CONCENTRIC ROTORS

It should be noted that the concentric rotors, viz., the LP rotor and the HP rotor, have a tendency to rub against each other. If the two rotors deform in such a way that the radial clearance between the two is consumed by the relative displacement of the two shafts, then the two will come in contact and will start rubbing. Figure 4.17 illustrates the effect of rub by developing contact stresses on the side of the fan shaft that comes into contact with the outer shaft rotating at about three times the rotational speed of the inner shaft. These contact stresses can create intense heat in the rub location, and at the end of the FBO event it is usually evident by the heat-generated discoloration of the metal.

4.9 SUMMARY OF BLADE CONTAINMENT AND WINDMILLING INTERACTION RESPONSE

In this chapter, we have covered the full detailed LS-DYNA® modeling methodology and the analytical results of full-engine FBO event simulations. The turbofan engine model included both the HP and LP rotors to produce an accurate physics-based model and to successfully predict the dynamic behavior of the engine in the case of an FBO event. It has been demonstrated that the dynamic loads predicted by the analytical model are in excellent agreement with the test data. It has been very well known in the aerospace industry that the cost of running an actual FBO test is an extremely expensive task, and as such only a very limited amount of test data are available to compare with the analytical results.

It should be noted that in the current modeling approach for rotordynamic analysis, it may be adequate to model all the fan blades with a relatively coarse mesh representation than what was described previously under a three-blade modeling procedure for containment analysis. However, the analytical simulation can achieve a much better accuracy from a single unified LS-DYNA® model by merging the two models together such that one replaces three coarse-mesh fan blades with the respective finely meshed blades in the fully bladed rotor model. It is recommended that the user should consider including some part of the wing of the aircraft and the pylon as well for better fidelity of the dynamic results, which in the latest versions of the code can also be represented as a superelement. More details of rotordynamic analysis and modeling issues will be discussed in the next chapter.

REFERENCES

1. S. K. Sinha and S. Dorbala (2008) "Dynamic Loads in the Fan Containment Structure of a Turbofan Engine", *ASCE 11th International Earth & Space Conference: Engineering, Science, Construction, and Operations in Challenging Environments 2008*, Long Beach, CA, March 3–5, 2008. https://ascelibrary.org/doi/10.1061/40988%28323%2977
2. S. K. Sinha and S. Dorbala (2009) "Dynamic Loads in the Fan Containment Structure of a Turbofan Engine", *ASCE Journal of Aerospace Engineering*, Vol. 22, No. 3, pp. 260–269. https://doi.org/10.1061/(ASCE)0893-1321(2009)22:3(260)
3. D. Arzina (2011) "Vibration Analysis of Compressor Blade Tip-Rubbing", Cranfield University, Cranfield, UK, School of Engineering, Thesis. https://dspace.lib.cranfield.ac.uk/handle/1826/7109
4. P. A. Kalinowski (2012) "Ein Beitrag zur Simulation und Messung von Blade Off Ereignissen", Doctoral Dissertation, Berlin, August 2012, (147 pages).
5. K. Sengoz, S. Kan and A. Eskandarian (2015) "Development of a Generic Gas Turbine Engine Fan Blade-out Full-Fan Rig Model", U.S. Department of Transportation, Federal Aviation Administration, Technical Report No. DOT/FAA/TC-14/43, August 2015, New England Regional Office—Aircraft Certification Service Engine & Propeller Directorate, Burlington, MA. https://www.tc.faa.gov/its/worldpac/techrpt/tc14-43.pdf
6. A. J. V. Klok (2016) "Experimental Impact Testing and Analysis of Composite Fan Cases", Ph.D. Thesis, Michigan State University, East Lansing. https://d.lib.msu.edu/etd/4319/datastream/OBJ/download/Experimental_Impact_Testing_and_Analysis_of_Composite_Fan_Cases.pdf
7. B. Ridley (2018) "Impact Studies of Containment Ring during Blade Shed in Turbofan Engines", Ph.D. Dissertation, University of Toronto, Toronto, Canada. https://tspace.library.utoronto.ca/handle/1807/91729
8. Б. С. Блинник, В. Ю. Мясников and И. И. Иванов (2018), "Исследование динамики системы ротор—корпус авиационного газотурбинного двигателя при обрыве лопатки вентилятора" [Investigation of the Dynamics of the Rotor-Housing System Aircraft Gas Turbine Engine When the Fan Blade Breaks"] (In Russian), *Engineering Journal: Science and Innovation* (In Russian), Vol. 62, No. 251, p. 534.1. https://doi.org/10.18698/2308-6033-2018-8-1797
9. G. Luo, C. Ma, W. Chen, L. Liu and Z. Zhao (2020) "Transient Analysis and Safety Assessment of Turbofan Engine Structures during Bird Ingestion", *International Journal of Aerospace Engineering*, Vol. 2020, No. 1, p. 7404587. https://doi.org/10.1155/2020/7404587
10. P. A. Roy and S. A. Meguid (2021) "Containment and Arrest of Blade Shedding in Gas Turbine Engines Using Novel Dual-Ring Design", *ASME Journal of Engineering for Gas Turbines and Power*, Vol.143, No. 7, p. 071015. https://doi.org/10.1115/1.4049489

11. M. Heidari, D.L. Carlson, S. Sinha, R. Sadegh, C. Heydari, H. Bayoumi and J. Son (2008) "An efficient multi-disciplinary simulation of Engine Fan-Blade out Event using MD Nastran", 49th AIAA/ASME/ASCE/AHS/ASC Structures, Structural Dynamics, and Materials Conference, Paper No. AIAA 2008-2333, 7–10 April 2008, Schaumburg, IL. http://dx.doi.org/10.2514/6.2008-2333
12. Y. N. Shmotin, D. V. Gabov, A. A. Ryabov, S. S. Kukanov and V. N. Rechkin (2006). "Numerical Analysis of Aircraft Engine Fan Blade-out", *Proceedings of the 42nd AIAA/ASME/SAE/ASEE Joint Propulsion Conference and Exhibit*, Sacramento, CA, Paper No. AIAA 2066–4620, American Institute of Aeronautics and Astronautics, Reston, VA. https://doi.org/10.2514/6.2006-4620
13. J. B. Husband (2007) " Developing an efficient FEM Structural Simulation of a Fan Blade Off test in a Turbofan jet engine", Ph.D. Thesis, October 2007, University of Saskatchewan, Saskatoon, Canada. http://www.scribd.com/doc/29847199/Jason-Husband-PhD-Thesis

5 Rotordynamic Response Under Shock Loading

5.1 INTRODUCTION OF IMPACT EVENTS IN ROTORDYNAMIC APPLICATIONS

In this chapter, we focus our attention on the transient dynamic response of the turbofan rotor under impact loading of shaft from the rotodynamic perspective. Depending upon the magnitude of impact load and also its duration of impulsive loading, the rotor can experience some significant transient vibratory response due to bird strike to severe damage to the engine due to a blade loss from the turbofan rotor. In the blade-loss event, it can also cause large oscillations of the entire engine during the windmilling action, which may continue up to 3 hours until the aircraft can safely land at the nearby airport.

Usually, the rotor dynamic analysis is concerned with items like minor unbalance response, whirl frequency, forward, and backward whirl of the rotor shaft, bearing loads, blade-tip rub, etc [see References 1–6]. Squeeze film dampers also play a very important role in turboengine rotors. In almost all turboengine applications, the rolling element bearings are invariably used between the rotor shaft and stator components.

5.1.1 Modeling of Rolling Element Bearings

Modeling of a ball or roller bearing in a rotordynamic model at multiple support location is an extremely challenging task. In the current modeling of the fan rotor, the following mathematical representation is used to simulate the dynamic characteristics of the three-bearing support system. In the current formulation, the axial stiffness of the ball bearing is neglected. However, the rotations of the inner race about the engine centerline are permitted. Axial stiffness of the ball bearing is neglected. It must be pointed out that the ball bearing carries axial and lateral loads, unlike the lateral only for roller bearings.

A typical ith rolling element bearing located at $Z = Z_i$ is mathematically represented by its dynamic stiffness matrix $[S]_b$ and damping matrix $[D]_b$ as

$$[S]_b = \begin{bmatrix} S^b_{XX} & S^b_{XY} & & \\ S^b_{YX} & S^b_{YY} & & \\ & & B^b_{XX} & \\ & & & B^b_{YY} \end{bmatrix}_i \text{ and, } [D]_b = \begin{bmatrix} D^b_{XX} & D^b_{XY} \\ D^b_{YX} & D^b_{YY} \end{bmatrix}_i \tag{5.1}$$

DOI: 10.1201/9781003749752-5

Here, $\left[S_{XX}^b\right]_i$, $\left[S_{XY}^b\right]_i$, $\left[S_{YX}^b\right]_i$, and $\left[S_{YY}^b\right]_i$ have the dimensions of linear stiffness as force per unit length, whereas $\left[B_{XX}^b\right]_i$ and $\left[B_{YY}^b\right]_i$ have the dimensions of bending stiffness as moment/unit radian. It should be noted that where ball bearings can take axial load with $\left[B_{XX}^b\right]_i$ and $\left[B_{YY}^b\right]_i$ terms being nonzero, whereas the roller bearings cannot take the axial load. Hence, in case of roller bearings, bending stiffness terms like $\left[B_{XX}^b\right]_i$ and $\left[B_{YY}^b\right]_i$ are invariably zero. Thus, the horizontal and vertical components of reaction forces, viz., $(F_X)_b$ and $(F_Y)_b$ at a typical bearing support location of $Z = Z_i$ due to stiffness terms can be readily computed as

$$(F_X)_b = \left[-S_{XX}^b\, u(Z_i,t) - B_{XX}^b\, u_{,ZZ}(Z_i,t) - S_{XY}^b\, v(Z_i,t)\right] \text{ at } Z = Z_i \tag{5.2}$$

$$(F_Y)_b = \left[-S_{YX}^b\, u(Z_i,t) - B_{YY}^b\, v_{,ZZ}(Z_i,t) - S_{YY}^b\, v(Z_i,t)\right] \text{at } Z = Z_i \tag{5.3}$$

Thus, the radial bearing load at a typical ith rolling element bearing located at $Z = Z_i$ will be

$$(F_{\text{Radial}})_i = \sqrt{(F_X)_b^{\,2} + (F_Y)_b^{\,2}} \tag{5.4}$$

Due to shaft being treated as a beam at the engine centerline with displacement (u, v), the movement of the inner race of the rolling element bearing must account for the slope with respect to the engine axis and the outer radius r_o of the shaft. The fundamental assumption of the beam bending theory applies in determining the shaft deformation as well. Plane sections of the shaft cross-section normal to the axis of the shaft (global Z-axis) remain plane as the shaft being treated as a beam is bent. Thus, for example, the horizontal movement of the outer surface of the shaft in the X-direction at Bearing No. 3 will be given as

$$\left[u + \left[\frac{\partial u}{\partial Z}\right] r_o\right]_{Z=Z_3} \tag{5.5}$$

Similarly, the vertical movement of the outer surface of the shaft in the Y-direction at Bearing No. 3 will be given as

$$\left[v + \left[\frac{\partial v}{\partial Z}\right] r_o\right]_{Z=Z_3} \tag{5.6}$$

During the impact event for typical bearing load shown in Figure 5.1, the differential movement of the outer and inner races of the rolling element bearing will have severe ovalization. Ovalization of the inner race results in local loading of only one sector of the bearing.

The loads of primary interest are at the ball bearing and the loads in the roller bearings are less significant. This is because rollers have greater lateral capability than the ball bearing. The peak transient loads on the bearing are directly affected by the nonlinearity in the dynamical system. The nonlinearity in the current set of equations is due to the following contributing factors:

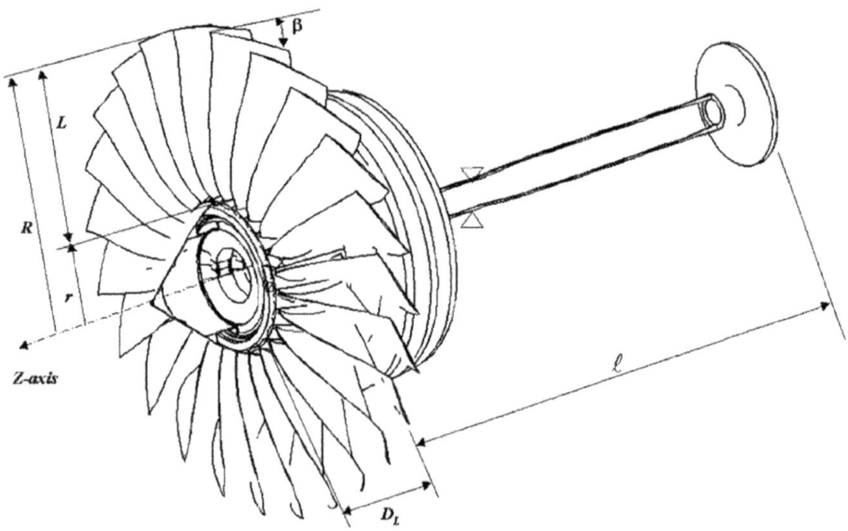

FIGURE 5.1 A typical fan rotor in a jet engine with hollow shaft: outside radius r_o; inside radius r_i. (Reprinted with permission from Reference [3].)

a. Contact-impact loading between the release blade and the first trail blade is always nonlinear. Contact loading is inherently nonlinear as well.
b. The fan blade airfoil has a pretwist in the span direction by design. The rotating condition generates centrifugal loading, which in combination with the impact load starts untwisting the blade in a centrifugal force field.
c. While the centrifugal loading creates stress-stiffening effect in the blade, the untwisting of a pretwisted blade causes stress softening. The combination of stress stiffening due to centrifugal loading and stress softening due to geometrical shape of the blade is highly nonlinear.
d. Large deformation of blades during impact. This introduces geometrical nonlinearity in the dynamical system.
e. The typical magnitude of impact load is in the range of 100–200 kN. This load eventually gets to the rotating shaft, and the shaft deforms in the lateral direction due to effective imbalance in the dynamical system. Furthermore, the effective stiffness of bearing due to the Hertzian contact between the rolling element and the outer and inner races produces another set of nonlinearities in the dynamical set of equations.
f. In a fixed frame of reference used in the current formulation (see Eqs. 5.27 and 5.28), the effective internal damping $\zeta_{\text{Effective}}$ is a function of the rotor speed. Here, in a rotor spinning with angular velocity Ω with its whirling frequency being ω_n, the effective internal or hysteretic damping $\zeta_{\text{Effective}}$ is given by

$$\zeta_{\text{Effective}} = \zeta\left(1 - \frac{\Omega}{\omega_n}\right) \tag{5.7}$$

For example, it can be seen that for subsynchronous operation of the rotating engine ($\Omega > \omega_n$), the cross-stiffness forces generated by the internal damping $\zeta_{\text{Effective}}$ are negative, which are destabilizing for forward whirl.

In short, the multiple complex set of dynamical parameters contributes to this class of problems to make as highly nonlinear.

5.1.2 Effect of Damping in a Rotating system

There are two types of damping present in any rotordynamic mechanical systems. Most of the damping can be provided by the damping in the bearing support system. Rolling element bearing in itself does not have much damping effect. Hence, squeeze film damper is the most common type of damper used in the aerospace applications. The shaft material viscoelastic damping is another form of damping; however, the material internal frictional damping introduces "whirling instability" in the fan rotor shaft. The effect of damping must be included in the mathematical formulation of rotordynamic applications for any time-domain solution with shock type of dynamic loading of the rotor shaft. In the mathematical derivation, the effect of material hysteretic damping is considered by Kimball's quadratic damping by expressing the elastic parameters the Young modulus E^* and the Shear modulus G^* as a complex number as follows:

$$E^* = E(1+i\delta) \tag{5.8}$$

$$G^* = G(1+i\delta) \tag{5.9}$$

where $i = \sqrt{-1}$ and δ is the internal damping factor of the structure. δ is also known as "Logarithmic decrement factor".

In a Jeffcott rotor, the critical viscous damping coefficient parameter is expressed as follows:

$$H = \frac{1}{2}\pi^2\ell^2\left[EI\rho A\right]^{1/2} \tag{5.10}$$

5.2 MATHEMATICAL FORMULATION OF IMPULSIVE LOADING IN ROTORDYNAMICS

In this section, we focus on mathematical forms of various types of forcing functions, which a spinning turbofan shaft rotor may experience. In any typical rotordynamic problem, the largest external forcing function is primarily caused by some type of imbalance introduced in the rotor system, which generates a rotating load vector on the spinning shaft due to the centrifugal action of the unbalanced mass. In a typical aerospace application, the turbofan engine rotor shaft encounters a sudden impulsive loading usually due to bird-strike or unexpected blade-loss event. These two types of dynamic events called bird-strike and fan blade out (FBO) have been discussed previously in Chapters 3 and 4, respectively. The transient dynamic response could also involve a severe plastic deformation of the impacted blade to a severe vibratory oscillation due to the repeated shock loading at the blade tip. The other type of intermittent pulse loading may also occur due to repeated blade tip rubbing against the inner surface of the containment ring.

The mathematical relationships for generating different forcing functions $F(t)$ acting on a rotating mechanical component to represent a periodic pulse at a rotating contact surface can have various pulse profiles. For example, with "p" as the pulse sequence number (i.e., $p = 1, 2, 3, \ldots$), the load application time for each pulse is always t_c. In the case of a single pulse, $F(t)$ is set equal to zero for all $p > 1$. The following are the mathematical relationships for generating a typical periodic pulse of magnitude $F_{\max}$ with an excitation frequency of $f_p - Hz$. In the special case of one pulse per revolution of the rotor spinning at an angular velocity of Ω, we have $f_p = \Omega/2\pi$.

a. *Sine-function pulse*: (based upon the excitation frequency $f_p - Hz$)

$$F(t) = 0 \quad \text{for} \quad \left[\left(\frac{(p-1)}{f_p} + t_c\right) \le t \le \frac{p}{f_p}\right] \tag{5.11}$$

and

$$F(t) = -F_{\max} \sin\left(\frac{\pi}{t_c}\left(t - \frac{(p-1)}{f_p}\right)\right) \quad \text{for} \quad \left[\left(\frac{(p-1)}{f_p}\right) \le t \le \left(\frac{(p-1)}{f_p} + t_c\right)\right] \tag{5.12}$$

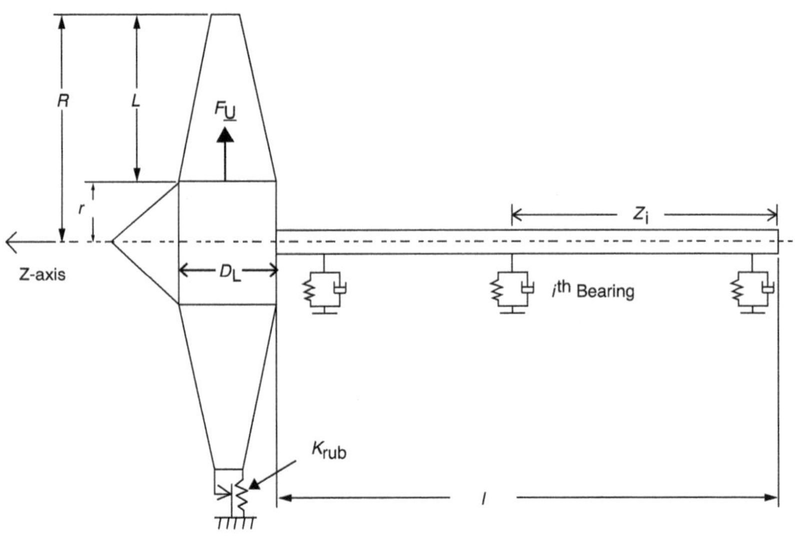

FIGURE 5.2 Fan rotor shaft supported on multiple bearings: number of blades = $\boldsymbol{N}$, mass density $= (\rho)_s$, cross-sectional area $= (\boldsymbol{A})_s = \pi\left(\boldsymbol{r}_o^2 - \boldsymbol{r}_i^2\right)$, flexural rigidity $= (\boldsymbol{EI})_s$, Coulomb friction force: $\boldsymbol{F_X} = \mu \boldsymbol{K_{rub}} \boldsymbol{v}(\ell, \boldsymbol{t})$, $\boldsymbol{F_Y} = -\mu \boldsymbol{K_{rub}} \boldsymbol{u}(\ell, \boldsymbol{t})$. For a typical $\boldsymbol{i}$ th bearing located at $\boldsymbol{Z} = \boldsymbol{Z_i}$, stiffness $= \left([S]_b\right)_i$ and damping $= \left([D]_b\right)_i$. In case of air gap between the blade tip and the outer case without any radial incursion in abradable, $\boldsymbol{K_{rub}} = 0$. (Reprinted with permission from Reference [3].)

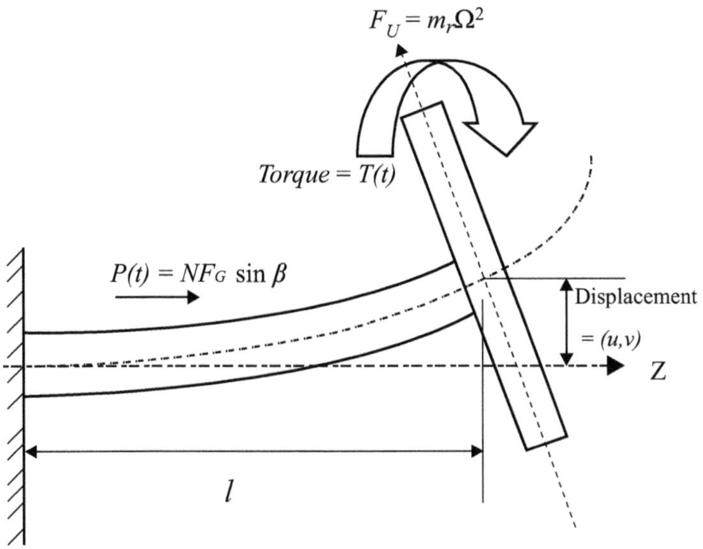

FIGURE 5.3 First flex deformation of the cantilever shaft with rigid disk: flexural rigidity of the shaft $(\boldsymbol{EI})$, mass/unit length of the shaft $(\boldsymbol{\rho A})$, mass of the disk $\boldsymbol{M_r}$, diametral mass moment of inertia of the disk $\boldsymbol{J_D} = (\boldsymbol{M_D r^2})$, $\boldsymbol{r}$ = outer radius of the disk. (Reprinted with permission from Reference [3].)

b. *Cosine-function pulse*: (based upon the excitation frequency f_p – Hz)

$$F(t) = 0 \quad \text{for} \quad \left[\left(\frac{(p-1)}{f_p} + t_c\right) \le t \le \frac{p}{f_p}\right] \tag{5.13}$$

and

$$F(t) = -\frac{F_{\max}}{2} + \frac{F_{\max}}{2}\cos\left(\frac{2\pi}{t_c}\left(t - \frac{(p-1)}{f_p}\right)\right) \text{for} \left[\left(\frac{(p-1)}{f_p}\right) \le t \le \left(\frac{(p-1)}{f_p} + t_c\right)\right] \tag{5.14}$$

c. *Symmetric triangular pulse*: (based upon the excitation frequency f_p – Hz)

$$F(t) = 0 \quad \text{for} \quad \left[\left(\frac{(p-1)}{f_p} + t_c\right) \le t \le \frac{p}{f_p}\right] \tag{5.15}$$

$$F(t) = -F_{\max}\left(\frac{2}{t_c}\left(t - \frac{(p-1)}{f_p}\right)\right) \quad \text{for} \quad \left[\left(\frac{(p-1)}{f_p}\right) \le t \le \left(\frac{(p-1)}{f_p} + \frac{t_c}{2}\right)\right] \tag{5.16}$$

and

$$F(t) = -F_{\max}\left(\frac{2}{t_c}\left(t_c - t - \frac{(p-1)}{f_p}\right)\right) \quad \text{for} \quad \left[\left(\frac{(p-1)}{f_p} + \frac{t_c}{2}\right) \le t \le \left(\frac{(p-1)}{f_p} + t_c\right)\right] \tag{5.17}$$

d. *Saw-tooth pulse*: (based upon the excitation frequency $f_p - \text{Hz}$):

$$F(t) = 0 \quad \text{for} \quad \left[\left(\frac{(p-1)}{f_p} + t_c\right) \le t \le \frac{p}{f_p}\right] \tag{5.18}$$

and

$$F(t) = -F_{\max}\left(\frac{1}{t_c}\left(t - \frac{(p-1)}{f_p}\right)\right) \quad \text{for} \quad \left[\left(\frac{(p-1)}{f_p}\right) \le t \le \left(\frac{(p-1)}{f_p} + t_c\right)\right] \tag{5.19}$$

e. *Pure Impulse function with varying magnitude of each pulse*: (based upon the excitation frequency $f_p - \text{Hz}$):

$$F(t) = -\sum_p (F_{\max})_p\, \delta\left(t - \frac{2\pi(p-1)}{\Omega}\right) \quad \text{for} \quad \left[0 \le t \le \frac{2\pi}{\Omega}\right] \tag{5.20}$$

f. *Rectangular pulse with step function*: (based upon the excitation frequency $f_p - \text{Hz}$):

$$F(t) = 0 \quad \text{for} \quad \left[\left(\frac{(p-1)}{f_p} + t_c\right) \le t \le \frac{p}{f_p}\right] \tag{5.21}$$

and

$$F(t) = -F_{\max} \quad \text{for} \quad \left[\left(\frac{(p-1)}{f_p}\right) \le t \le \left(\frac{(p-1)}{f_p} + t_c\right)\right] \tag{5.22}$$

The time-history plot of $F(t)$ for a special case of $f_p = \Omega/2\pi$ is shown in Figure 7.6. This pulse profile can be expanded in the form of a cosine series as

$$F(t) = c_0 + \sum_{k=1}^{\infty} c_k \cos(k\Omega t) \tag{5.23}$$

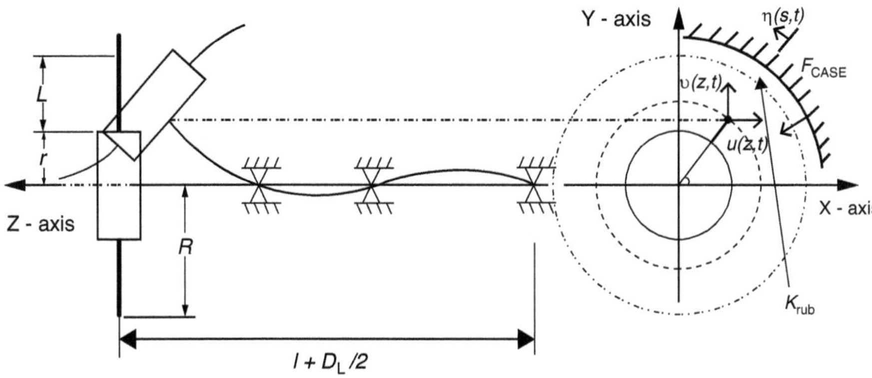

FIGURE 5.4 A typical dynamic mode shape of a bladed rotor with three-bearing supports. (Reprinted with permission from Reference [3].)

Where the Fourier coefficients c_0 and c_k are

$$c_k = \frac{\int_0^{2\pi/\Omega} F(t)\cos(k\Omega t)\,dt}{\int_0^{2\pi/\Omega} \cos^2(k\Omega t)\,dt} = \frac{\int_0^{2\pi/\Omega} F_{max}\cos(k\Omega t)\,dt}{\int_0^{2\pi/\Omega} \cos^2(k\Omega t)\,dt} = \frac{\sin(k\Omega t_c)}{k\pi}F_{max} \tag{5.24}$$

and

$$c_0 = -\left(\frac{\Omega t_c}{2\pi}\right)F_{max} \tag{5.25}$$

Thus, in a rectangular shape pulse-profile of magnitude Fmax and contact load-time t_c with one pulse per revolution, we can write

$$F(t) = -\,F_{max}\left[\left(\frac{\Omega t_c}{2\pi}\right) + \sum_{k=1}^{\infty}\frac{\sin(k\Omega t_c)}{k\pi}\cos(k\Omega t)\right] \tag{5.26}$$

The above series representation of $F(t)$ enables us to solve the rotor dynamic stability problem with periodic pulse impact in the frequency domain using the procedure of harmonic balance. It should be noted that each term of the series ($k = 1, 2, 3, 4, \ldots$, etc.) will yield a different set of solution for the stability zones; and the total solution will be in the form of a sum of individual sets obtained for each value of "k".

5.2.1 Governing Rotordynamic Equations of a Turbofan Rotor Shaft

Let us consider the dynamic response of an overhung rotor supported by a set of N_b number of bearings at multiple axial locations and deformable blades rubbing against a case at the other end (see Figure 5.1). The rotor shaft may be under a periodic sinusoidal axial force of magnitude P_t superimposed on a constant static axial load

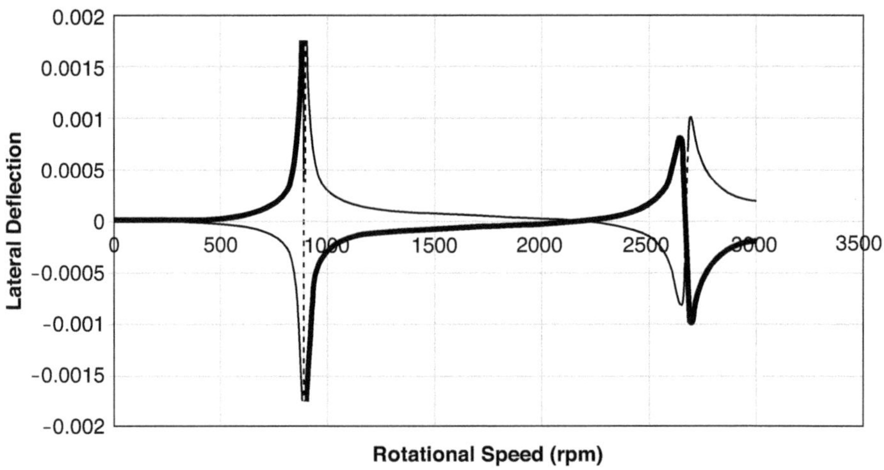

FIGURE 5.5 Harmonic response $(\boldsymbol{u}, \boldsymbol{v})$ of the rotor at Bearing No. 3 support for a nominal unbalance (—— *u*; —— *v*). (Reprinted with permission from Reference [3].)

P_0 and torque T_0 on the system. In rotating machinery, such pulsating axial forces are usually caused by misalignment, aerodynamic forces, and inlet distortion. The rotor system considered in this chapter is composed of a flexible hollow shaft, a rigid disk, and a full set of elastically deformable blades, a sector of which may be rubbing against the inside surface of an outer cylindrical containment structure. In this discussion, we have assumed that the containment structure behaves like a rigid wall and the spinning shaft disk with bladed rotor as an impactor hitting its inside concave surface at a shallow oblique angle. The dynamics of rotating blade tips hitting a rigid cylindrical wall was considered in full detail in author's previous investigations. The inner surface of the fan case has a lining of a relatively soft material like honeycomb, foam, or an abradable coating against which the blade tip is allowed to rub with stiffness in the radial direction represented as K_{Rub}. The tip-rub forces can deform the blade in all the three directions of the fan rotor, viz., tangential, axial, and radial. The radial force at the blade tip F_{Rub} calculated in this fashion would also result in a tangential load of magnitude μF_{Rub} acting in the direction opposite to the direction of shaft rotation. The effective coefficient of friction μ at the blade-tip/abradable contact interface surface is usually established in an empirical way by measuring the instantaneous speed drop of the rotor shaft during the rub event.

It should be recognized that in determining the dynamic characteristics of an overhung rotor shaft, the gyroscopic moments play a significant role. This effect is further magnified in an asymmetric rotor with a large-size fan blade missing from one of the disk sectors caused by a blade loss. One can use the energy method approach to derive the mathematical relationship to determine the gyroscopic effects in a bladed nonsymmetric fan rotor. As it can be seen that this effect not only makes the mass matrix time-dependent varying at twice the frequency of running speed rather gyroscopic matrix also becomes a sinusoidal function of time.

The Coulomb friction between the blade tip and the outer surface introduces skew-symmetric terms in the stiffness matrix for the rotor shaft which, depending upon the amount of effective damping present in the dynamical system, can initiate instability in the bladed-rotor shaft. In this derivation, the shaft is assumed to be supported at multiple axial locations z_i along the length of the shaft, which may be either ball-bearing or roller-bearing, with their respective stiffness $[S]_b$ and damping $[D]_b$ parameters. As a general case, these support bearing may not necessarily be symmetrical in the $X-Y$ plane and as such for any typical ith bearing located at $Z = z_i$ their effective stiffness and damping in the matrix form are written as shown earlier in Eq. (5.1).

To derive the governing equations of motion of the fan rotor, we start with a rectangular Cartesian coordinate (X, Y, Z) system attached to the engine structure with the centerline coinciding with the global Z-axis and its origin located at the back end of the engine (see Figure 5.1). Referring to Figure 5.1, it should be noted that although a typical fan shaft can have a varying inner radius as shown here schematically, in our current discussion we inherently assume a slightly simpler case where both outer and inner radii of the shaft r_o and r_i are constant for the entire length of the shaft $(0 \leq Z \leq \ell)$. Here, the time-dependent shaft deformation $\left(u(Z,t), v(Z,t)\right)$ at the engine centerline for a typical point is a function of its axial location (Z), measured from the aft-end of the engine. In our current formulation of typical beam type of differential equation for engine centerline, we use Dirac delta functions δ to capture the point loading and discrete masses, such as $\delta(Z - Z_i)$ for bearing locations and $\delta(Z - \ell)$ for disk mass-related dynamic parameters. Using this derivation methodology, even the boundary conditions become part of the differential equation.

The full 3-D analytical model developed here in the global (X, Y, Z) coordinate system will help in determining the dynamic load of the bearing support system, as the shaft deforms due to bending and torsion loading caused mainly by the blade loss during the FBO event. To account for the total mass of a fully bladed rotor, we assume that the fan disk mass located at $(Z = \ell)$ is M_D with the number of radial airfoil blades being N with each of its mass equal to M_b. The remainder of the fan disk and shaft are treated as an overhung hollow circular beam supported on a three-bearing system, typically used in almost all large commercial jet engines. The advantage of three-bearing configuration and its FEA modeling scheme for a two-spool rotor in aeroengine has been discussed elsewhere. It is worth mentioning that while overall rotodynamic problem is 3-D, the model captures the axial aspect of the impact in the fan model and carries the axial component to the ball bearing. The lateral components in the (X, Y) directions are resolved from the 3-D fan impact then carried on to the shaft to assess the shaft and bearing dynamic response. The corresponding rotor shaft equation in the horizontal global X-direction of the fixed inertial frame of reference is as follows. These equations use commonly used notions; however, for readers' convenience, at the end of this chapter, a list of notations is also provided.

$$(EI)_s u,_{ZZZZ} + 2\zeta H[\dot{u},_{ZZZZ} + \Omega v,_{ZZZZ}] - 2I_d\Omega[\dot{v},_{ZZ} - \Omega u,_{ZZ}] + \rho_s A_s \ddot{u} - I_d \ddot{u},_{ZZ} +$$

$$T(t)v,_{ZZZ} - P(t)u,_{ZZ} + \sum_{i=1}^{N_b}\left[S_{XX}^b u + S_{XY}^b v + B_{XX}^b u,_{ZZ} + D_{XX}^b \dot{u} + D_{XY}^b \dot{v}\right]\delta(Z - Z_i) +$$

$$\left[(M_D + N M_b)\ddot{u} - J_{D-YY}\ddot{u},_{ZZ} - (J_{D-XX} + J_{D-YY})\Omega[\dot{v},_{ZZ} - \Omega u,_{ZZ}]\right]\delta(Z - \ell)$$

$$+ K_{\text{rub}}\left[\left(u - R(u,_Z)^2\right) + \mu\left(v - R(v,_Z)^2\right)\right]\delta(Z - \ell) = q_X(Z,t)$$

$$= [F_U \cos(\theta_0 + \Omega t)]\,\delta(Z - \ell) \tag{5.27}$$

Similarly, the corresponding fan rotor shaft equation of motion in the vertical global Y-direction is as follows:

$$(EI)_s v,_{ZZZZ} + 2\zeta H[\dot{v},_{ZZZZ} - \Omega u,_{ZZZZ}] + 2I_d\Omega[\dot{u},_{ZZ} + \Omega v,_{ZZ}] + \rho_s A_s \ddot{v} - I_d \ddot{v},_{ZZ}$$

$$-T(t)u,_{ZZZ} - P(t)v,_{ZZ} + \sum_{i=1}^{N_b}\left[S_{YX}^b u + S_{YY}^b v + B_{YY}^b v,_{ZZ} + D_{YX}^b \dot{u} + D_{YY}^b \dot{v}\right]\delta(Z - Z_i)$$

$$+\left[(M_D + N M_b)\ddot{v} - J_{D-XX}\ddot{v},_{ZZ} + (J_{D-XX} + J_{D-YY})\Omega[\dot{u},_{ZZ} + \Omega v,_{ZZ}]\right]\delta(Z - \ell)$$

$$+ K_{\text{rub}}\left[\left(v - R(v,_Z)^2\right) - \mu\left(u - R(u,_Z)^2\right)\right]\delta(Z - \ell) = q_Y(Z,t)$$

$$= [F_U \sin(\theta_0 + \Omega t)]\,\delta(Z - \ell) \tag{5.28}$$

where I_d is the diametral mass moment of inertia of the shaft (per unit length). For the detailed list of mathematical symbols and other terminology symbols, the reader is referred to the Nomenclatures section at the end of this chapter. It should be noted that in a fully bladed fan rotor disk with the usual even number of inserted blades, one always satisfies the condition: $(J_{D-XX} = J_{D-YY})$. However, in a specific case of aftermath of FBO event due to the damage of the rotor disk caused by its missing blade, the transient dynamic response of the fan rotor shaft encounters the new diametral inertia terms about the two axes being asymmetrical in the $X-Y$ plane: $(J_{D-XX} \neq J_{D-YY})$. Additionally, in the post-FBO scenario, due to spin of the fan-rotor shaft about the global Z-axis, the effective diametral moment of inertia terms J_{D-XX} and J_{D-YY} of the fan disk at any instant becomes time dependent. For any particular instant of time "t" with angular velocity Ω, the change in diametral mass moment of inertia terms of the fan disk with one missing blade of mass M_b originally starting at time $t = 0$ from a fully bladed fan disk can be determined as follows:

$$[J_{D-XX}]_{\text{time}=t} = [J_{D-XX}]_{\text{time}=0} - M_b\left(r + \frac{L}{2}\right)^2 \sin^2(\Omega t) \tag{5.28a}$$

$$[J_{D-YY}]_{\text{time}=t} = [J_{D-YY}]_{\text{time}=0} - M_b\left(r+\frac{L}{2}\right)^2 \cos^2(\Omega t) \tag{5.28b}$$

$$\left[(M_D+(N-1)\,M_b)\right]_{\text{time}=t} = \left[(M_D+N\,M_b)\right]_{\text{time}=0} - M_b \tag{5.28c}$$

Furthermore, the presence of skew-symmetric stiffness terms due to Coulomb friction in Eqs. (5.27) and (5.28) highlights the destabilizing effect of the rub. To determine whether a particular rub is actually destabilizing to the rotor or if it can ride-through the rub action requires that the blade tip movement be computed very precisely, which forces us to include the nonlinear terms in the blade-tip movement calculation. The rub takes place, when the rotor deformation at the fan-disk location is large enough to consume the available tip clearance between the blade tips and the surrounding containment ring stator structure. This can happen in any 360° circumferential space around containment at the fan-disk location of $Z = \ell$. Due to bending of the shaft and the resulting rotation of the disk and blades, the term containing the Coulomb friction effect at the blade tip, namely, $R(u,_Z)^2$ and $R(v,_Z)^2$, are truly nonlinear and are included for an accurate determination of the rub condition. However, under small rotation assumption, they can be reduced into linear terms with a rotation factor "e" as shown below:

$$R(u,_Z)^2 = eRu,_Z \quad \text{and,} \quad R(v,_Z)^2 = eRv,_Z \tag{5.29}$$

Depending upon the fan blade-tip radius "R," the typical value of the rotation factor "e" usually varies anywhere from 0.01 to 0.1. In the linearization process under small rotation assumptions, for the direct rub stiffness terms, one can further simplify by approximating it as follows:

$$(u,_Z)^2 \approx 0 \quad \text{and,} \quad (v,_Z)^2 \approx 0 \tag{5.30}$$

The other important terms affecting the shaft dynamics modeling scheme for a typical rolling element bearing support system for the rotor shaft have been discussed in Section 5.1.1.

Figure 5.2 shows the transverse view of the same fully bladed turbofan rotor system. In this figure, the global origin ($Z = 0$) of the fixed frame of reference (X, Y, Z) is located at the aft-end of the fan shaft. Here, with the unbalanced mass M_U acting on the pressure side of the fan airfoil blade, the effective unbalance force F_U on the rotor shaft will be

$$F_U = M_U\Omega^2(r+s_{CG}) \tag{5.31}$$

Although the most common external load in any rotordynamic problem is the rotating load vector due to unbalance, there are also additional loads acting on the rotating system. During the normal operation of an aero-engine fan at a given speed, the gas

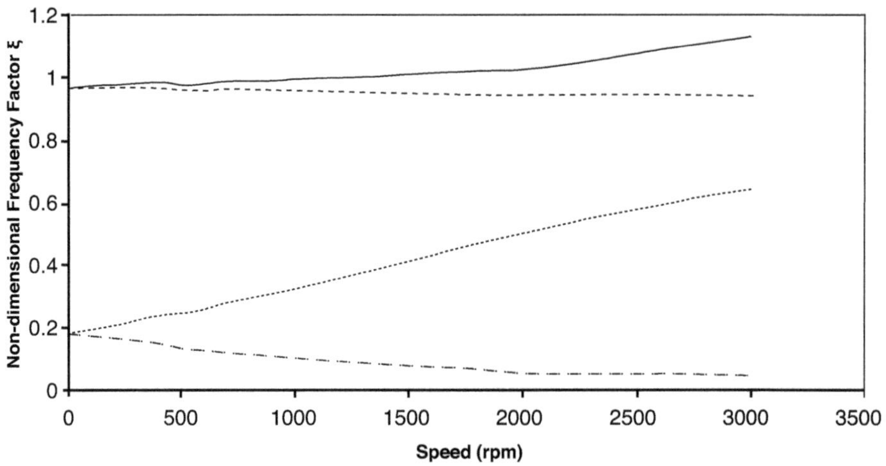

FIGURE 5.6 Campbell diagram showing the gyroscopic effect of the polar mass moment of inertia of the disk and blades on the rotor shaft natural frequency for a typical case of $M_D / M_S = 15$ with ratio of the disk/shaft polar inertia = 50 [- - - -, forward whirl (first flex); — — —, backward whirl (first flex); ——, forward whirl (second flex); — — —, backward whirl (second flex)]. (Reprinted with permission from Reference [3].)

loads F_G on the blades apply a steady-state load torque on the shaft. For any other speed, this load torque is proportional to the square of the speed. If the fan blade-tip rub takes place under these conditions, then both the torque due to F_G and Coulomb friction are opposite to the direction of rotation. However, during the wind-milling operation, the gas load F_G becomes the driving torque and the Coulomb friction like always is acting opposite to the direction of motion trying to straighten the blades.

The blade-tip rub introduces additional loads making the loading on the shaft much more complex than a simple rotating unbalance. In the rotordynamic equations (5.27) and (5.28), all the loads transferred to the shaft through the blades can be treated like external loads. When the tips of the blades start rubbing against the inner surface of the containment structure, then depending upon the magnitude of the radial load (F_a) acting along the blade stacking axis in the longitudinal span direction, the blade tips can either skate on the inner surface if it is sufficiently hard or they can cut through the inner surface if it is relatively soft. If the ultimate shear strength of the inner surface of abradable-like material is τ in psi units; and the radial incursion of the blade tip or the interference with the inner surface is "ε" then for a blade of the chord length "C" at the tip, the condition for either skating or cutting-through can be established as follows: (A) skating will occur if $F_a < 2\tau\varepsilon C$ and (B) cutting-through will occur if $F_a > 2\tau\varepsilon C$.

If the inner surface material has been cut-through, then as the tip-rub progresses, the cut material will be pushed away from the path of the rotating blade tip like a plowing action. As a result of that, the blade-tip rub produces the following loads on the rotor shaft:

A. During skating of the blade tip on the inner surface with the coefficient of friction "μ" and the blade tip pretwist angle, also sometimes called stagger angle "β", the radial shear load on the shaft by a typical jth blade, the bending load in the direction perpendicular to the blade stacking axis on a typical jth blade, and the torque on the shaft by a typical jth blade are given, respectively. In a local blade coordinate system attached to any typical jth radial blade's "span-chord-thickness" directions, the external force vector during the tip-rub event is expressed as $\left(-F_a\hat{e}_s + \mu F_a \sin\beta\hat{e}_c + \mu F_a \cos\beta\hat{e}_t\right)$. The thickness direction component of the blade tip-rub load generates a torque on the rotor shaft in a direction opposite to the direction of rotation of the shaft, the magnitude of which is equal to $\mu F_a \cos\beta(r+L)$.

B. During the plowing action of the blade tip of the inner surface of the filler abradable material with density "ρ_c," and blade tip tangential velocity "ΩR," the radial shear load on the shaft by a typical radial incursion of blade tip inside the abradable surface "ε"; the corresponding bending loads on the shaft at the fan-disk location of $Z = \ell$ are determined in the following steps. Here, the radial blade's longitudinal direction load is $F_a = 2\tau\varepsilon C$. The corresponding load vector at the blade tip in a local blade coordinate system is attached to any typical jth radial blade's "span-chord-thickness" directions, and the external force vector during the tip-rub event is expressed as $\left(-F_a\hat{e}_s + \mu F_a \sin\beta\hat{e}_c + \mu F_a \cos\beta\hat{e}_t\right)$. The thickness direction component of the blade tip-rub load generates a torque on the rotor shaft opposite to the direction of rotation of the shaft, the magnitude of which is equal to $\mu F_a \cos\beta(r+L)$. This yields the blade tip-rub torque on the shaft under the abradable removal condition as equal to $2\mu\tau\varepsilon C\cos\beta(r+L)$. It should be noted that the torque on the fan rotor shaft by a typical j th blade during its plowing action of the abradable affects the dynamic response of the rotor shaft as follows. In this situation, the blade tip tries to accelerate the removed abradable material in front of it and then attempts to push it forcibly with its tangential velocity reaching up to the blade-tip tangential velocity. Using this assumption, the torque generated on the fan rotor shaft during the plowing action of the blade tip can be used to approximately estimate the amount of abradable material being removed by the rubbing action of blade tip.

$$\text{Rub torque on the shaft} = \frac{1}{2}\rho_c\left(\Omega R\cos\beta\right)^2 \varepsilon C\cos\beta(r+L) = \frac{F_a}{4\tau}\rho_c\,\Omega^2\left(R\cos\beta\right)^3 \tag{5.32}$$

It should be obvious to the reader that the magnitude of the rub torque on the fan shaft is proportional to the square of the rotational speed of the rotor and cubic to the fan size. The axial load on the shaft by a typical j th rub blade is as follows. Since any external torque has a destabilizing effect on the rotor, even with the same blade-tip tangential velocity, a larger engine with the blade tips rubbing against the containment ring can become unstable much quicker than a smaller engine. It should be obvious in the above equation that the blade's longitudinal direction spanwise load $\left(F_a\right)_j$ is a function of time "t". Thus, the fan rotor shaft sees only steady axial pull due to the gas load on the blades called F_G; however, it has both steady-state and time-varying torque terms in the left-hand side of Eqs. (5.27) and (5.28):

$$P(t) = P_0 + P_t = NF_G \sin\beta \tag{5.33}$$

$$T(t) = T_0 + T_t = NF_G\left(r + \frac{L}{2}\right)\cos\beta + \sum_{j=1}^{N} \mu\,(F_a)_j \cos\beta\,(r+L) \tag{5.34}$$

In a rigorous sense, the above relationship for $P(t)$ and $T(t)$ shown in Eqs. (5.33) and (5.34) is correct only at a constant speed. During acceleration and deceleration of the rotor, Ω is no longer constant and F_G as well as $(F_a)_j$ are also time-dependent, which make the entire dynamic even highly nonlinear. The additional load during the plowing action of the blade tip against the relatively soft filler material in the case imparts a pulse load on the shaft and should be treated accordingly in the dynamical equation for the shaft.

5.2.2 Galerkin's Method and Assumed Displacement Function

In Eqs. (5.27) and (5.28), the time-dependent shaft deformations in the X and Y directions are assumed as follows:

$$u(Z,t) = \sum_{m=0}^{\infty} U_m(t)\,W_m(Z) = U_0(t) + \sum_{m=1}^{\infty} U_m(t)\left[1-\cos(\propto_m Z)\right] \tag{5.34a}$$

$$v(Z,t) = \sum_{m=0}^{\infty} V_m(t)\,W_m(Z) = V_0(t) + \sum_{m=1}^{\infty} V_m(t)\left[1-\cos(\propto_m Z)\right] \tag{5.34b}$$

where

$$\propto_m = \frac{(2m-1)\pi}{2\ell}.\ [m = 1,2,3....] \tag{5.34c}$$

This set of deformation shape function arbitrarily satisfies the boundary conditions of a typical long cantilever beam type rotor with an overhung disk at $Z = \ell$. At the free end of the rotor, both the displacement and the slope are maximum at the same time, which this particular sinusoidal displacement function captures very well. Here, U_m and V_m are the peak deflections of the shaft centerline in the global X and Y directions, respectively, at $Z = \ell$. Here, using the assumed shaft displacement shape function as $W_m(Z)$, we have the following relationships for the derivatives of the shape functions in the transverse direction $W_m'(Z)$, $W_m''(Z)$, etc. as

$$W_0(Z) = 1\ (\text{rigid body mode of the fan rotor shaft deformation}) \tag{5.35}$$

$$W_m(Z) = 1 - \cos(\propto_m Z) \tag{5.36}$$

$$W_m'(Z) = \propto_m \sin(\propto_m Z) \tag{5.37}$$

$$W_m''(Z) = (\propto_m)^2 \cos(\propto_m Z) \tag{5.38}$$

$$W_m'''(Z) = -(\propto_m)^3 \sin(\propto_m Z) \tag{5.39}$$

$$W_m''''(Z) = -(\propto_m)^4 \cos(\propto_m Z) \tag{5.40}$$

At the end of the shaft for the overhung fan disk location at $Z = \ell$, the above shaft deformation shape functions and their derivatives simplify as follows:

$$W_m(\ell) = 1 - \cos(\propto_m \ell) = 1 \tag{5.41}$$

$$\sin(\propto_m \ell) = (-1)^{m+1} \tag{5.42}$$

$$\cos(\propto_m \ell) = 0 \tag{5.43}$$

$$W_m'(\ell) = \propto_m \sin(\propto_m \ell) = \propto_m (-1)^{m+1} \tag{5.44}$$

$$W_m''(\ell) = (\propto_m)^2 \cos(\propto_m \ell) = 0 \tag{5.45}$$

$$W_m'''(\ell) = -(\propto_m)^3 \sin(\propto_m \ell) = -(\propto_m)^3 (-1)^{m+1} \tag{5.46}$$

$$W_m''''(\ell) = -(\propto_m)^4 \cos(\propto_m \ell) = 0 \tag{5.47}$$

Using the assumed deformation function for the shaft and its spatial derivatives outlined in Eqs. (5.30a–5.47), Galerkin's method can be used to convert the coupled partial differential equations of motion (refer to Eqs. 5.27 and 5.28) into matrix form. To that end, on applying Galerkin's method with the series of assumed deformation shape function terms $W_k(Z)$, one obtains a set of second order ordinary differential equations of motion of which a typical kth equation can be expressed as follows using symbolically $\mathfrak{R}(u,v)$ as a partial differential operator:

$$\int_0^\ell \mathfrak{R}(u,v) W_k(Z)\, dZ = 0 \tag{5.48}$$

Thus, Galerkin's method simplifies the coupled partial differential equations of motion of the fan rotor shaft in the global (X,Y) directions (Eqs. 5.27 and 5.28), respectively. On applying Galerkin's method in the overhung fully bladed fan-disk rotor shaft supported like a cantilever beam, the natural boundary conditions at the fan-disk end are not automatically satisfied by the assumed deformed shape function. Since the natural boundary conditions of the shear force equal to zero at the disk end are not satisfied for $Z = \ell$, the typical n th algebraic equation of motion in the global X-direction of the fan rotor shaft for the complete set of simultaneous equation is written as follows:

$$
\begin{aligned}
&\sum_{m=0}^{\infty}\left[(EI)_s U_m \int_0^{\ell} W_m'''' W_n dZ + 2\zeta H\left[\dot{U}_m \int_0^{\ell} W_m'''' W_n dZ + \Omega V_m \int_0^{\ell} W_m'''' W_n dZ\right]\right.\\
&+(\rho A)_s \ddot{U}_m \int_0^{\ell} W_m W_n dZ - I_d \ddot{U}_m \int_0^{\ell} W_m'' W_n dZ - 2\, I_d \Omega\left[\dot{V}_m \int_0^{\ell} W_m'' W_n dZ - \Omega U_m \int_0^{\ell} W_m'' W_n dZ\right]\\
&+ T(t) V_m \int_0^{\ell} W_m''' W_n dZ - P(t) U_m \int_0^{\ell} W_m'' W_n dZ +\\
&\sum_{i=1}^{N_b}\left[S_{XX}^b\, U_m + S_{XY}^b V_m + D_{XX}^b \dot{U}_m + D_{XY}^b \dot{V}_m\right] \int_0^{\ell} W_m W_n \delta(Z - Z_i) dZ\\
&+\sum_{i=1}^{N_b}\left[B_{XX}^b U_m\right] \int_0^{\ell} W_m'' W_n \delta(Z - Z_i) dZ +\\
&\left[\begin{gathered}(M_D + N\, M_b)\ddot{U}_m\, W_m(\ell)\, W_n(\ell) + J_{D\text{-}YY}\ddot{U}_m W_m'(\ell) W_n'(\ell) + (J_{D\text{-}XX} + J_{D\text{-}YY})\\
\Omega\left[\dot{V}_m W_m'(\ell) W_n'(\ell) - \Omega U_m\, W_m''(\ell)\, W_n(\ell)\right]\end{gathered}\right]\\
&+K_{\text{rub}}\left[U_m W_m(\ell) W_n(\ell) + \mu\left(V_m W_m(\ell) W_n(\ell) - eRV_m W_m'(\ell) W_n(\ell)\right)\right]\Bigg]\\
&\quad + P(t) U_m W_m'(\ell) W_n(\ell) - T(t) V_m W_m'(\ell) W_n'(\ell)\\
&-(EI)_s U_m\, W_m'''(\ell)\, W_n(\ell) + (EI)_s U_m\, W_m''(\ell)\, W_n'(\ell) - F_U \cos(\theta_0 + \Omega t)\\
&\quad W_n(\ell) = \int_0^{\ell} \Re_X(Z) W_n(Z) dZ = 0
\end{aligned}
\tag{5.49}
$$

Similarly, after the substitution of assumed deformation shape function $W_m(Z)$, the corresponding equation (Eq. 5.28) in the vertical global Y-direction, and applying Galerkin's method, the typical n th algebraic equation of motion of the complete set become as follows:

$$\sum_{m=0}^{\infty}\Bigg[(EI)_s V_m \int_0^{\ell} W_m'''' W_n\, dZ + 2\zeta H\left[\dot{V}_m \int_0^{\ell} W_m'''' W_n dZ - \Omega U_m \int_0^{\ell} W_m'''' W_n dZ\right]$$

$$+(\rho A)_s \ddot{V}_m \int_0^{\ell} W_m W_n dZ - I_d \ddot{V}_m \int_0^{\ell} W_m'' W_n dZ + 2\, I_d\Omega\left[\dot{U}_m \int_0^{\ell} W_m'' W_n dZ - \Omega V_m \int_0^{\ell} W_m'' W_n dZ\right]$$

$$-T(t) U_m \int_0^{\ell} W_m''' W_n dZ - P(t) V_m \int_0^{\ell} W_m'' W_n dZ +$$

$$\sum_{i=1}^{N_b}\left[S_{YX}^b\, U_m + S_{YY}^b V_m + D_{YX}^b \dot{U}_m + D_{YY}^b \dot{V}_m\right]\int_0^{\ell} W_m W_n \delta(Z - Z_i) dZ$$

$$+\sum_{i=1}^{N_b}\left[\, B_{YY}^b V_m\right]\int_0^{\ell} W_m'' W_n \delta(Z - Z_i) dZ +$$

$$\Bigg[(M_D + N\, M_b)\ddot{V}_m\, W_m(\ell)\, W_n(\ell) + J_{D-XX}\ddot{V}_m W_m'(\ell) W_n'(\ell) - (J_{D-XX} + J_{D-YY})\Omega$$

$$\left[\dot{U}_m W_m'(\ell) W_n'(\ell) + \Omega V_m\, W_m''(\ell)\, W_n(\ell)\right]\Bigg]$$

$$+K_{\text{rub}}\left[V_m W_m(\ell) W_n(\ell) - \mu\Big(U_m W_m(\ell) W_n(\ell) - eRU_m W_m'(\ell) W_n(\ell)\Big)\right]\Bigg]$$

$$+P(t) V_m W_m'(\ell) W_n(\ell) + T(t) U_m W_m'(\ell) W_n'(\ell) - (EI)_s V_m\, W_m'''(\ell)\, W_n(\ell) +$$

$$(EI)_s V_m\, W_m''(\ell)\, W_n'(\ell) - F_U \sin(\theta_0 + \Omega t)\, W_n(\ell) = \int_0^{\ell} \Re_Y(Z) W_n(Z) dZ = 0 \tag{5.50}$$

To determine the $U_0(t)$ and $V_0(t)$ terms in the assumed displacement functions, the rigid body equation is obtained by setting $n = 0$ and using $W_0(Z) = 1$ in the above set of equations, which yields the following sets of second-order ordinary differential equations of motion. The rigid body motion of the rotor is governed by the following two equations:

$$\left[\int_0^{\ell}(\rho A)_s\,dZ + (M_D + N\,M_b)\right]\ddot{U}_0 + \sum_{i=1}^{N_b}\left[D_{XX}^b\right]\dot{U}_0 + \sum_{i=1}^{N_b}\left[D_{XY}^b\right]\dot{V}_0$$

$$+\sum_{m=1}^{\text{Modes}}\left[\sum_{i=1}^{N_b}\left[D_{XX}^b W_m(Z_i)\right]\dot{U}_m + \sum_{i=1}^{N_b}\left[D_{XY}^b W_m(Z_i)\right]\dot{V}_m\right] + \sum_{i=1}^{N_b}\left[S_{XX}^b + K_{\text{rub}}\right]U_0$$

$$+\sum_{i=1}^{N_b}\left[S_{XY}^b + \mu K_{\text{rub}}\right]V_0 + \sum_{m=1}^{\text{Modes}}\left[\sum_{i=1}^{N_b}\left[\begin{matrix} S_{XX}^b W_m(Z_i) + B_{XX}^b W_m''(Z_i) + K_{\text{rub}}W_m(\ell) + \\ P(t)W_m'(\ell)\end{matrix}\right]U_m\right.$$

$$\left.+\sum_{i=1}^{N_b}\left[S_{XY}^b W_m(Z_i) + \mu K_{\text{rub}}W_m(\ell)\right]V_m\right] - F_U\cos(\theta_0 + \Omega t) = 0$$

(5.51)

$$\left[\int_0^{\ell}(\rho A)_s\,dZ + (M_D + N\,M_b)\right]\ddot{V}_0 + \sum_{i=1}^{N_b}\left[D_{YY}^b\right]\dot{V}_0 + \sum_{i=1}^{N_b}\left[D_{YX}^b\right]\dot{U}_0 +$$

$$\sum_{m=1}^{\text{Modes}}\left[\sum_{i=1}^{N_b}\left[D_{YY}^b W_m(Z_i)\right]\dot{V}_m + \sum_{i=1}^{N_b}\left[D_{YX}^b W_m(Z_i)\right]\dot{U}_m\right] + \sum_{i=1}^{N_b}\left[S_{YY}^b + K_{\text{rub}}\right]V_0$$

$$+\sum_{i=1}^{N_b}\left[S_{YX}^b - \mu K_{\text{rub}}\right]U_0 +$$

$$\sum_{m=1}^{\text{Modes}}\left[\begin{matrix}\sum_{i=1}^{N_b}\left[S_{YY}^b W_m(Z_i) + B_{YY}^b W_m''(Z_i) + K_{\text{rub}}W_m(\ell) + P(t)W_m'(\ell)\right]V_m + \\ \sum_{i=1}^{N_b}\left[S_{YX}^b W_m(Z_i) - \mu K_{\text{rub}}W_m(\ell)\right]U_m\end{matrix}\right]$$

$$-F_U\sin(\theta_0 + \Omega t) = 0$$

(5.52)

Similarly, the shaft deformational modes causing the whirling of the rotor shaft are governed by the following sets of the second-order ordinary differential equations.

$$\sum_{i=1}^{N_b}\left[S^b_{XX}W_m(Z_i)+B^b_{XX}W_m''(Z_i)+K_{\text{rub}}W_m(\ell)+P(t)W_m'(\ell)\right]U_0$$

$$+\sum_{i=1}^{N_b}\left[S^b_{XY}W_m(Z_i)+\mu K_{\text{rub}}W_m(\ell)\right]V_0$$

$$+\sum_{i=1}^{N_b}\left[D^b_{XX}W_m(Z_i)\right]\dot{U}_0+\sum_{i=1}^{N_b}\left[D^b_{XY}W_m(Z_i)\right]\dot{V}_0+\sum_{m=1}^{\text{Modes}}\left[(EI)_s U_m\int_0^{\ell}W_m''''W_n dZ+2\zeta H\right.$$

$$\left[\dot{U}_m\int_0^{\ell}W_m''''W_n dZ+\Omega V_m\int_0^{\ell}W_m''''W_n dZ\right]+(\rho A)_s\ddot{U}_m\int_0^{\ell}W_mW_n dZ-I_d\ddot{U}_m$$

$$\int_0^{\ell}W_m''W_n dZ-2I_d\Omega\left[\dot{V}_m\int_0^{\ell}W_m''W_n dZ-\Omega U_m\int_0^{\ell}W_m''W_n dZ\right]+T(t)V_m\int_0^{\ell}W_m'''W_n dZ$$

$$-P(t)U_m\int_0^{\ell}W_m''W_n dZ+\sum_{i=1}^{N_b}\left[S^b_{XX}U_m+S^b_{XY}V_m+D^b_{XX}\dot{U}_m+D^b_{XY}\dot{V}_m\right]$$

$$\int_0^{\ell}W_mW_n\delta(Z-Z_i)dZ+\sum_{i=1}^{N_b}\left[B^b_{XX}U_m\right]\int_0^{\ell}W_m''W_n\delta(Z-Z_i)dZ$$

$$+\begin{bmatrix}(M_D+N\,M_b)\ddot{U}_m\,W_m(\ell)\,W_n(\ell)+J_{D\text{-}YY}\ddot{U}_mW_m'(\ell)W_n'(\ell)+(J_{D\text{-}XX}+J_{D\text{-}YY})\\ \Omega\left[\dot{V}_mW_m'(\ell)W_m'(\ell)-\Omega U_m\,W_m''(\ell)\,W_n(\ell)\right]\end{bmatrix}$$

$$+K_{\text{rub}}\left[U_mW_m(\ell)W_n(\ell)+\mu\left(V_mW_m(\ell)W_n(\ell)-eRV_mW_m'(\ell)W_n(\ell)\right)\right]\Bigg]+$$

$$P(t)U_mW_m'(\ell)W_n(\ell)-T(t)V_mW_m'(\ell)W_n'(\ell)$$

$$-(EI)_sU_m\,W_m'''(\ell)\,W_n(\ell)+(EI)_sU_m\,W_m''(\ell)\,W_n'(\ell)-F_U\cos(\theta_0+\Omega t)$$

$$W_n(\ell)=\int_0^{\ell}\Re_X(Z)W_n(Z)dZ=0 \tag{5.53}$$

Similarly, after the substitution of assumed deformation shape function $W_m(Z)$, the corresponding equation (Eq. 5.28) in the vertical global Y-direction, and applying Galerkin's method, the typical nth algebraic equation of motion of the complete set become as follows:

$$\sum_{i=1}^{N_b}\left[S_{YX}^b W_m(Z_i)-\mu K_{\text{rub}} W_m(\ell)\right]U_0 +$$

$$\sum_{i=1}^{N_b}\left[S_{YY}^b W_m(Z_i)+B_{YY}^b W_m''(Z_i)+K_{\text{rub}} W_m(\ell)+P(t)W_m'(\ell)\right]V_0 +$$

$$\sum_{i=1}^{N_b}\left[D_{YX}^b W_m(Z_i)\right]\dot{U}_0 + \sum_{i=1}^{N_b}\left[D_{YY}^b W_m(Z_i)\right]\dot{V}_0$$

$$+\sum_{m=1}^{\text{Modes}}\begin{bmatrix}(EI)_s V_m \int_0^\ell W_m'''' W_n dZ + 2\zeta H\left[\dot{V}_m \int_0^\ell W_m'''' W_n dZ - \Omega U_m \int_0^\ell W_m'''' W_n dZ\right] + \\ (\rho A)_s \ddot{V}_m \int_0^\ell W_m W_n dZ - I_d \ddot{V}_m \int_0^\ell W_m'' W_n dZ\end{bmatrix}$$

$$+2\, I_d\Omega\left[\dot{U}_m \int_0^\ell W_m'' W_n dZ - \Omega V_m \int_0^\ell W_m'' W_n dZ\right] - T(t)U_m \int_0^\ell W_m''' W_n dZ - P(t)V_m \int_0^\ell W_m'' W_n dZ +$$

$$\sum_{i=1}^{N_b}\left[S_{YX}^b\ U_m + S_{YY}^b V_m + D_{YX}^b \dot{U}_m + \left[S_{YX}^b\ U_m + S_{YY}^b V_m + D_{YX}^b \dot{U}_m + D_{YY}^b \dot{V}_m\right]\right.$$

$$\int_0^\ell W_m W_n \delta(Z-Z_i)dZ + \sum_{i=1}^{N_b}\left[B_{YY}^b V_m\right]\int_0^\ell W_m'' W_n \delta(Z-Z_i)dZ +$$

$$\begin{bmatrix}(M_D + N\ M_b)\ddot{V}_m\ W_m(\ell)\ W_n(\ell) + J_{D-XX}\ddot{V}_m W_m'(\ell)W_n'(\ell) - (J_{D-XX}+J_{D-YY})\Omega \\ \left[\dot{U}_m W_m'(\ell)W_n'(\ell) + \Omega V_m\ W_m''(\ell)\ W_n(\ell)\right]\end{bmatrix}$$

$$+K_{\text{rub}}\left[V_m W_m(\ell)W_n(\ell) - \mu\left(U_m W_m(\ell)W_n(\ell) - eRU_m W_m'(\ell)W_n(\ell)\right)\right]\Bigg] +$$

$$P(t)V_m W_m'(\ell)W_n(\ell) + T(t)U_m W_m'(\ell)W_n'(\ell)$$

$$-(EI)_s V_m\ W_m'''(\ell)\ W_n(\ell) + (EI)_s V_m\ W_m''(\ell)\ W_n'(\ell) - F_U \sin(\theta_0 + \Omega t)$$

$$W_n(\ell) = \int_0^\ell \Re_Y(Z)W_n(Z)dZ = 0 \tag{5.54}$$

The above set of second-order ordinary coupled equations of motion outlined in Eqs. (5.51–5.54) is expressed in a matrix form as follows:

$$[\mathrm{M}]\{\ddot{f}\} + \left[[\mathrm{C}_D] + [\mathrm{C}_G]\right]\{\dot{f}\} + \left[[\mathrm{K}_S] + [\mathrm{K}_I] - \left[[\mathrm{B}_1] + [\mathrm{B}_2]\psi(t)\right]\right]$$

$$\{f\} = \{F(t)\} \tag{5.55}$$

where

[**M**] Mass matrix (symmetrical). It should be noted that in the case of post-FBO event, the terms of the mass matrix are not only asymmetric rather due to missing release blade in one of the fan disk sectors, it also becomes time-dependent as a function of time "*t*". The reader is referred to Eqs. (5.29–5.31) to implement the time-dependent updating in the numerical scheme.

[$\mathbf{C}_D$] Damping matrix (symmetrical or nonsymmetrical depending upon the bearing)

[$\mathbf{C}_G$] Gyroscopic matrix (skew-symmetric, causes forward and backward whirl in the shaft)

[$\mathbf{K}_S$] Stiffness matrix (generally symmetrical, but may be nonsymmetrical due to nonsymmetrical bearing stiffness in the *X* and *Y* directions)

[$\mathbf{K}_I$] Instability matrix (skew-symmetric caused by the internal damping in the shaft and also torque in the system)

[$\mathbf{B}_1$] Time-dependent stiffness matrix (generally symmetrical and causes parametric resonance due to axial load on the shaft)

[$\mathbf{B}_2$] Time-dependent stiffness matrix (skew-symmetric and causes parametric instability in the shaft due to fluctuating torque in the system)

$\psi(t)$ Time-dependent function

$\{F(\text{t})\}$ Right-hand side column vector containing external forces on the dynamical system.

It can be seen that in Eq. (5.55), the terms containing generalized coordinates in the column vector $\{f(t)\}$ have dimensions of length. Suppose, we consider "*m*" number of deformation modes on the shaft. Then the size of above matrices in two-dimensional (*X–Y*) space will be $(2+2m)\times(2+2m)$. Here, for brevity, we can introduce the following notations for the generalized coordinates $\{f(t)\}$:

Rotor deflections:

$$f_1(t) = U_0(t),\ f_3(t) = U_1(t),\ f_5(t) = U_2(t),\ f_{1+2m}(t) = U_m(t). \tag{5.56}$$

$$f_2(t) = V_0(t),\ f_4(t) = V_1(t),\ f_6(t) = V_2(t),\ f_{2+2m}(t) = V_m(t). \tag{5.57}$$

Specifically, the key non-zero terms of the matrices [**M**], [**C**], [**K**], and [**B**] of Eq. (5.55) for the overhung rotor-shaft stiffness matrix [**K**], with 2*m* rows and 2*n* columns are as follows:

$$[M_{1,1}] = \left[\int_0^{\ell} (\rho A)_s\, dZ + (M_D + N\ M_b)\right] \tag{5.58}$$

$$[M_{2,2}] = \left[\int_0^{\ell} (\rho A)_s\, dZ + (M_D + N\ M_b)\right] \tag{5.59}$$

$$\left[M_{2m+1,2n+1}\right] = \int_0^{\ell} (\rho A)_s W_m W_n dZ - \int_0^{\ell} I_d \; W_m \; W_n'' dZ \; + (M_D + N \; M_b) \; W_m(\ell) \; W_n(\ell) + J_{D-YY} W_m'(\ell) W_n'(\ell) \tag{5.60}$$

$$\left[M_{2m+2,2n+2}\right] = \int_0^{\ell} (\rho A)_s W_m W_n dZ - \int_0^{\ell} I_d \; W_m \; W_n'' dZ \; + (M_D + N \; M_b) \; W_m(\ell) \; W_n(\ell) + J_{D-XX} W_m'(\ell) W_n'(\ell) \tag{5.61}$$

$$\left[K_{1,1}\right] = \sum_{i=1}^{N_b} \left[\left(S_{XX}^b\right)_i + K_{\text{rub}} \right] \tag{5.62}$$

$$\left[K_{1,2}\right] = \sum_{i=1}^{N_b} \left[\left(S_{XY}^b\right)_i + \mu K_{\text{rub}} \right] \tag{5.63}$$

$$\left[K_{2,1}\right] = \sum_{i=1}^{N_b} \left[\left(S_{YX}^b\right)_i - \mu K_{\text{rub}} \right] \tag{5.64}$$

$$\left[K_{2,2}\right] = \sum_{i=1}^{N_b} \left[\left(S_{YY}^b\right)_i + K_{\text{rub}} \right] \tag{5.65}$$

$$\left[C_{1,1}\right] = \sum_{i=1}^{N_b} \left[\left(D_{XX}^b\right)_i \right] \tag{5.66}$$

$$\left[C_{1,2}\right] = \sum_{i=1}^{N_b} \left[\left(D_{XY}^b\right)_i \right] \tag{5.67}$$

$$\left[C_{2,1}\right] = \sum_{i=1}^{N_b} \left[\left(D_{YX}^b\right)_i \right] \tag{5.68}$$

$$\left[C_{2,2}\right] = \sum_{i=1}^{N_b} \left[\left(D_{YY}^b\right)_i \right] \tag{5.69}$$

overhung rotor-shaft stiffness matrix [**K**], with 2*m* rows and 2*n* columns are

$$\left[K_{2m+1,1}\right]=\left[K_{1,2m+1}\right]=\sum_{i=1}^{N_b}\left[\left(S_{XX}^b\right)_i W_m(Z_i)+\left(B_{XX}^b\right)_i W_m''(Z_i)\right]+ K_{\text{rub}}W_m(\ell)+P(t)W_m'(\ell) \tag{5.70}$$

$$\left[K_{2m+1,2}\right]=\left[K_{2,2m+1}\right]=\sum_{i=1}^{N_b}\left[\left(S_{XY}^b\right)_i W_m(Z_i)\right]+\mu K_{\text{rub}}W_m(\ell) \tag{5.71}$$

$$\left[K_{2m+2,1}\right]=\left[K_{1,2m+2}\right]=\sum_{i=1}^{N_b}\left[\left(S_{YX}^b\right)_i W_m(Z_i)\right]-\mu K_{\text{rub}}W_m(\ell) \tag{5.72}$$

$$\left[K_{2m+2,2}\right]=\left[K_{2,2m+2}\right]=\sum_{i=1}^{N_b}\left[\left(S_{YY}^b\right)_i W_m(Z_i)+\left(B_{YY}^b\right)_i W_m''(Z_i)\right]+ K_{\text{rub}}W_m(\ell)+P(t)W_m'(\ell) \tag{5.73}$$

$$\left[C_{2m+1,1}\right]=\left[C_{1,2m+1}\right]=\sum_{i=1}^{N_b}\left[\left(D_{XX}^b\right)_i W_m(Z_i)\right] \tag{5.74}$$

$$\left[C_{2m+1,2}\right]=\left[C_{2,2m+1}\right]=\sum_{i=1}^{N_b}\left[\left(D_{XY}^b\right)_i W_m(Z_i)\right] \tag{5.75}$$

$$\left[C_{2m+2,1}\right]=\left[C_{1,2m+2}\right]=\sum_{i=1}^{N_b}\left[\left(D_{YX}^b\right)_i W_m(Z_i)\right] \tag{5.76}$$

$$\left[C_{2m+2,2}\right]=\left[C_{2,2m+2}\right]=\sum_{i=1}^{N_b}\left[\left(D_{YY}^b\right)_i W_m(Z_i)\right] \tag{5.77}$$

$$\begin{aligned}\left[K_{2m+1,2n+1}\right] &= (EI)_S\left[\int_0^{\ell} W_m W_n'''' dZ - W_m(\ell)\, W_n'''(\ell)\right]+2I_d\Omega^2 \\ &\left[\int_0^{\ell} W_m\, W_n'' dZ - W_m(\ell)\, W_n'(\ell)\right] \\ &+2J_d\Omega^2\left[W_m'(\ell)W_n'(\ell)\right]-P(t)\left[\int_0^{\ell} W_m W_n'' dZ - W_m(\ell)\, W_n'(\ell)\right] \\ &+\sum_{i=1}^{N_b}\left[\left(S_{XX}^b\right)_i W_m(Z_i)W_n(Z_i)+\left(B_{XX}^b\right)_i W_m(Z_i)\, W_n''(Z_i)\right]\end{aligned} \tag{5.78}$$

$$\left[K_{2m+1,2n+2}\right] = 2\zeta H\Omega\left[\int_0^{\ell} W_m W_n''''dZ - W_m(\ell)\, W_n'''(\ell)\right] + T(t)$$

$$\left[\int_0^{\ell} W_m W_n'''dZ - W_m'(\ell) W_n'(\ell)\right] + \sum_{i=1}^{N_b}\left[\left(S_{XY}^b\right)_i W_m(Z_i) W_n(Z_i)\right] \tag{5.79}$$

$$\left[K_{2m+2,2n+1}\right] = -2\zeta H\Omega\left[\int_0^{\ell} W_m W_n''''dZ - W_m(\ell)\, W_n'''(\ell)\right] - T(t)$$

$$\left[\int_0^{\ell} W_m W_n'''dZ - W_m'(\ell) W_n'(\ell)\right] + \sum_{i=1}^{N_b}\left[\left(S_{YX}^b\right)_i W_m(Z_i) W_n(Z_i)\right] \tag{5.80}$$

$$\left[K_{(2m+2,2n+2)}\right] = (EI)_S\left[\int_0^{l} W_m W_n'''' \, dZ - W_m(l) W_n'''(l)\right] + 2I_d\Omega^2$$

$$\left[\int_0^{l} W_m W_n'' \, dZ - W_m(l) W_n'(l)\right] + 2J_d\Omega^2[W_m'(l) W_n'(l)] - P(t)$$

$$\left[\int_0^{l} W_m W_n'' \, dZ - W_m(l) W_n'(l)\right] + \sum_{(i=1)}^{(N_b)}\left[(S_{YY}^b)_i W_m(Z_i) W_n(Z_i) + (B_{YY}^b)_i W_m(Z_i) W_n''(Z_i)\right] \tag{5.81}$$

Similarly, the corresponding terms for the velocity-dependent matrix [**C**] are

$$\left[C_{2m+1,2n+1}\right] = 2\zeta H\left[\int_0^{\ell} W_m W_n''''dZ - W_m(\ell)\, W_n'''(\ell)\right] + \sum_{i=1}^{N_b}\left[\left(D_{XX}^b\right)_i W_m(Z_i) W_n(Z_i)\right] \tag{5.82}$$

$$\left[C_{2m+1,2n+2}\right] = -2\Omega\left[\int_0^{\ell} I_d W_m W_n''dZ\right] + \sum_{i=1}^{N_b}\left[\left(D_{XY}^b\right)_i W_m(Z_i) W_n(Z_i)\right] + \tag{5.83}$$

$$\left(J_{D-XX} + J_{D-YY}\right)\Omega\left[W_m'(\ell) W_n'(\ell)\right]$$

$$\left[C_{2m+2,2n+1}\right] = 2\Omega\left[\int_0^\ell I_d W_m W_n'' dZ\right] + \sum_{i=1}^{N_b}\left[\left(D_{YX}^b\right)_i W_m(Z_i)W_n(Z_i)\right] - (J_{D-XX}+J_{D-YY})\Omega\left[W_m'(\ell)W_n'(\ell)\right] \tag{5.84}$$

$$\left[C_{2m+2,2n+2}\right] = 2\zeta H\left[\int_0^\ell W_m W_n''' dZ - W_m(\ell)\, W_n'''(\ell)\right] + \sum_{i=1}^{N_b}\left[\left(D_{YY}^b\right)_i W_m(Z_i)W_n(Z_i)\right] \tag{5.85}$$

5.2.3 Time-Domain Response of a Fan Rotor with High Unbalance

The transient response of a decelerating or accelerating rotor is of great interest in any rotordynamics problem. The two most common techniques for integrating the equations of motion and computing the displacement and velocity response of a dynamical system are (1) *Newmark's β-method:* it is an unconditionally stable implicit numerical scheme, which is commonly used for determining the long-term transient response in the area of structure dynamics; (2) *fourth-order Runge–Kutta Method*: it is an explicit technique and is commonly used for computing short-term transient response of nonlinear dynamical problems, usually with pulse and impact types of loading.

The system of coupled nonlinear transient equations of motion outlined in Eq. (5.55) in a matrix form is solved in time domain using an accurate sixth-order Runge–Kutta time marching forward the numerical integration scheme. In this numerical scheme in order to handle the nonlinearity in its proper detail, the deformed shape of the rotor shaft and its loading conditions are updated continuously at every time step including at the mid-time points as well. The application of this method is demonstrated for a rotor spinning at a constant speed Ω, and only external load on the dynamical system is due to nominal unbalance load of magnitude F_U acting as a rotating load vector at the fan disk location of $Z = \ell$. To apply this method, the system of dynamical equations shown in a matrix form by Eq. (5.55) is simplified by dropping many unnecessary complexities due to gysrocopic effects, etc., which can be written as shown below in a familiar form of equation of motion:

$$[\mathrm{M}]\{\ddot{f}\}+[\mathrm{C}_D]\{\dot{f}\}+[\mathrm{K}_S]\{f\}=\{F(t)\}_{\mathrm{Rotating\_Unbalannce\_load}} \tag{5.86}$$

It should be noted that the size (Rows X Columns) of the mass matrix $[\mathrm{M}]$ and the stiffness matrix $[\mathrm{K}_S]$ are $(2+2m)\times(2+2m)$. The number of rows in the column vector of $\{f\}$ representing displacement terms like $\left[\{f\}^{-1} = \{\, U_0\ V_0\ U_1\ V_1\ U_2\ V_2\ \ldots\ldots\ldots\ldots\ldots\ \}\right]$ is $(2+2m)$.

Equation (5.56) yields the acceleration terms of the generalized coordinates like the column vector $\{\ddot{f}\}$ as follows:

$$\{\ddot{f}\} = [\mathrm{M}]^{-1}\left[\{F(t)\} - [\mathrm{C}_D]\{\dot{f}\} - [\mathrm{K}_S]\{f\}\right] \tag{5.87}$$

This enables one to compute the acceleration vector terms like $\{\ddot{f}\}_{t+\Delta t}$ at the new time step as follows:

$$\{\ddot{f}\}_{t+\Delta t} = [\mathrm{M}]^{-1}\left[\{F\}_{t+\Delta t} - \{F\}_t - [\mathrm{C}_D]\{\dot{f}\}_t - [\mathrm{K}_S]\{f\}_t\right] \tag{5.88}$$

In this equation, $[\mathrm{M}]^{-1}$ represents the terms in the inverse of the mass matrix $[\mathrm{M}]$, and the column vectors $\{F\}_{t+\Delta t}$ and $\{F\}_t$ represent the cumulative sum of all the external forces at the new and old time steps, respectively, acting on the dynamical system made up of the fan shaft and a fully bladed disk. In addition to the centrifugal load effects of unbalance mass, the external force vector $\{F(t)\}$ may also contain the effect of contact forces determined by the penalty method, where all the radial deformations are in the blade tips due to rub stiffness effect represented by the equivalent term K_{rub}, and the surrounding fan containment ring acts like a rigid wall. In the numerical integration algorithm of the nonlinear equations of motion, the number of blades rubbing the case "N_r", the radial load on these blades $\{F_{\text{rub}}\}_{t+\Delta t}$, and the corresponding coefficient of friction m are updated according to the direction of instantaneous velocity of the tip of the respective blades. During the time-marching forward solution, the computer program also keeps track of the bearing loads, and if any of the bearings generate higher loads than its load-carrying capacity, it is considered failed at that instant and is taken out of the analysis from then onwards. At the new time step of $(t+\Delta t)$, the corresponding velocity column vector terms $\{\dot{f}\}_{t+\Delta t}$ and displacement column vector terms $\{f\}_{t+\Delta t}$, respectively, in a finite-difference form may be written as follows:

$$\{\dot{f}\}_{t+\Delta t} = \{\dot{f}\}_t + \Delta t\,\{\ddot{f}\}_{t+\Delta t} \tag{5.89}$$

$$\{f\}_{t+\Delta t} = \{f\}_t + \Delta t\,\frac{\{\dot{f}\}_t + \{\dot{f}\}_{t+\Delta t}}{2} + \frac{1}{2}(\Delta t)^2\{\ddot{f}\}_{t+\Delta t} \tag{5.90}$$

5.2.3.1 Rotor Mode Frequencies and Initial Condition for Transients

Here, first we introduce a new term m_r, which is called "mass moment of unbalance". The dimensions of this term are mass units times radial distance units. Then, the initial conditions for the time-marching forward integration process are established by determining the steady-state forced harmonic response for a nominal unbalance mass of m_r. It is rotating with an angular velocity Ω generating a rotating unbalance radial load vector of magnitude F_U such that in a complex number notation: $F_U e^{i\Omega t} = \left(m_r\Omega^2\right)e^{i\Omega t}$. Referring to Eq. (5.88), the steady-state harmonic response shaft deformation column vector $\{\delta_{\max}\}$ provides the initial condition of displacement vector represented as $\{f\}_{t=0}$ in the equation of motion such that we can write as follows:

a. Rotating load vector components at constant angular velocity Ω:

$$F_X = F_U \cos(\Omega t) = \left(m_r \Omega^2\right)\cos(\Omega t) \text{ and, } F_Y = F_U \sin(\Omega t) = \left(m_r \Omega^2\right)\sin(\Omega t). \tag{5.91}$$

b. Steady-state rotor deformation vector components:

$$\delta_X = \delta_{\max}\cos(\Omega t) \quad \text{and} \quad \delta_Y = \delta_{\max}\sin(\Omega t). \tag{5.92}$$

Thus, in the expanded matrix form, on substituting Eqs. (5.91) and (5.92) in Eq. (5.86) and after some lengthy algebraic manipulations, the following equation is obtained:

$$\begin{bmatrix} -[\mathrm{M}]\Omega^2 + [\mathrm{K}_S] & -[\mathrm{C}_D]\Omega \\ [\mathrm{C}_D]\Omega & -[\mathrm{M}]\Omega^2 + [\mathrm{K}_S] \end{bmatrix} \begin{Bmatrix} \{\delta_X\} \\ \{\delta_Y\} \end{Bmatrix} = \begin{Bmatrix} \left\{\left(m_r\Omega^2\right)\right\} \\ \left\{-\left(m_r\Omega^2\right)\right\} \end{Bmatrix} \tag{5.93}$$

The phase difference for each mode with respect to the rotating unbalance load vector and the displacement peaking out "ψ" are written as follows:

$$\psi = \tan^{-1}\left[\frac{\delta_Y}{\delta_X}\right]. \tag{5.94}$$

The numerical scheme for computing the mass and stiffness matrices is verified and checked by determining the eigenvalues of a cantilever hollow cylindrical shaft with a rigid disk mounted at the free end (see Figure 5.3). Equations (5.53) and (5.54) are easily reduced to the more familiar form for a cantilever shaft, rotating at a constant angular velocity of Ω of span length ℓ mounted with a free-end rigid disk of mass M_D and diametral mass moment of inertia as J_D. For determining the eigenvalues after

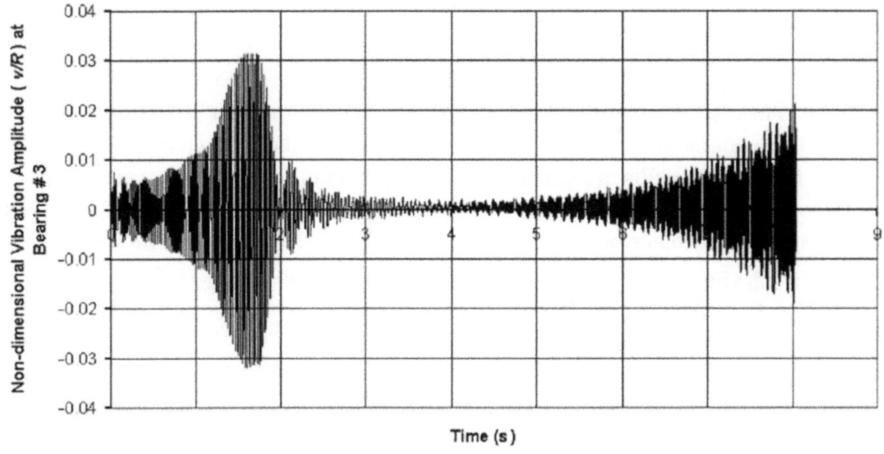

FIGURE 5.7 Analytically computed transient response of a decelerating fan rotor after the blade loss. (Reprinted with permission from Reference [3].)

dropping the shaft damping terms and also the rub stiffness terms related to K_{rub}, the respective simplified equations in the global X and Y directions are as follows:

$$\sum_{m=1}^{\text{Modes}}\left[(\rho A)_s\int_0^{\ell} W_m W_n dZ - I_d\int_0^{\ell} W_m'' W_n dZ + M_D\, W_m(\ell)\, W_n(\ell) + J_D W_m'(\ell) W_n'(\ell)\right]\ddot{U}_m$$

$$+\left[(EI)_s\int_0^{\ell} W_m'''' W_n dZ - (EI)_s\, W_m'''(\ell)\, W_n(\ell)\right]U_m$$

$$-2\Omega\left[I_d\left[\int_0^{\ell} W_m'' W_n dZ\right] - J_D\left[W_m'(\ell) W_n'(\ell)\right]\right]\dot{V}_m$$

$$= F_U W_n(\ell)\cos(\theta_0 + \Omega t) \tag{5.95}$$

$$\sum_{m=1}^{\text{Modes}}\left[(\rho A)_s\int_0^{\ell} W_m W_n dZ - I_d\int_0^{\ell} W_m'' W_n dZ + M_D\, W_m(\ell)\, W_n(\ell) + J_D W_m'(\ell) W_n'(\ell)\right]\ddot{V}_m$$

$$+\left[(EI)_s\int_0^{\ell} W_m'''' W_n dZ - (EI)_s\, W_m'''(\ell)\, W_n(\ell)\right]V_m \tag{5.96}$$

$$+2\Omega\left[I_d\left[\int_0^{\ell} W_m'' W_n dZ\right] - J_D\left[W_m'(\ell) W_n'(\ell)\right]\right]\dot{U}_m$$

$$= F_U W_n(\ell)\sin(\theta_0 + \Omega t)$$

Using the above set of equations outlined in Eqs. (5.95) and 5.96), the frequency characteristics of a simple overhung cantilevered type of rotor shaft are computed for a typical large engine rotor with its geometrical and mass data shown in Table 5.1. The sample numerical results are generated for a typical fan rotor with parameters listed in Table 5.1. For the numerical results, all linear dimensions are nondimensionalized with respect to the outer radius of R, and the computed frequency coefficient parameter ξ for varying ratios of $\left(M_D / M_s \right)$ is normalized as shown here. The nondimensional frequency parameter ξ is defined by treating the shaft as a simple beam and using its flexural rigidity $(EI)_s$ and mass per unit length $(\rho A)_s$ as follows:

$$\omega_n = \frac{\xi}{\ell^2}\sqrt{\frac{(EI)_s}{(\rho A)_s}}\ \text{radians/s} = \frac{1}{2\pi}\frac{\xi}{\ell^2}\sqrt{\frac{(EI)_s}{(\rho A)_s}}\ \text{Hz} \tag{5.97}$$

TABLE 5.1
Details of the Parameters Considered for the Sample Rotor Used in the Discussion

M_b	Mass of one blade	10 kg
M_D	Mass of the disk	150 kg
D_L	Axial length of the rigid disk	25 cm
R	Blade-tip radius	150 cm
r	Blade root radius = disk outer radius	50 cm
L	Blade length	100 cm
$(A)_b$	Blade cross-sectional area	5000 mm^2
$(I)_b$	Blade area moment of inertia	0.5E6 mm^4
K_{rub}	Case filler material stiffness	10 MN/m
ℓ	Length of the shaft	350 cm
r_0	Outside radius of the shaft	7.0 cm
r_i	Inside radius of the shaft	5.5 cm
$(\rho)_S$	Mass density of the shaft material	7.833 g/cm^3
β	Blade stagger angle at the tip	60
Speed		3000 rpm
Ω	Rotor initial steady-state spin velocity (rad/s)	314.159 rad/s
m_r	Mass moment of unbalance about engine axis	1E6 g-cm
θ_0	Circumferential location of the unbalance at time	$t = 0$ 180°
N	Number of blades on the rotor	28
N_b	Number of discrete bearings supporting the shaft	3
Z_i	Bearing support locations	0, 300 cm, 325 cm
μ_{static}	Static coefficient of friction between the blade tip and the outer case inner surface	0.1
μ_{dyn}	Dynamic coefficient of friction between the blade tip and the outer case inner surface	0.01
$(E)_S$	Elastic Young's modulus of the rotor shaft material	200 GPa
$(E)_b$	Elastic Young's modulus of the blade material	117 GPa
C	Blade chord	35 cm

The computed nondimensional natural frequency parameter ξ for varying disk mass M_D is summarized in Table 5.2. The computed values of ξ for the static condition match with the closed-form solution given in many textbooks. The tabular data illustrates the fact that lower flex frequencies move down rapidly with the increasing mass; however, for higher flex modes the change is not significant. The effect of increasing mass moment of inertia of the disk "J_D" is shown in Table 5.3. It is obvious

TABLE 5.2
Computed Nondimensional Frequencies ξ of a Cantilever Shaft with Rigid Overhung Disk of Concentrated Point Mass M_D

Disk-Mass $\left(M_D/M_s \right)$	1st Flex	2nd Flex	3rd Flex	4th Flex	5th Flex	6th Flex
0.00	3.5160	22.0351	61.7052	120.9317	199.9649	298.7559
0.10	2.9678	19.3563	55.5245	110.7328	185.4243	279.7140
0.25	2.4766	17.8520	53.0221	107.5703	181.7910	275.7414
0.50	2.0163	16.9018	51.7063	106.0816	180.1953	274.0769
0.75	1.7431	16.4846	51.1823	105.5178	179.6080	273.4753
1.00	1.5573	16.2505	50.9011	105.2217	179.3031	273.1653
2.00	1.1582	15.8613	50.4528	104.7583	178.8309	272.6872
3.00	0.9628	15.7201	50.2958	104.5986	178.6695	272.5239
4.00	0.8415	15.6472	50.2159	104.5176	178.5882	272.4415
5.00	0.7569	15.6026	50.1674	104.4687	178.5393	272.3917
6.00	0.6936	15.5727	50.1349	104.4359	178.5065	272.3579
7.00	0.6439	15.5511	50.1116	104.4124	178.4830	272.3334
8.00	0.6035	15.5348	50.0940	104.3948	178.4656	272.3154
9.00	0.5699	15.5220	50.0804	104.3810	178.4520	272.3008
10.00	0.5414	15.5119	50.0694	104.3700	178.4411	272.2888
15.00	0.4438	15.4810	50.0365	104.3368	178.4088	272.2525
35.00	0.2918	15.4455	49.9988	104.2980	178.3744	272.2000

TABLE 5.3
Effect of Mass Moment of Inertia of the Disk on the Nondimensional Static Frequencies ξ of a Cantilever Shaft with Rigid Disk of Mass M_D (For a Typical Value of $M_D/M_S = 15$, J_D = Diametral Mass Moment of Inertia of the Disk = $\left(M_D r^2\right)/4$, r = Outer Radius of the Disk)

$\dfrac{(J_D)}{(\rho A\,\ell^3)_S}$	1st Flex	2nd Flex	3rd Flex	4th Flex	5th Flex	6th Flex
0.000	0.4438	15.4810	50.0365	104.3368	178.4088	272.2525
0.025	0.4429	10.5528	27.3565	65.3865	125.7987	207.5618
0.050	0.4421	8.2670	25.2366	64.5455	125.3496	207.2700
0.500	0.4297	2.9697	23.3562	63.8173	124.9521	207.0110
2.500	0.3753	1.5359	23.1966	63.7546	124.9166	206.9938
5.000	0.3284	1.2499	23.1767	63.7465	124.9152	206.9867
10.000	0.2654	1.0755	23.1668	63.7419	124.9148	206.9845
25.000	0.1906	0.9767	23.1603	63.7388	124.8836	206.9536
50.000	0.1456	0.9428	23.1596	63.7357	124.8525	206.9135

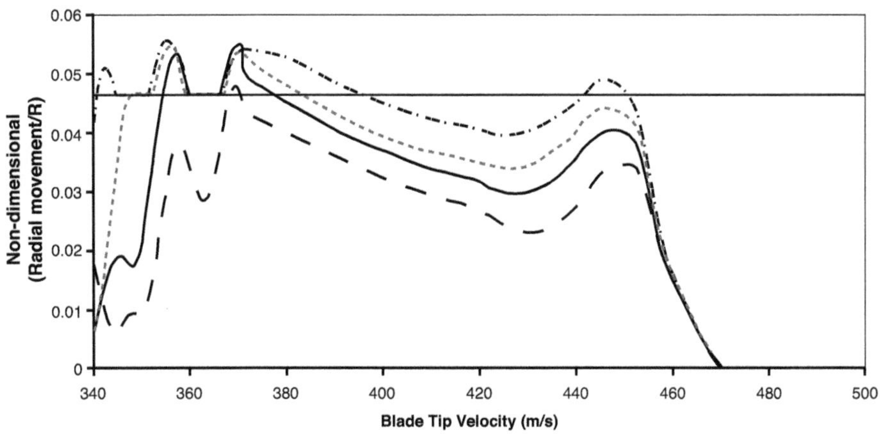

FIGURE 5.8 Non-dimensional radial movement of the center of fan disk after the blade loss (case filler material thickness/blade-tip radius = 0.0467) [— — —, K_{rub} = 0.0 MN/m (air gap); K_{rub} = MN/m; ———, K_{rub} = 10 MN/m;— — — —, K_{rub} = MN/m; ——, line for blade tips deforming under hard rub against the rigid fan case]. (Reprinted with permission from Reference [3].)

that the effect of increase in "J_D" for the lower mode is again more pronounced than for the higher modes, where the nondimensional natural frequency parameter ξ becomes asymptotic very quickly.

The two-system modes identified for the above rotor are 14.86 Hz (891.5 rpm) and 44.49 Hz (2669.5 rpm). The typical mode shape is shown in Figure 5.4. The harmonic lateral response at the Bearing No. 3 support for a nominal unbalance with the rigid disk is illustrated in Figure 5.5. The effects of gyroscopic terms are plotted in Figure 5.6 by listing the forward and backward whirl frequencies for the first and second flex bending modes at different rotational speeds Ω.

Before analyzing the transient response of the rotor, we have also investigated the steady-state dynamic response of individual blades on this rotor during the rotating and nonrotating conditions. The dynamic buckling characteristics of each blade with its root clamped at the top of the disk in terms of its frequency and the critical compressive buckling load is considered in Table 5.4. Here, we see that the speed has a significant effect on the first and second flex mode frequencies of the blade, but relatively minimal effect on the higher modes.

5.2.3.2 Rotor Dynamic Response Through Resonance

The transient response of this rotor is considered under an assumed condition of an unbalance created due to the sudden blade loss from the disk during normal operation of 3000 rpm. After the blade is lost, the engine stalls and the rotor starts decelerating until it comes to a complete stop. During the deceleration, as the rotor passes through its resonance the disk goes through large radial excursion causing the blades to rub momentarily against the outer case. The transient response of this rotor for a typical blade-loss condition is analyzed by numerically solving the coupled equations of motion (Eq. 5.88). The computed lateral dynamic response of Bearing No. 3 is shown in Figure 5.7.

TABLE 5.4
Dynamic Buckling Characteristics of Each Blade with Its Root Clamped at the Top of the Disk

Flexural Mode	Frequency				Critical Buckling Load (MN)	
	(a) Stationary		(b) 3000 rpm		(a) Stationary	(b) 3000 rpm
	Frequency Parameter ξ	Frequency (Hz)	Frequency Parameter ξ	Frequency (Hz)		
1st flex	3.5160	42.927	6.7116	81.942	0.1446	0.4892
2nd flex	22.0351	269.028	25.8109	315.127	1.3011	1.8937
3rd flex	61.7052	753.362	65.6918	802.034	3.6142	4.2267
4th flex	120.9317	1476.460	125.1833	1528.369	7.0834	7.7416
5th flex	199.9649	2441.380	204.5480	2497.336	11.7098	12.3959
6th flex	298.7559	3647.524	302.7786	3696.637	17.4925	18.2613

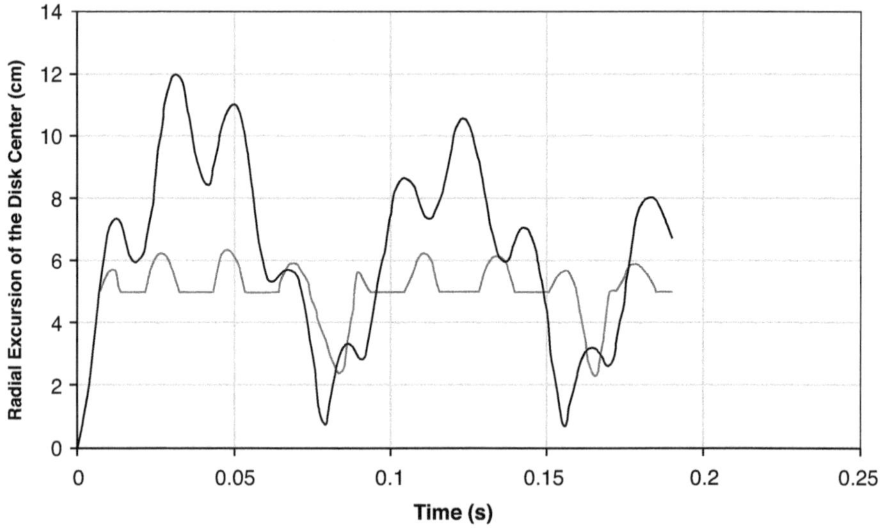

FIGURE 5.9 Transient radial response of the disk center (movement of rotor center beyond 5 cm due to blade-tip pulse buckling) with and without the blade rubbing against the fan case after the blade loss [——, tip clearance = 12 cm; ——, tip clearance = 5 cm]. (Reprinted with permission from Reference [3].)

It can be seen that the peak dynamic response during deceleration does not occur at the same speed corresponding to the system mode frequencies, rather it takes place at a lower speed, which depends upon the rate of slowing down of the rotor. The opposite happens during acceleration, where the peak response takes place at a higher speed. This is a generic characteristic of an accelerating and decelerating system.

The rotor model described in Table 5.1 of this chapter is used to study the effect of various parameters affecting the rub load. The effect of varying filler stiffness K_{rub} is highlighted in Figure 5.8, where the ratio of the tip clearance of the hard surface to the blade-tip radius is kept constant as 0.0467 (rigid case gap = 7 cm; R = 150 cm). The rotor radial movement exhibits two distinctly different peaks, the first at 450 m/s tip velocity and the second at 370 m/s. The transient radial movement of the rotor is computed for three different values of filler material stiffness, namely, 5, 10, and 20 MN/m in addition to the first case with a free air gap and without any filler material in the fan case. The hard rub of blade tip against the outer case would occur at a filler stiffness less than 2 MN/m. During the hard rub, the blade tips deform and the disk center still keep on moving toward the case. Depending upon the ductility of the blade material, the blade tips may plastically curl, or if it is brittle then it may be machined under grinding action. The effect of changing tip clearance is shown in Figure 5.9 by plotting the radial movement of the fan rotor after the blade loss for two different tip clearance values of 5 and 12 cm: The free unimpeded radial movement of the rotor at the disk center with this much amount of unbalance wants to be 12 cm: Obviously, a tip clearance less than 12 cm would result in the blade tips rubbing against the case, resulting in the dynamic pulse buckling at some critical value of the radial load from the case (Figure 5.10). For example, with 5 cm of tip clearance the rotor disk-center moves up to 6.5 cm from the center of rotation due to tip buckling of the blade resulting in radial movement of the blade tip by 1.5 cm. During this movement of the blade,

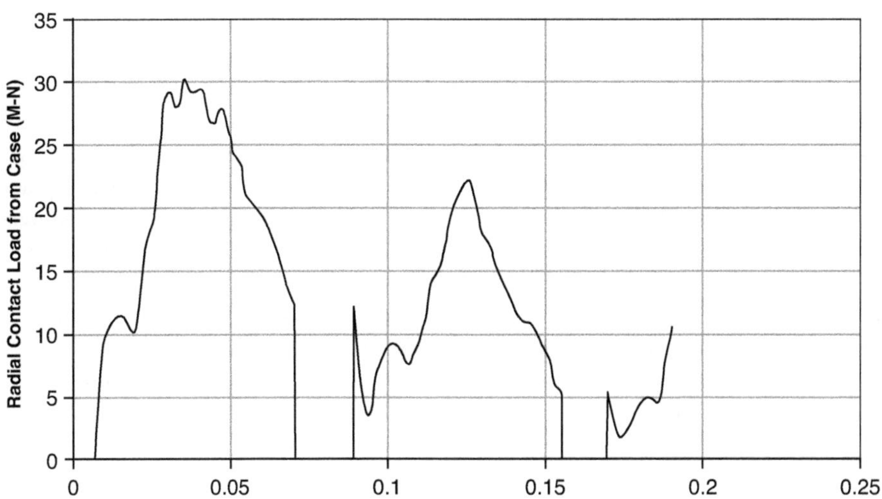

FIGURE 5.10 Transient radial response of the disk center with and without the blade rubbing against the fan case after the blade loss. (Reprinted with permission from Reference [3].)

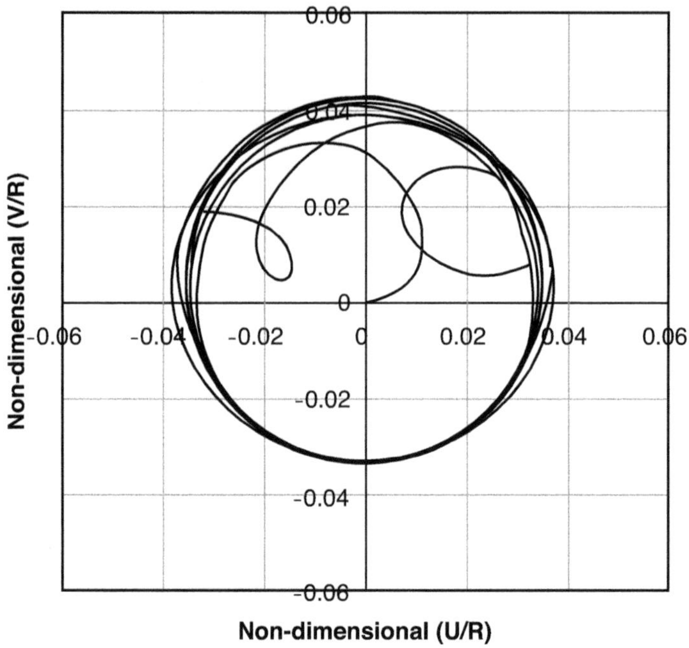

FIGURE 5.11 Non-dimensional orbit of the fan disk center after the unbalance created due to blade loss at 9 o'clock position. (Reprinted with permission from Reference [3].)

the sum of the radial load from the fan case to the blade tips goes up to 30 MN (see Figure 5.10) with first contact lasting for about 0.06 second. It should be noted that this is the analytically computed highest possible radial load on the rotor due to the simplifying assumption of elastic buckling and the blade geometry being completely radial. In reality, the blade tips will deform plastically, once the load is large enough to yield the material and will bend backward easily due to the blade geometry not being completely radial and as such the peak dynamic load in most rub situations will be somewhat lower, but would spread over a longer period than 0.06 second. The instantaneous dynamic stiffness of the rotor during rub changes, but the first elastic contact results in an equivalent blade segment elastic stiffness of 1749 MN/m. However, if the free radial movement of the rotor through the air gap is also taken into account for the effective stiffness calculation, then the numerical value of the effective blade segment elastic stiffness reduces to 190 MN/m. The transient orbit of the fan disk center after the blade loss is shown in Figure 5.11.

For understanding the effect of changing blade-tip clearance on the contact load, the same rotor model was run for the air gap values of 1, 2, 4, 5, 7, 10, and 12 cm: The computed values of peak contact load and the peak radial movement of the rotor in a nondimensional form as a function of nondimensional tip clearances are shown in Figure 5.12. The plot shows that as the clearance is increased, for a given unbalance of a blade loss, the radial movement of the rotor increases almost linearly and the contact load decreases with a change in the slope of the curve. From these data, one

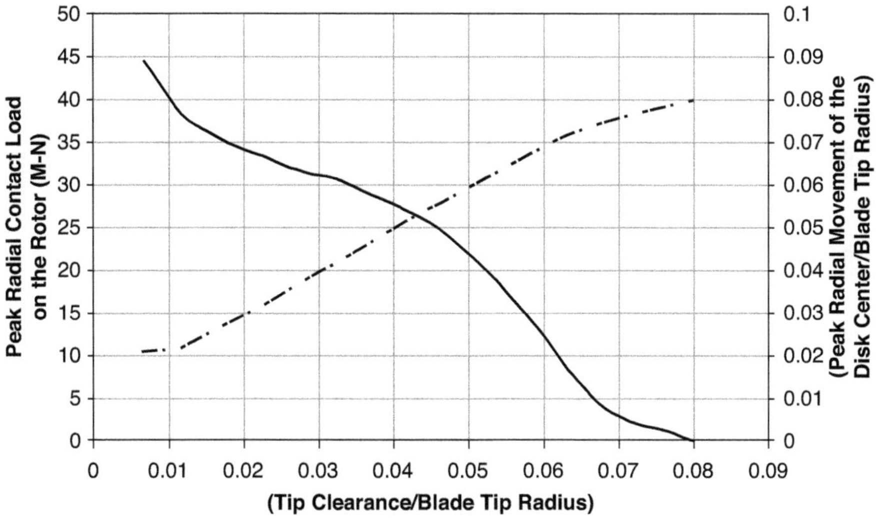

FIGURE 5.12 Peak radial contact load on the rotor and the disk center radial movement as a function of the tip clearance (no filler material) [— —, radial movement; ——, contact load]. (Reprinted with permission from Reference [3].)

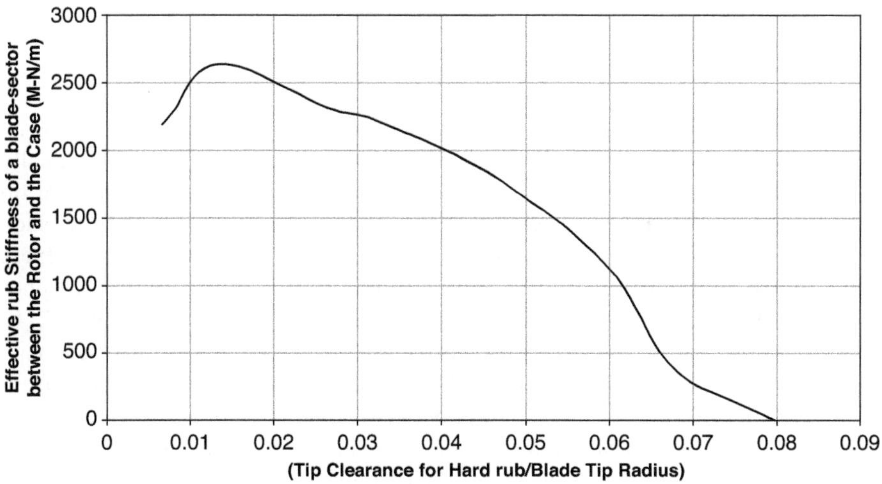

FIGURE 5.13 Effective blade tip rub stiffness between the rotor and the case as a function of tip clearance (no casing filler material). (Reprinted with permission from Reference [3].)

can compute the effective dynamic stiffness of a segment of rub blades on the rotor during the hard rub as follows:

$$\text{Effective stiffness} = \frac{(\text{Contact load})}{(\text{Total radial movement of the rotor center} - \text{tip clearance})} \tag{5.97}$$

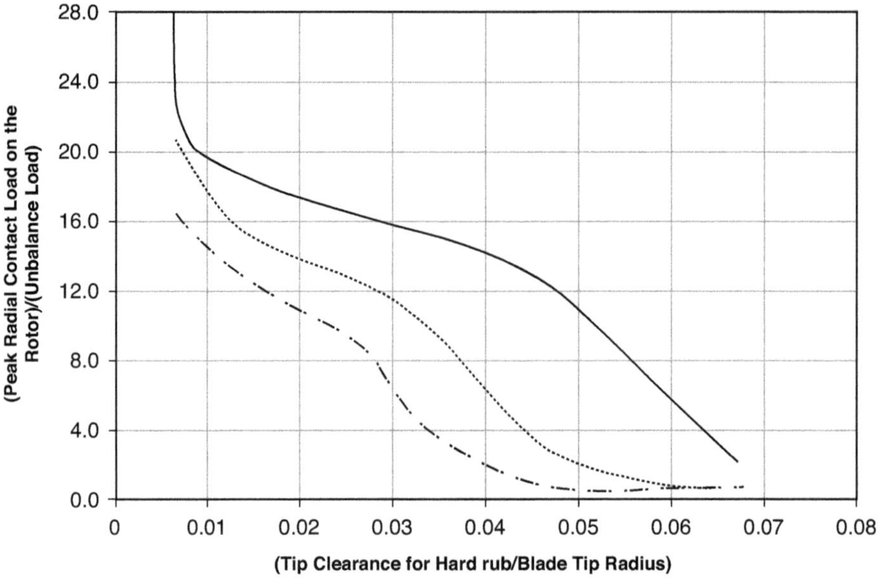

FIGURE 5.14 Non-dimensional peak radial tip rub load on the rotor from the case during hard rub as a function of tip clearance and filler material stiffness [— — — —, air gap $K_{rub} = 0$; filler material $K_{rub} = 10$ MN/m; — — — —, filler material $K_{rub} = 20$ MN/m]. (Reprinted with permission from Reference [3].)

The above relationship is used to plot the curve shown in Figure 5.13 from the contact load data of Figure 5.12. For most part, the blade segment is dynamically more flexible with increasing clearance. There appears to be a nondimensional tip clearance value of 0.013, at which the dynamic elastic stiffness value of the blade segment peaks. The peak radial contact load is also analyzed for changing stiffness values of the filler material K_{rub}. We have studied its effect for two different values of $K_{rub} = 10$ and 20 MN/m in addition to free air gap, i.e., $K_{rub} = 0$ (see Figure 5.14). As expected, the contact load decreases as the filler material stiffness increases (e.g., for nondimensional tip clearance of 0.03 the nondimensional force decreases from 11.5 to 6.5). It is caused because the rotor outward radial velocity during the hard rub decreases, or the rotor is slowed down faster as the filler material gets stiffer.

An occasional hard local rub against the case may not be destabilizing. In these situations, the rotor can momentarily withstand frictional torque even with a high coefficient of friction. The numerical solution has shown that for an impulse type of torque, the rotor experiences relatively large radial excursion; however, it recovers within a few milliseconds and goes to normal steady-state rotation. The presence of damping in the bearing support $[D]_b$ matrix plays a significant role in making the system dynamically stable. But, in 360° rub, even a very (Figure 5.13) effective blade-tip rub stiffness between rotor and the case as a function of tip clearance (no filler material) nominal unbalance and a low dynamic coefficient of friction $\mu_{dynamic} = 0.01$, radial excursion quickly grows into large magnitude. On generating the contact load data at the blade tip, shown in Figure 5.14, it can be seen that the

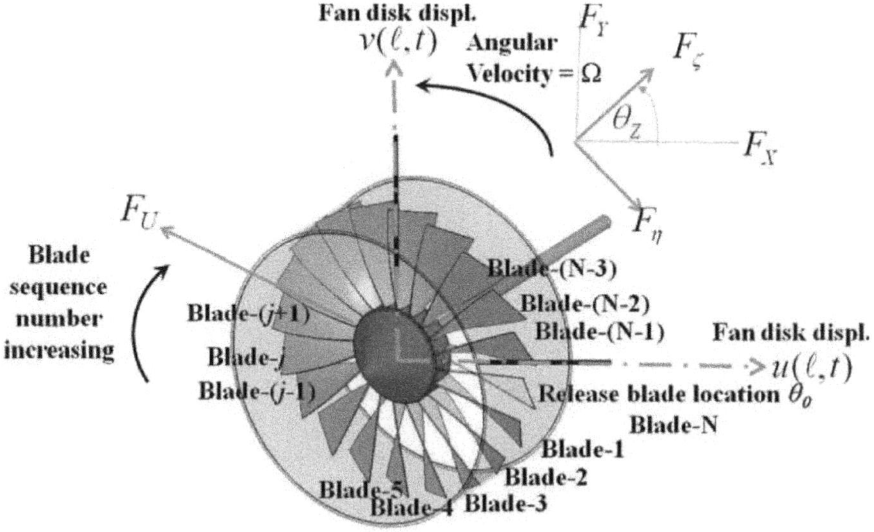

FIGURE 5.15 Schematic front view of the bladed-fan disk with asymmetric rotor due to blade loss during a typical fan-blade-out (FBO) event with blade release at 0° with respect to the horizontal X-axis. (Reprinted with permission from Reference [5].)

contact load decreases as filler material stiffness is increased. It is assumed that the driving torque is maintaining a constant speed of the rotor. For a given spinning rpm of the rotor, it becomes less stable with increasing torque and more stable with increasing damping.

One of the most interesting observations is that the load torque has much more stabilizing effect at higher speeds on a rotor than a driving torque. As a matter of fact, the rotor becomes stable with increasing load torque and less stable with increasing driving torque. An increasing load torque will decelerate the rotor whereas an increasing driving torque will accelerate it. After the blade loss, the frictional torque with the case can cause a very rapid speed drop of the rotor shaft. A correct estimate of transient dynamic load is necessary for the design of the supporting structure.

5.2.3.3 Turbofan Rotor Behavior due to FBO Event with Blade Deformation

In this section, we will focus our discussion specifically on the dynamic characteristics of a fully bladed fan rotor, and the blade airfoils are highly deformable. In the previous section, this item was touched only very briefly in reference to the mode shape of the fan rotor, shown in Figure 5.4, and the interaction of rotor shaft and airfoil blade deformation together. It should be noted that in the previous section, the entire flexibility at the blade casing the flexibility was only in the abradable. Additionally, during the FBO event, the rotor unbalance load F_U is acting exactly in the opposite direction of the blade loss location on the disk sector. In Figure 5.15, the blade is released at 0° with respect to the horizontal X-axis. The number of blades on this rotor under consideration is $N = 20$.

When the blades are considered rigid, the eigenvalue solution yields first two fundamental modes of shaft vibration as 63.8 and 70.11 Hz (see Figure 5.16).

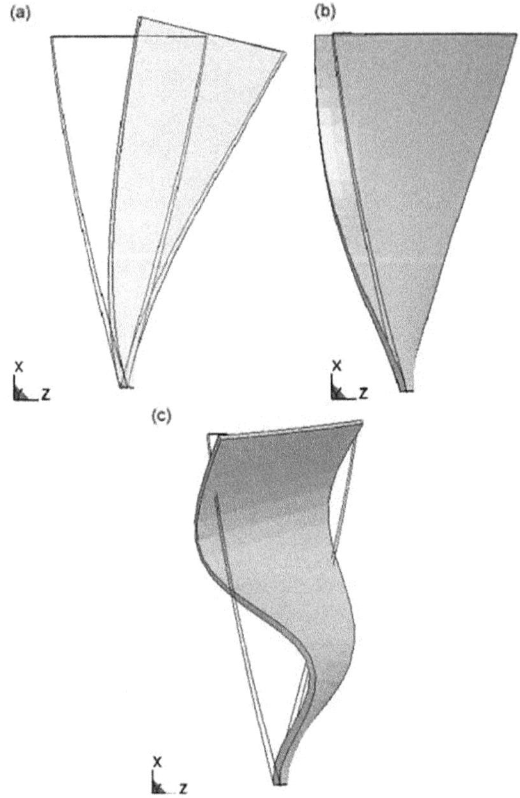

FIGURE 5.16 Fan blade frequencies and respective mode shapes in stationary and running conditions: (a) 1F-mode (45.89 Hz stationary blade, 51.65 Hz at 1000 rev/min, 84.08 Hz at 3000 rev/min), (b) 1T-mode (182.78 Hz stationary blade, 183.44 Hz at 1000 rev/min, 198.89 Hz at 3000 rev/min), and (c) 2T-mode (552.50 Hz stationary blade, 550.82 Hz at 1000 rev/min, 549.86 Hz at 3000 rev/min). (Reprinted with permission from Reference [5].)

However, when the blades are considered flexible, with its representative elastic Young's modulus value, the first shaft bending mode frequency drops by 8.4% to 58.45 Hz. As shown in Figure 5.17, there is a very clear participation of the blade first flexural mode shapes with the deformation of the shaft.

Interestingly, for this rotor, the bending mode of the shaft in the rotating condition at the top speed of 3000 rev/min again moves up to 62.73 Hz, which is very close to the rotor frequency with the rigid blade assumption. Thus, it shows that in case of an axisymmetric rotor the assumption of blade being rigid is good enough for determining the critical speed of the rotor. However, when the same rotor is analyzed with a missing blade making it asymmetric, the shaft mode at 58.45 Hz bifurcates in two vibrational modes with the lowest rotor modes showing at 58.35 Hz and the next one slightly higher at 59.97 Hz. Similarly, the other modes previously showing up at 70.26 Hz shift along 70.26 and 70.28 Hz, respectively. In the rotating condition with spinning speed as 1000 and 3000 rev/min, the corresponding natural frequencies of this rotor are computed as follows:

a. 61.31 Hz and 70.30 Hz at 1000 rev/min (rotating symmetric rotor)
   62.85 Hz and 70.35 Hz at 1000 rev/min (rotating asymmetric by FBO blade)
b. 63.23 Hz and 70.20 Hz at 3000 rev/min (rotating symmetric rotor)
   64.44 Hz and 70.21 Hz at 3000 rev/min (rotating asymmetric by FBO blade).

It should be noted that the above frequencies still inherently assume that all the bearing supports are fully active. However, in general by the time the windmilling of the asymmetric rotor with a blade missing occurs, Bearing No. 3, closest to the overhung fan disk may also be damaged or fused due to huge imbalance in the system. It should be noted that the above frequencies still inherently assume that all the bearing supports are fully active. The design philosophy behind the failure of Bearing No. 3 allowed to fail is discussed in quite detail by VanDuyn [see Reference 7] in his patent disclosure. For this type of asymmetric rotor with a large imbalance, the critical speeds are best established by determining the forced response of the rotor at varying rotational speed of the shaft. Figure 5.7 shows the forced harmonic response of such a damaged rotor shaft by computing its amplitudes of horizontal and vertical motions at the bearing support location separately for each speed (0–3000 rev/min) being changed at the increment of 1 rev/min. The response plot highlights the existence of two resonant speeds for this type of decelerating damaged rotor, with the first one occurring at 2380 rev/min (39.67 Hz) and the second one at 100 rev/min (1.67 Hz). In general, a damaged rotor shaft due to blade loss invariably decelerates due to aerodynamic stalling of the turbofan. That is the most common observation after a typical blade loss. A rapidly decelerating rotor shaft may exhibit its peak vibrational response within a band of speed slightly lower than its resonant speed computed in this manner (Figures 5.17 and 5.18).

However, in a hypothetical scenario of a fan blade-loss event coupled with the simultaneous mid-shaft failure, the turbine-end of the rotor shaft may accelerate after the blade-out occurs. The original shaft mode frequency (corresponding to 58.45 Hz stationary and 62.73 Hz in a rotating condition of 3000 rev/min with the three undamaged bearing support system), shown in Figure 5.9, drops to 39.67 Hz with only Bearings-1 and -2 remaining active as a result of the fusing of Bearing-3 soon after encountering the large imbalance caused by the fan blade loss. The shaft mode frequencies of 39.67 Hz, sometimes known as "decoupled mode", and the lower one at 1.67 Hz can also be easily identified by solving eigenvalue problem of the fan rotor shaft supported only on two remaining bearings, i.e., Bearing No. 1 and Bearing No. 2, respectively.

The fan rotor starts slowing down immediately after the blade loss as the rotor shaft decelerates and its rotational speed drops from the peak speed of 3000 rev/min to the state of rest. In most of the cases also per FAA (Federal Aviation Administration) requirements for safe operation of the crippled aircraft, this happens in 15 seconds or less. The resonant speed of 2380 rev/min of the damaged rotor-bearing system exhibits the fan blades momentarily rubbing against the fan case/abradable during the transient event as the rotor shaft during the spool-down passes through the 39.67 Hz excitation frequency. As some of the blade tips may experience intermittent rubs, the rotor–shaft response frequency moves up slightly due to the additional load path in the dynamical system. Thus, effectively the fan-rotor response frequency during the transient event of the spool-down may have slight shift due to nonlinearity of the system. These nonlinear contact rub effects will definitely influence the dynamic

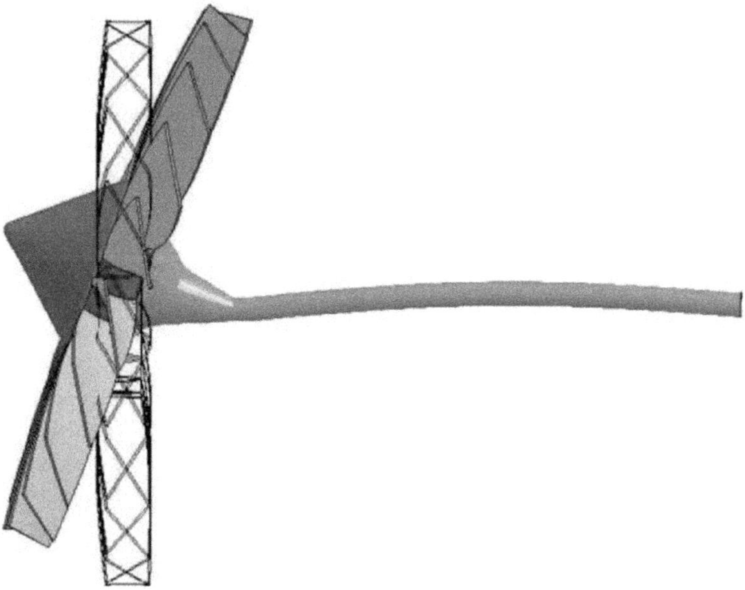

FIGURE 5.17 Symmetric rotor natural modes with rigid blades (no blade participation). (Reprinted with permission from Reference [5].)

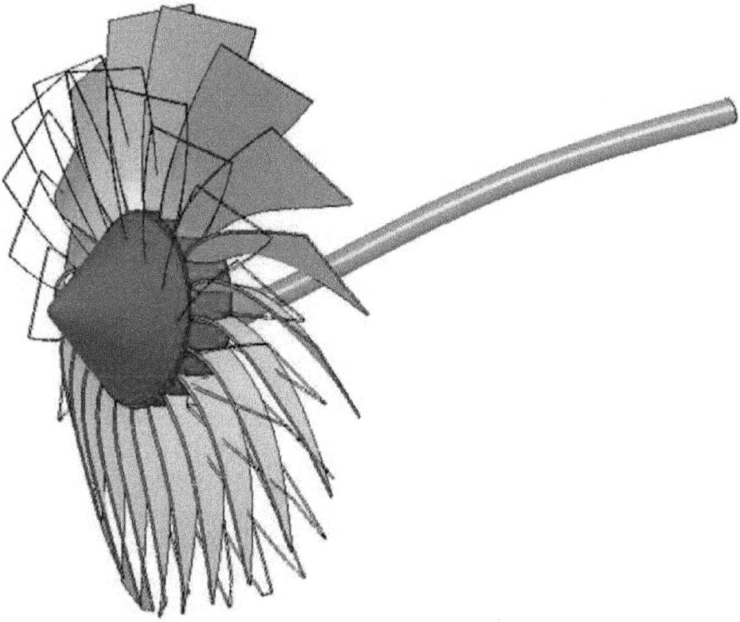

FIGURE 5.18 Symmetric rotor natural modes with flexible blades (blades deforming in its own flexural vibrational mode). (a) 58.45 and 70.26 Hz at zero speed (stationary condition) and (b) 62.73 and 70.20 Hz at 3000 rev/min (rotating condition). (Reprinted with permission from Reference [5].)

response of the damaged rotor during the windmilling phase in a significant way. The dynamic characteristics of such a nonlinear system can only be predicted with any confidence by using a time-marching numerical scheme for integrating the equations of motion. Section 3.2 discusses the transient response of such a rotor.

The lower resonant speed of 100 rev/min may become important in the windmilling of the damaged rotor during the fly-home mission. At this time since the rotor is very lightly damped and the natural frequencies in the two directions are slightly separated, the rotor can wildly oscillate around 100 rev/min and also at multiples of the windmilling speed.

5.2.3.4 Dynamic Response of Fan Rotor Impacted by Bird Strike

For determining the fan rotor transient vibratory response after a bird strike is discussed in this section. To solve the rotordynamic problem, one needs to transform the bird impact load from the airfoil coordinate system to the fan rotor global coordinate system. In this case referring to Figure (5.4), the unbalance load on the fan rotor shaft F_U caused by the bird slice of mass B_m at the blade airfoil attached to the fan disk located at $Z = \ell$ is determined based upon the span height "s" of bird-strike impact. Here, the horizontal and vertical components of the unbalance rotating radial load vector F_U acting on the fan rotor shaft in the global $(X - Y)$ plane are as follows, with the angle of the impacted blade being $\theta_0 = 90°$ with respect to the X-axis:

$$(F_X)_2 = B_m\Omega^2(r+s)\cos(\theta_0 + \Omega t) \tag{5.98}$$

$$(F_Y)_2 = B_m\Omega^2(r+s)\sin(\theta_0 + \Omega t) \tag{5.99}$$

To that end referring to Figure (5.19), we start with the following coordinate transformation equation. As shown in the previous figure (see Figure 5.4), the setting angle β_r is also the angle of pretwist of the blade at the airfoil root and at the top of the fan disk, which are radially coincident. For the rotor model, the polar coordinate system used in the analysis is shown in Figure 5.19a. Thus, the coordinate transformation equations for the $(\hat{e}_T, \hat{e}_A, \hat{e}_R)$ polar coordinate system in the matrivx form are written as follows (Figure 5.20):

$$\begin{Bmatrix} \hat{e}_T \\ \hat{e}_A \\ \hat{e}_R \end{Bmatrix} = \begin{bmatrix} -\sin(\Omega t) & \cos(\Omega t) & 0 \\ 0 & 0 & 1 \\ \cos(\Omega t) & \sin(\Omega t) & 0 \end{bmatrix} \begin{Bmatrix} \hat{i} \\ \hat{j} \\ \hat{k} \end{Bmatrix} \tag{5.100}$$

Similarly, the blade "Chord-Normal-Span" coordinate system is

$$\begin{Bmatrix} \hat{e}_c \\ \hat{e}_n \\ \hat{e}_s \end{Bmatrix} = \begin{bmatrix} -\sin\beta & \cos\beta & 0 \\ -\cos\beta & -\sin\beta & 0 \\ 0 & 0 & 1 \end{bmatrix} \begin{Bmatrix} \hat{e}_T \\ \hat{e}_A \\ \hat{e}_R \end{Bmatrix} \tag{5.101}$$

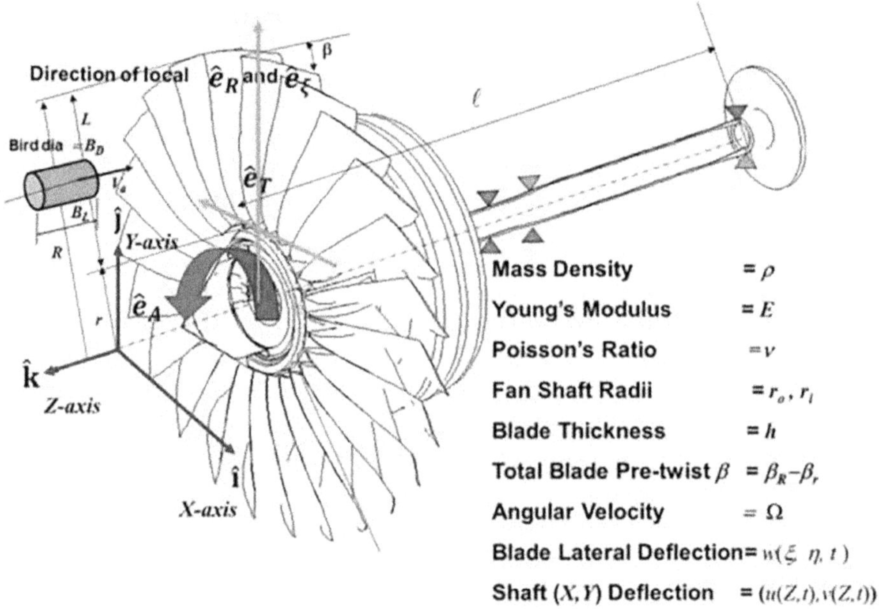

FIGURE 5.19 A typical three-bearing supported turbofan bladed-rotor hollow shaft (length = ℓ)with pretwisted fan blades of thickness $\boldsymbol{h}$ impacted by a cylindrical hydrodynamic bird projectile of diameter $\boldsymbol{B_D}$ and length $\boldsymbol{B_L}$, respectively. (Reprinted with permission from Reference [6].)

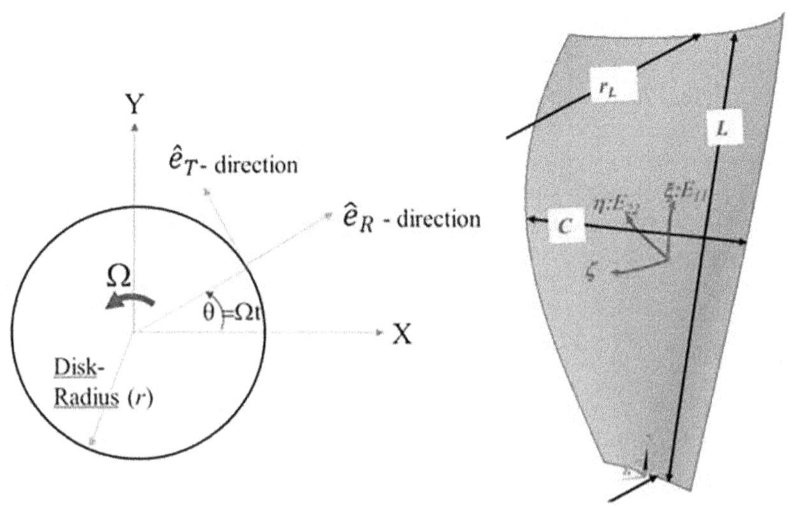

FIGURE 5.20 Two different curvilinear systems used in the mathematical formulation: (a) polar coordinate system used in the shaft dynamic formulation (forward-looking aft view of engine in $\hat{e}_A$ direction coinciding with global **Z**-axis) and (b) (ξ,η,ζ) thin-shell local Lagrangian 3D coordinate system, showing view from the concave side of the fan blade airfoil. (Reprinted with permission from Reference [6].)

Thus, with a fan blade with airfoil root setting angle being <ı>β_r</ı>, we have

$$\begin{Bmatrix} \hat{e}_c \\ \hat{e}_n \\ \hat{e}_s \end{Bmatrix} = \begin{bmatrix} -\sin(\beta_r + \beta' s) & \cos(\beta_r + \beta' s) & 0 \\ -\cos(\beta_r + \beta' s) & -\sin(\beta_r + \beta' s) & 0 \\ 0 & 0 & 1 \end{bmatrix} \begin{Bmatrix} \hat{e}_T \\ \hat{e}_A \\ \hat{e}_R \end{Bmatrix} \tag{5.102}$$

Thus, with the rotational angular velocity Ω, the "Chord-Normal-Span" coordinate system relationship becomes

$$\begin{Bmatrix} \hat{e}_c \\ \hat{e}_n \\ \hat{e}_s \end{Bmatrix} = \begin{bmatrix} \sin\beta\sin(\Omega t) & -\sin\beta\cos(\Omega t) & \cos\beta \\ \cos\beta\sin(\Omega t) & -\cos\beta\cos(\Omega t) & -\sin\beta \\ \cos(\Omega t) & \sin(\Omega t) & 0 \end{bmatrix} \begin{Bmatrix} \hat{i} \\ \hat{j} \\ \hat{k} \end{Bmatrix} \tag{5.103}$$

Thus,

$$\begin{Bmatrix} \hat{i} \\ \hat{j} \\ \hat{k} \end{Bmatrix} = \begin{bmatrix} \sin\beta\sin(\Omega t) & \cos\beta\sin(\Omega t) & \cos(\Omega t) \\ -\sin\beta\cos(\Omega t) & -\cos\beta\cos(\Omega t) & \sin(\Omega t) \\ \cos\beta & -\sin\beta & 0 \end{bmatrix} \begin{Bmatrix} \hat{e}_c \\ \hat{e}_n \\ \hat{e}_s \end{Bmatrix} \tag{5.104}$$

where the angle of pretwist β in the airfoil at any typical span-height of s is given by (see Figure 5.4)

$$\beta = (\beta_r + \beta' s). \tag{5.105}$$

In the numerical scheme, it is assumed that the pretwist angle $\beta(s)$ of the fan blade changes linearly from the root of the airfoil toward the tip. As such, the twist angle rate change parameter β' is constant. This is a perfectly valid assumption. In all fan blade designs, each airfoil cross-section is linearly rotated about its stacking axis, as we move radially outward in the span direction. Thus, the spanwise twist angle of blade moves from β_r being at the root to twist angle of β_R at the tip.

For determining the external forces on the fan shaft due to impact forces of the bird strike, we will assume the fan blade deformation to be negligibly small such that any local deformation will not result in any geometry change, which will affect the time history of impact forces called $F_{\text{Impact}}(t)$. This assumption is perfectly reasonable in large size composite fan blades. Thus, the net external forces on the fan shaft are as follows:

$$F_X = (F_X)_1 + (F_X)_2 = -F_{\text{Impact}}(t)\left[\cos(\beta_r + \beta' s)\sin(\theta_0 + \Omega t)\right] + B_m\Omega^2(r+s)\cos(\theta_0 + \Omega t) \tag{5.106}$$

$$F_Y = (F_Y)_1 + (F_Y)_2 = F_{\text{Impact}}(t)\left[\cos(\beta_r + \beta' s)\cos(\theta_0 + \Omega t)\right] + B_m\Omega^2(r+s)\sin(\theta_0 + \Omega t) \tag{5.107}$$

$$F_Z = F_{\text{Impact}}(t)\left[\sin(\beta_r + \beta' s)\right] \tag{5.108}$$

In the rotordynamic coupled shaft equations, we substitute the external loads determined as shown in Eqs. (5.37)–(5.48). Next, the transverse rate of loadings in two dimensions on the shaft $q_X(Z,t)$ and $q_Y(Z,t)$ modeled as a beam by using Dirac's delta function $\delta(Z-\ell)$ in the shaft longitudinal direction at $Z = \ell$ are expressed as

$$q_X(Z,t) = F_X\delta(Z-\ell) \tag{5.109}$$

$$q_Y(Z,t) = F_Y\delta(Z-\ell) \tag{5.110}$$

It should be noted that although the external loads F_X and F_Y are boundary conditions on the fan shaft acting at the disk located at $Z = \ell$, they can be treated as a continuous function by the use of Dirac delta function $\delta(Z-\ell)$, which is how Eqs. (5.49) and (5.50) are obtained.

The axial loading $P(t)$ on the shaft considered a beam column is:

$$P(t) = F_Z \tag{5.111}$$

The corresponding dynamic torque $T(t)$ on the rotor shaft caused by the tangential component $F_T(t)$ of the bird impact load on the blade airfoil is

$$T(t) = F_T(t)(r+s) \tag{5.112}$$

Using these external forces on the rotor shaft, a direct time-integration scheme is used to solve transient oscillatory dynamic response of the fan rotor impacted by a bird with the following characteristics data.

It should be noted that the bearing ID numbers used in this paper are intentionally reversed to keep the derivation more general and are opposite to common practice used in the jet engine industry. Thus, in terms of typical jet engine industry notation, the ball bearing, nearest to the fan disk, will be Bearing No. 1, and the other two roller bearings will be called Bearing No. 2 and Bearing No. 3, respectively. The reader is referred to another common typical bearing number system used in a recent reference.

Using the flexible blade model, the first bending mode frequency of the fan shaft, described above, is obtained as 58.45 Hz. However, in the rotating condition at the operating speed of 3000 rpm, the shaft first bending mode frequency moves up to 62.73 Hz [see Reference 5]. In the forced response analysis of the fan rotor, we also consider the small amount of damping in the bearing support system for Bearing No. 2 and No. 3. Bearing damping is usually small and is affected by the friction and the relative deformation between the rolling element and the outer and the inner races. The loads of primary interest are at the ball bearing and the loads at the roller bearing are less significant. This is because rollers have greater lateral capability. There are two damping sources in a typical ball bearing, which are represented by the damping matrix called $[D]_b$ as discussed earlier in Section 5.1.1, first the Hertzian contact between the ball and the cylindrical race, and the second the interface friction between the bearing and the housing structure. Other relevant impact parameters considered for the current sample example are assumed as listed in Table 5.5. The transient analysis for this hypothetical bird-strike scenario has been carried out for one full revolution of the fan-rotor shaft, which happens to be about 20 ms. As an example to demonstrate the numerical characteristics of the transient vibratory response of a typical turbofan engine rotor, a 2-kg bird-strike input data outlined in Tables 5.5 and 5.6 will be considered in the following section.

5.2.3.5 Rotor Orbit and Bearing Loading

In this section as shown in Table 5.7, the peak magnitude values of key dynamic response parameters and their respective time lines are computed numerically and have been summarized here for a better insight in the transient vibratory characteristics of a large engine turbofan rotor. From Table 5.7 of summarized results, it can be seen that most of the bird impact-related dynamic parameters peak around 3.0–3.8 ms from the time of the initial contact. However, the shaft dynamic parameters peak around 0.2 second from the time of impact. After the initial impact shock response has dissipated, the shaft responds at the vibratory frequency of around 84 Hz, which happens to the blade first flex bending mode frequency at 3000 rpm, instead of first overhung shaft mode natural frequency of the rotor shaft being at 62.73 Hz. It should be noted that in a separate analysis [see Reference 5], the corresponding blade first flex bending mode bench frequency (1F mode in stationary condition of blade) is 45.85 Hz.

The radial load on the bearing is almost three times of the impact load on the blade airfoil caused by the bird impact. The response is quite significant, as the load on the Bearing No. 3 for a large size bird can be enough to damage the ball Bearing No. 3 in the front of the engine, which is closest to the fan disk.

For wider application of the present numerical results, the rotor orbit at the fan disk after the bird–strike is shown in a nondimensional form with respect to the fan diameter of the turbofan engine (Figure 5.21). It should be noted that the radial movement of rotor orbit is highest in the second quadrant due to bird hitting the fan at $\theta = 0°$.

The transient dynamic response of the fan disk in the horizontal and vertical directions is illustrated in Figures 5.22 and 5.23, respectively. Fan disk response peaked about 0.2 second after the initial impact of the bird. Another key parameter of interest in any rotordynamic analysis is the magnitude of the radial load on the bearing due to dynamic magnification, which is peaked at 372 kN at 0.2047 second (Figure 5.24) after the initial impact of the bird on the engine. The bird-slice unbalance load F_U on the rotor system was only 110 kN, which was less than one-third of the Bearing No. 3 peak radial load (372 kN). Even a small amount of bearing damping is essential and beneficial to withstand a sudden dynamic shock on the fan rotor in the aftermath of bird-strike impact load of $F_{\text{Impact}}(t)$ on the blade airfoil.

TABLE 5.5
Details of the Parameters Considered for the Sample Rotor Used in Discussion

Mass of one blade	$M_b = 10$ kg
Blade-tip radius	$R = 150$ cm
Blade root radius = Disk outer radius	$r = 50$ cm
Blade span length	$L = 100$ cm
Average (mean) blade chord	$C = 35$ cm
Blade chord at the root	$C_0 = 20$ cm
Blade chord at the tip	$C_L = 50$ cm
Blade average aspect ratio	$L/C = 2.0$
Blade cross-sectional area	$A_b = 5000$ mm^2
Blade area moment of inertia	$(I)_b = 0.5E6$ mm^4
Mass of the overhung fan disk	$M_D = 150$ kg
Mass of additional rotor components (spinner cone, booster spool, etc.)	$M_{\text{Additional}} = 280$ kg
Length of the shaft	$\ell = 350$ cm
Outside radius of the shaft	$r_o = 7.0$ cm
Inside radius of the shaft	$r_i = 5.5$ cm
Mass density of the shaft material	$(\rho)_s = 7.833$ gm/cc
Blade stagger angle at the tip	$\beta_R = -60°$
Rotational speed of the fan rotor	$rpm = 3000$
Rotor Initial steady-state spin velocity (radian/s)	$\Omega = 314.159$ radians/s
Mass moment of unbalance about engine axis	$m_r = B_m\,(r + s)$
Circumferential location of the unbalance at time "t" = 0 with respect to horizontal	$\theta_0 = 0°$
Number of blades on the fan disk	$N = 20$
Number of discrete bearings supporting the shaft	$N_b = 3$
Elastic Young's modulus of the rotor shaft material	$(E)_s = 200$ GPa
Elastic Young's modulus of the blade material	$(E)_b = 117$ GPa
Percentage of critical damping in shaft	$\zeta = 0.5\% = 0.005$
ith Bearing support location from aft-end of fan shaft:	Z_i
a. Roller bearing-1: at $Z = 0.3$ cm	$S^1{}_{XX} = S^1{}_{YY} = 2000.0$ MN/m
b. Roller bearing-2: at $Z = 265.0$ cm	$S^2{}_{XX} = S^2{}_{YY} = 715.0$ MN/m
c. Ball bearing-3: at $Z = 325.0$ cm	$S^3{}_{XX} = S^3{}_{YY} = 565.0$ MN/m
Rotational stiffness	$B^3{}_{XX} = B^3{}_{YY} = 7.500E6$ N.m/radian
Ball bearing-3: Radial load capability	1 MN

TABLE 5.6
Bird Impact Parameters

Mass of the bird cylinder	$B_M = 2.0$ kg
Bird material density	$(\rho)_B = 0.9$ g/cm^3
Bird cylinder projectile length	$B_L = 22.4$ cm
Bird cylinder projectile diameter	$B_D = 11.2$ cm
Bird slice mass	$B_m = 0.89$ kg
Bird spanwise impact location on blade	$s = 75$ cm
Aircraft velocity	$V_a = 100$ m/s
Angular velocity of fan rotor	$\Omega = 314.159$ radians/s
Blade tangential velocity at contact	$V_t = 392.7$ m/s
Bird normal impact velocity	$V_{\text{Impact}} = 178.1$ m/s
Twist in the blade at impact (75% span)	$\beta = -45°$
Blade airfoil maximum thickness	$h = 2$ cm

TABLE 5.7
Time-Line and Peak Amplitudes of Transient Rotordynamic Response of the Fan Rotor Shaft with 1-m Long Airfoil Blades Hit by a 2-kg Bird

Parameters	Peak Magnitude	Time from Initial Impact
1. Horizontal displacement	−3.28 mm	0.2032 second
2. Vertical displacement	+3.16 mm	0.2000 second
3. Radial displacement of the Fan Disk	3.28 mm	0.2030 second
4. Peak radial load on Bearing No. 3	372 kN	0.2047 second
5. Bearing No. 3 vibratory amplitude	0.641 mm	0.2050 second
6. Peak bird impact load on the blade airfoil	171 KN	0.00031 second
7. Spanwise strain in the blade airfoil	0.716%	0.00076 second
8. Transverse deflection at the LE of the blade airfoil	5.34 cm	0.00375 second

5.3 FAR FIELD DAMAGE IN AEROENGINE DUE TO FAN BLADE LOSS (THREE PHASES OF ANALYSIS)

The most common external load in any rotordynamic problem is the rotating load vector due to unbalance. During the FBO event, the sudden unbalance load on the rotor shaft is extremely severe, and as a result of that the fan disk on the heavy side of the rotor moves off-center by a large amount. A detailed full FBO analysis requires three consecutive steps which are as follows:

1. *Prestress step*: Fan blade prestresses due to thrust, gravity, and gyroscopic effects are computed. This step is typically performed over a detailed mesh with an implicit scheme like ANSYS®.

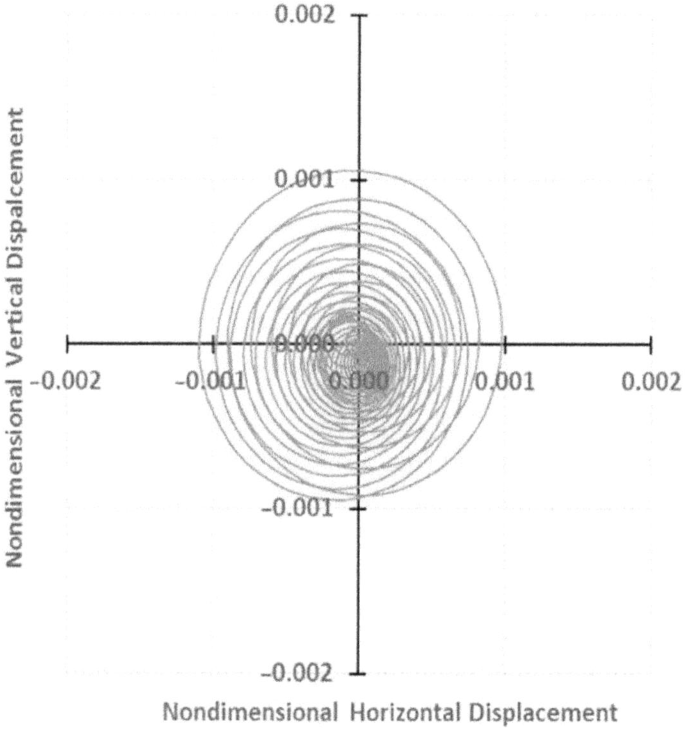

FIGURE 5.21 Nondimensional rotor orbit with respect to fan diameter after a 2-kg bird strike at 75% span (peak radial deflection 3.28 mm at 0.203 second after the initial impact). (Reprinted with permission from Reference [6].)

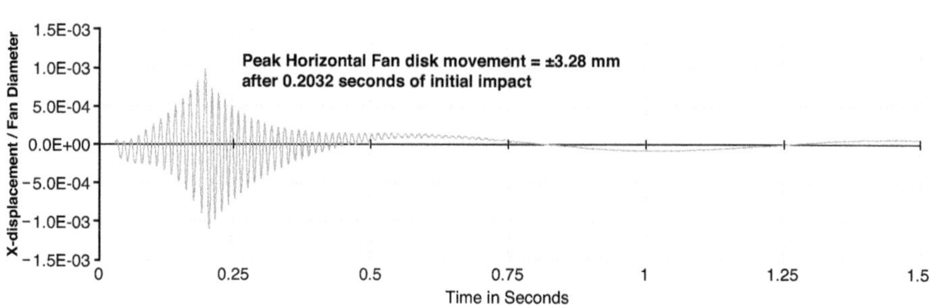

FIGURE 5.22 Transient horizontal fan disk movement = 3.28 mm at 0.2032 second after initial impact. (Reprinted with permission from Reference [6].)

2. *Fan blade off step*: The release of the blade is explicitly simulated together with damage of the blade, rubbing with the engine case, damage to the engine containment. The prestress state computed at Step 1 serves as initial condition of this step and this simulation is usually performed over a small-time interval with small time steps, usually using LS-DYNA®.

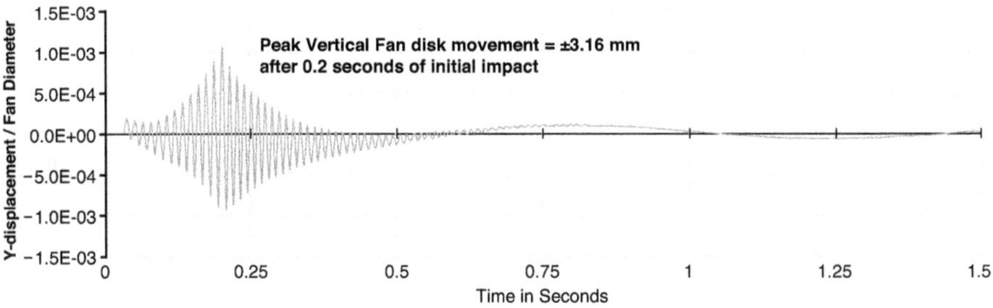

FIGURE 5.23 Transient vertical fan disk movement = 3.16 mm at 0.2 second after initial impact. (Reprinted with permission from Reference [6].)

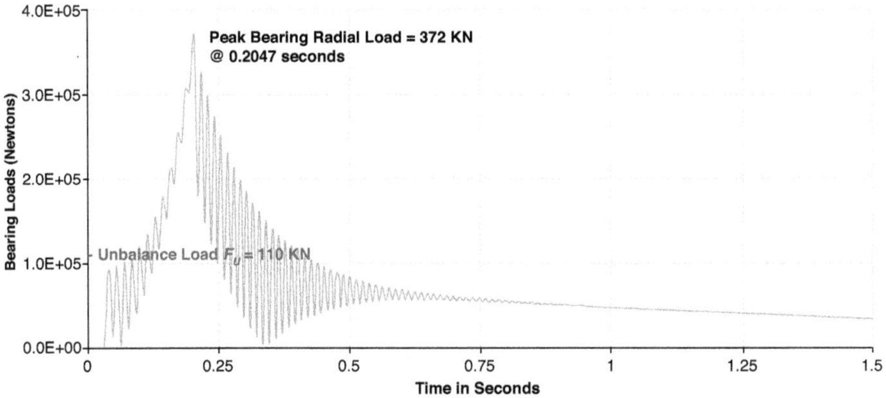

FIGURE 5.24 Dynamic radial load on Ball bearing 3 from the aft end of the engine (peak radial load = 372 kN at 0.2047 second). (Reprinted with permission from Reference [6].)

3. *Postimpact loads step:* The forces calculated during Step 2 allow a transient analysis to be performed of the moments after the blade has impacted the casing. These forces are used as loads for a much longer analysis usually accomplished through an implicit scheme using a coarser mesh. This analysis covers the event until the engine run-down (rotor windmilling).

   The first two steps are usually performed by the aeroengine manufacturer and usually involve proprietary data and models. These are computationally very expensive (especially Step 2 while Step 1 implies linear elasticity analysis, rotordynamics effects and possibly, depending on the geometry of the blade, geometric nonlinearities) since they are based on fine meshes (3D elements). The latter step is usually performed by the airframe manufacturer to get an estimate of FBO loads for pylon sizing. The airframe manufacturer is given forces from Step 2 by the aero-engine manufacturer. These forces are as follows:

   - Blade impact loads on the engine case
   - Seizure or engine torque due to rubbing loads.
   - Unbalance loads coming from the mass of the missing blade material.

The postimpact analysis needs the fan rotor speed time history of rotor slowing down as well. At this point, the methodology should distinguish between the different analyses.

- Steps 1 and 2 where most of the nonlinearity is predicted by the simulation. This is done by the aeroengine manufacturer during the design stage of the containment structure and aims to ensure that the engine containment will resist the high impact loads from blade off. In this analysis, rotor speed unbalance and impact loads, and seizure torque are only by-products at the engine level of the analysis. These two steps and particularly Step 2 remain challenging computational modeling tasks, which needs very refined 3-D FE mesh of the containment system.
- Step 3 where most of the nonlinearity is previously simulated (or gained from experiments, e.g., rotor speed) is transferred to the airframe manufacturer through "engine level" data, such as rotor speed. In this analysis, most of the nonlinearities are imposed as either boundary (or initial) conditions and occur as forced functions on the right-hand side.

It should be noted that Step 3 is not an FBO simulation but a rotordynamics simulation. This is the reason that this part of the containment analysis is included in Chapter 5 of this book. It is essential not only for the airframe manufacturer but also for the aeroengine manufacturer since it provides a prediction of the performance of the engine after the event. In particular, when simulating the postimpact behavior, the aeroengine manufacturer will determine a windmill speed that may cause a high level of vibrations. For more details to bridge the gap between the two types of analysis by mixing different models and numerical schemes, the reader should refer to another paper [see Reference 7], where the current author has collaborated with different experts across multiple industries.

5.4 WINDMILLING RESPONSE OF DAMAGED ROTOR AFTER BLADE LOSS

During the normal operation of an aeroengine fan at a given speed, the gas loads F_G on the blades apply a steady-state load torque on the shaft. A typical turbofan rotor starts behaving like a turbine rotor, when it is subjected to a head-wind, which feeds energy to the dynamical system. The safety objective of FAA regulation 33.74 is to ensure that an engine that continues to rotate after shutdown will not create a hazard to the aircraft. Continued rotation refers to a condition in which any part of the main rotating system in an engine continues to rotate after the engine has been shut down. Continued rotation can be caused by windmilling or mechanical effects or a combination of both. Windmilling is the rotation of a nonoperating engine due to the airflow-induced forces on the blades caused by the forward motion of the aircraft. Mechanical effects include, for example, drive shaft clutch drag in some multiengine rotorcraft installations, which may result in continued rotation of the engine after it has been shut down. The FAA document [see Reference 8] describes, "The vibratory loads resulting from the failure of a fan blade have traditionally been regarded as insignificant relative to other portions of the design load spectrum for the airplane.

However, the progression to larger fan diameters and fewer blades with larger chords has changed the significance of engine structural failures that result in an imbalanced rotating assembly. This condition is further exacerbated by the fact that fans will continue to windmill in the imbalance condition following engine shut down. Current rules require provisions to stop the windmilling rotor where continued rotation could jeopardize the safety of the airplane. However, large high bypass ratio fans are practically impossible to stop in flight.

In a damaged fan rotor after losing a blade, intermittent rub can produce a very large vibrational response, which may contain various types of subharmonic characteristics. In an aircraft engine after a fan blade-out event, the windmilling of an asymmetric rotor takes place during the fly-home period. A typical turbofan rotor usually will windmill at 1/3 of the full-thrust speed of the rotor. Using the same transient response technique described in Section 3.2 of this book, we have analyzed the dynamic response of the damaged rotor, if it were windmilling at 1000 rev/min. The time-domain dynamic response of the damaged nonaxisymmetric rotor at 1000 rev/min is summarized as follows:

a. Horizontal and vertical movement of the fan disk.
b. Blade tip-rub load time history.
c. Rotor response frequency.

The above time-domain results are analyzed at 1000 rev/min for 8 seconds. Here, the forcing function is the unbalanced force caused by the missing blade as well as the solution scheme updates the mass and stiffness matrices as the rotor spins. As shown in Figure 5.12, the peak horizontal movement of the rotor at the center of the fan disk oscillates between 0.5 and 2.0 cm. The zoomed-in dynamic response of this rotor, shown in Figure 5.13, exhibits various types of frequency content. This time-domain response includes the effect of both oscillations in the mass matrix at twice the rotational speed of the rotor as well as the time-dependent terms in the stiffness matrix due to intermittent blade tip-rub against abradable inner surface (Figures 5.25–5.27).

Similar time-domain response of the rotor in the vertical direction has been plotted and is shown in Figure 5.14. The asymmetricity of the rotor is quite evident by the fact that the horizontal direction response is significantly different from its vertical direction response. As shown in Figure 5.14, the peak vertical movement of the rotor at the center of the fan disk oscillates between 1.0 and 1.7 cm.

Figure 5.15 shows the zoomed-in view of the vertical response in a quasi-steady state of the rotor windmilling at 1000 rev/min. The time-domain response of the asymmetric rotor during windmilling shown in Figure 5.15 clearly highlights the fact that in its dynamic behavior, there are multiple frequencies superimposed on one another. We also recover the rub response of the running asymmetric rotor during windmilling in the aftermath of the blade loss. At this point, we will determine the frequency content of the dynamic response by using a fast-Fourier transform of the time-domain data displacement data. The actual time-domain behavior of the rotor–shaft comprises responses at several different frequencies, and the instantaneous whirl frequency of this type of asymmetric fan–rotor can also be determined using Eq. (5.58).

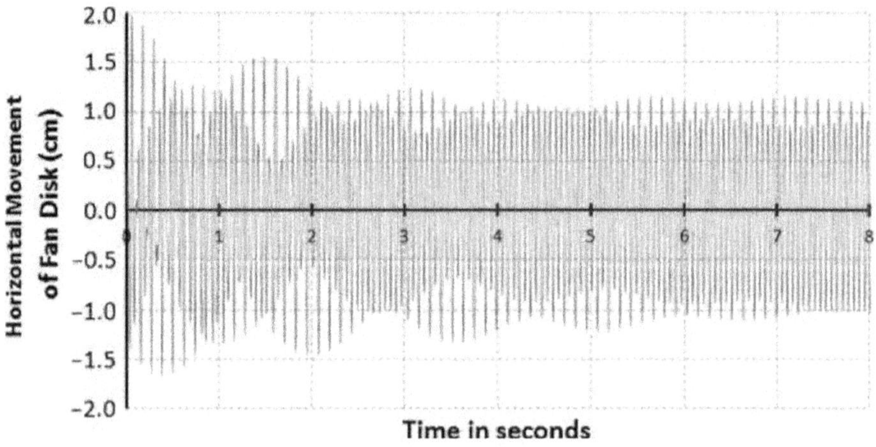

FIGURE 5.25 Time-domain horizontal response of a nonsymmetric shaft at the fan disk-center during windmilling at 1000 rev/min after a blade loss. (Reprinted with permission from Reference [5].)

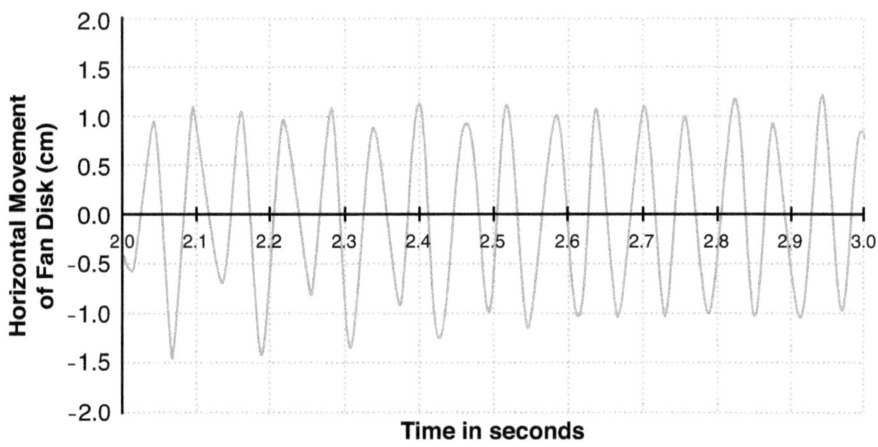

FIGURE 5.26 Zoomed-in view of a typical 1 second window of the horizontal response of the nonsymmetric shaft windmilling at 1000 rev/min after a blade loss. (Reprinted with permission from Reference [5].)

The above time-domain results also help us in determining the rub-load on the rotor coming from the blade tips rubbing against the abradable. Figure 5.16 shows the time-domain instantaneous magnitudes of the analytically determined rub loads on the fan rotor. The plot shows that the magnitude of the contact-impact rub load in the radial direction of this rotor at the fan disk oscillates about the mean value of 100,000 N with its amplitude peaking up to 190,000 N. A simple hand calculation shows that an impulsive load of this magnitude in the longitudinal direction of the airfoil will produce a compressive membrane strain in the vicinity of rub zone at the blade tip of the order of 0.033% (Figure 5.28–5.30).

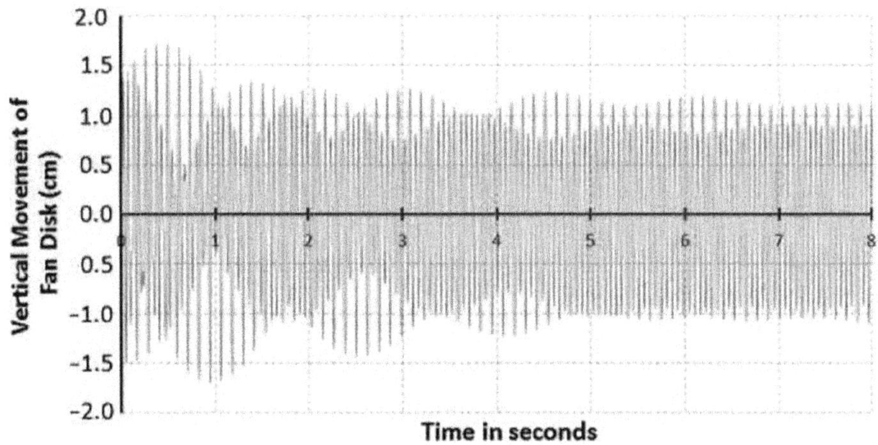

FIGURE 5.27 Time-domain vertical response of a nonsymmetric shaft at the fan disk-center during windmilling at 1000 rev/min after a blade loss. (Reprinted with permission from Reference [5].)

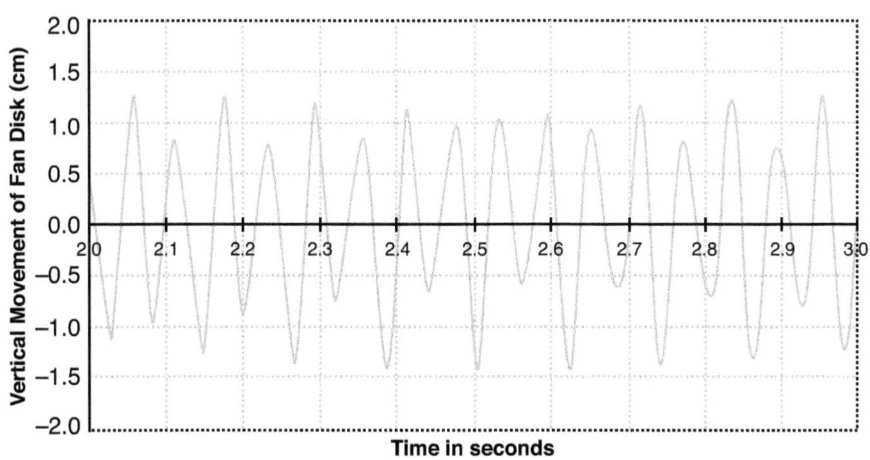

FIGURE 5.28 Zoomed-in view of a typical 1 second window of the vertical response of the nonsymmetric shaft windmilling at 1000 rev/min after a blade loss. (Reprinted with permission from Reference [5].)

Figure 5.17 shows that during windmilling a damaged nonsymmetric rotor would respond at a subharmonic frequency. For example, the natural frequencies of this rotor at 1000 rev/min are at 70.35 and 62.85 Hz, but the rotor shows high magnitude oscillation not only at its excitation frequency of 16.66 Hz, but it also exhibits subharmonic response at 8.25 Hz. The subharmonic frequency coincides with half of the excitation frequency O/2, and the system also responds at its sum frequency of 25.4 Hz.

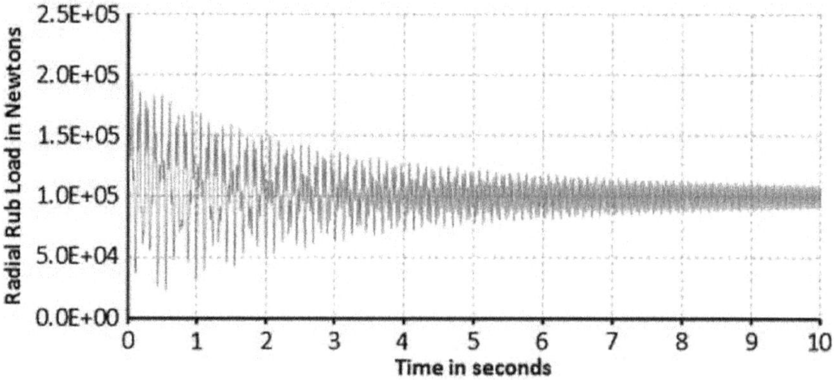

FIGURE 5.29 Time-domain response of the radial rub load on the rotor of the asymmetric shaft at the fan disk-center during windmilling after a blade loss. (Reprinted with permission from Reference [5].)

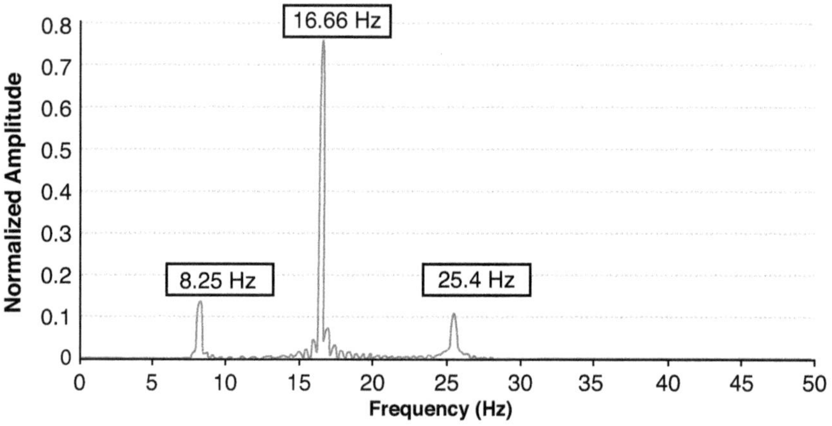

FIGURE 5.30 Frequency content of vibratory response of a windmilling nonsymmetric fan rotor windmilling at 1000 rev/min (16.66 Hz) at the bearing support. (Reprinted with permission from Reference [5].)

Figure 5.31 is plotted for a steady-state rotational speed of 1000 rev/min (16.67 Hz) as a result of unbalance caused by the missing blade. Any whirl response of the rotor at a frequency higher than the running excitation frequency will result in a forward-whirl response of the rotor. Similarly, the dynamic response occurring at less than the running excitation frequency will exhibit backward whirl characteristics of the rotor shaft. The forward and backward whirl response of the fan rotor can very easily be visualized by creating an animation of the fan-disk orbit from the transient response plotted in Figures 5.25 and 5.27. In addition, Figure 5.31 shows that this damaged asymmetric rotor during windmilling at 1000 rev/min will go into distinct backward-whirl behavior at 6.1 Hz and other two minor ones at 10.9 and 12.2 Hz. In addition, it will show a complex forward-whirl behavior with its whirling frequencies

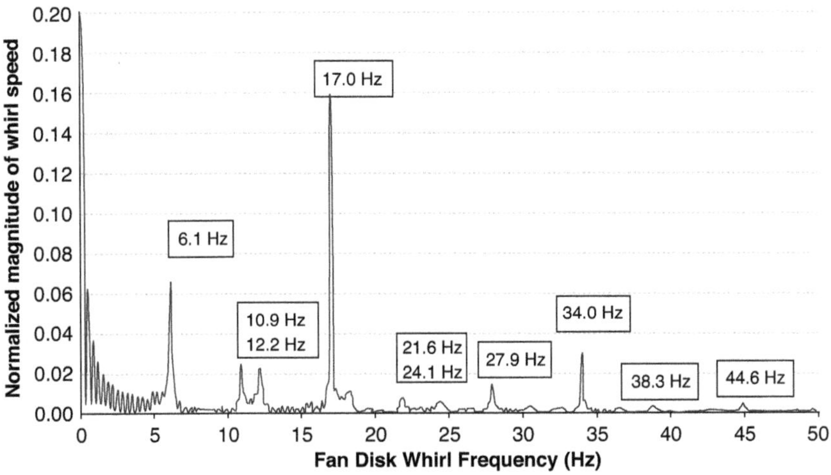

FIGURE 5.31 Whirling frequencies of a windmilling nonsymmetric fan rotor at 1000 rev/min (16.66 Hz) with a blade missing. (Reprinted with permission from Reference [5].)

at 27.9 and 34.0 Hz. There are several other minor blips in the whirl frequency plot, indicating that this shaft may also exhibit a very complex forward-whirl behavior at other frequencies such as 21.6, 24.1, 38.3, and 44.6 Hz. Depending upon the magnitude of the shaft oscillation during whirling, the radial rub load on the fan rotor will be oscillating between 25,000 and 190,000 N (see Figure 5.29). The most dominant whirl characteristic for this long rotor will be a synchronous whirl in the vicinity of excitation frequency of 17 Hz. The existence of multiple whirl frequencies indicates that the different sections of the fan–rotor will be whirling at varying frequencies. Considering its critical speed at 100 rev/min, the low-frequency whirl may result in large-magnitude oscillations of the shaft during windmilling.

This rotor depending upon the blade-tip clearance γ will have occasional rub in some isolated circumferential locations of the fan disk. During the solution of the governing equations numerically, we keep track of the particular blade tips' coordinate locations to determine whether any parts of the blade tips are in contact with the abradable on the containment casing. The time-dependent circumferential location of a typical blade identified as "Blade $-\, j$"which might be rubbing at any time "t" is expressed as $\left(\vartheta_j + \Omega t\right)$. The intermittent rubbing with blades being flexible in the longitudinal direction introduces a coupling between the longitudinal and transverse deformations along its span. This coupling is caused by the Coriolis forces as a result of longitudinal wave of the blade traveling up-and-down along the span due to the blade rotating in the centrifugal force field [see Reference 4].

5.5 BLADE BENDING DURING TIP-RUB AND ITS EFFECT ON ROTOR SHAFT MOVEMENT

The effect of blade tip-rub loading was briefly discussed earlier in Section 5.2.1. In Section 5.2.3.3, we also discussed the effect of blade deformation due to various flexural and torsional mode shapes of the blade airfoil and its interaction effects with the

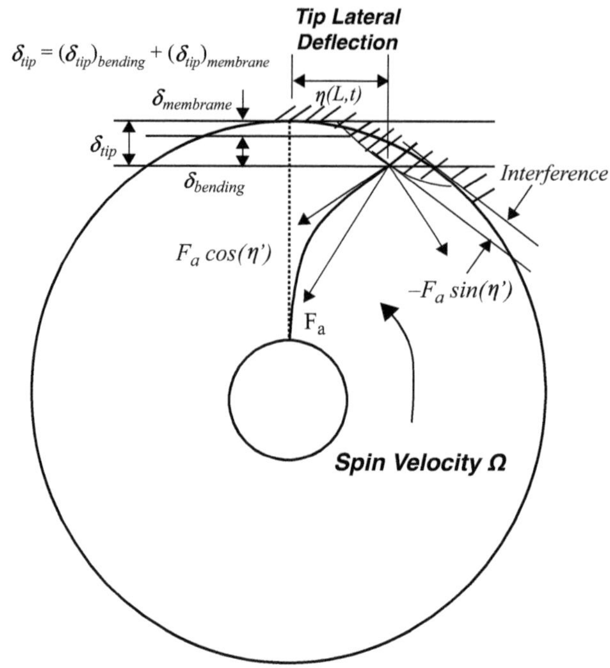

FIGURE 5.32 Blade deformation during the interference-induced rub at the tip due to clearance closure with the outer case. (Reprinted with permission from Reference [4].)

shaft mode frequencies. However, that discussion was limited to the free vibrational modes of the blade airfoil deformation only. In this section, we will discuss as how to numerically determine the radial movement of fan disk center, when the blade tips are rubbing against the inner surface of the containment ring or the abradable surface. This effect of blade deformation due to bending in the direction opposite to the direction of rotation, observed usually during the heavy tip rub, is schematically shown in Figure 5.32. Such "hard spots" on the fan casing structure in the circumferential direction of the containment ring occur, where the local tip-rubbing takes place, which are caused due to "case ovalization" and the resulting interference.

In this section, we are presenting a numerical scheme, which determines all the dynamic forces on the rotating system with enforced time-dependent displacements of the blade tip caused by the interference with the outer case. The interference in the dynamical system takes place when the running (steady-state) tip clearance "γ"is used up during the radial movement of the tip, henceforth denoted as "δ_{tip}". During the sudden interference, the contact load at the tip surface pushes the blade tip inward toward the center of spin and the total tip deflection along the beam-axis δ_{tip} in the radial direction of the rotor can be considered made up of two parts (see Figure 5.32), which are written as follows:

$$\delta_{\text{tip}} = \left(\delta_{\text{tip}}\right)_{\text{bending}} + \left(\delta_{\text{tip}}\right)_{\text{membrane}} \tag{5.113}$$

Suppose the blade being considered a rotating cantilevered radial beam of span length of $(s = L\}$ with clamped at one end of $s = 0$, the transverse deflection $\eta(s,t)$ due to flexural bending and the longitudinal deflection $\zeta(s,t)$ due to membrane stretching are represented as follows:

$$\eta(s,t)=\sum_{n=1}^{\infty}W_n(s)Y_n(t)=\sum_{n=1}^{\infty}\frac{[1-\cos\alpha_n s]}{\alpha_n}Y_n(t) \tag{5.114}$$

$$\zeta(s,t)=\sum_{n=1}^{\infty}V_n(s)Z_n(t)=\sum_{n=1}^{\infty}\frac{[\sin\alpha_n s]}{\alpha_n}Z_n(t) \tag{5.115}$$

The radial movement of the blade tip "δ_{tip}" is computed as follows. Here, the blade-tip longitudinal deflection $(\delta_{\text{tip}})_{\text{bending}}$ due to beam bending effect can be expressed as shown below. It should be noted that for the rotating disk, one needs to add the following term for accurate value of the blade-tip location with respect to the engine axis.

$$(\delta_{\text{blade-root}})_{\text{membrane}} = \text{Fan disk's radial growth due to centrifugal effect}$$

$$\begin{aligned}(\delta_{\text{tip}})_{\text{bending-computed}} &= -\int_0^L [1-\cos(\eta')]ds\\ &= -\int_0^L\left[1-\cos\left(\sum_{k=1}^{N}W'_k\,Y_k\right)\right]ds\\ &= -\int_0^L\left[1-\cos\left(\sum_{k=1}^{N}\sin(\alpha_k s)\,Y_k\right)\right]ds\end{aligned} \tag{5.116}$$

Similarly, the blade-tip longitudinal deflection $(\delta_{\text{tip}})_{\text{membrane}}$ due to membrane-stretching effect can be computed as shown below:

$$\begin{aligned}(\delta_{\text{tip}})_{\text{membrane-computed}} &= (\zeta(L,t))_{\text{computed}} = \sum_{k=1}^{N}V_k(L)\,Z_k\\ &= \sum_{k=1}^{N}\left[\frac{\sin\alpha_k L}{\alpha_k}\right]Z_k\end{aligned} \tag{5.117}$$

If the specified tip clearance (gap) is γ, then for all $\delta_{\text{tip}} > \gamma$, we will set $\delta_{\text{tip}} = \gamma$, $\zeta(L,t)=0$, i.e.,

$$(\delta_{\text{tip}})_{\text{bending-computed}} + (\delta_{\text{tip}})_{\text{membrane-corrected}} = \gamma \tag{5.118}$$

or,

$$(\delta_{\text{tip}})_{\text{bending-computed}} = \gamma - (\delta_{\text{tip}})_{\text{membrane-corrected}} \tag{5.119}$$

or,

$$\{Z_k(t)\}_{\text{corrected}} = \frac{(\delta_{\text{tip}})_{\text{membrane-corrected}}}{(\delta_{\text{tip}})_{\text{membrane-computed}}} \{Z_k(t)\}_{\text{computed}} \tag{5.120}$$

or,

$$\{\Delta Z_k(t)\} = \{Z_k(t)\}_{\text{computed}} - \{Z_k(t)\}_{\text{corrected}} \tag{5.121}$$

And

$$\text{Interference} = \left[(\delta_{\text{tip}})_{\text{computed}} - \gamma\right] \tag{5.122}$$

Thus, during the blade-tip rub with the blade deformation being included, the actual fan disk rotor movement at $Z = \ell$ in the global X and Y directions per Eqs. (5.32) and (5.33) are as follows:

$$u(\ell,t) = \sum_{m=1}^{\infty} U_m(t)\, W_m(\ell) \tag{5.123}$$

$$v(\ell,t) = \sum_{m=1}^{\infty} V_m(t)\, W_m(\ell) \tag{5.124}$$

Referring to Figure 5.16 at any instant of time of time "t", suppose a typical jth blade on the fan disk is rubbing against the containment structures' inner surface. Then, the circumferential location of this typical jth rub-blade at time "t" from the instant of the blade release time is established as follows. This blade will be in the circumferential location of $\left((2\pi j/N) + \Omega t\right)$ radians, measured with respect to the horizontal global X-axis in the direction of fan rotation. Due to the elastic deformation of the blade tip during tip rub per Figure 5.32, the effective radial deflection of the fan rotor shaft will be additional amount of which the radial value will be equal to "δ_{tip}" as given in Eq. 5.113. Thus, the effective deformation of fan shaft at the fan disk location of $Z = \ell$, called $(u(\ell,t))_{\text{Effective}}$ and $(v(\ell,t))_{\text{Effective}}$ can be determined as follows:

$$(u(\ell,t))_{\text{Effective}} = u(\ell,t) + \delta_{\text{tip}} \cos\left((2\pi j / N) + \Omega t\right) \tag{5.125}$$

$$(v(\ell,t))_{\text{Effective}} = v(\ell,t) + \delta_{\text{tip}} \sin\left((2\pi j / N) + \Omega t\right) \tag{5.126}$$

Thus, the numerical methodology described here can be used to determine the fan disk movement for the entire FBO event, which takes into account of all the three phases of the time marching forward scheme, outlined in Section 5.3 on rotordynamics. Mode detailed in-depth methodology will be further discussed in Chapter 7 of this book.

NOMENCLATURE

$(A)_s$	cross-sectional area of shaft $= \pi\left(r_o^2 - r_i^2\right)$
$(A)_b$	cross-sectional area of the blade
$[\mathbf{B}_1]$	time-dependent stiffness matrix (symmetrical due to axial load on the shaft)
$[\mathbf{B}_2]$	time-dependent stiffness matrix (skew-symmetric due to fluctuating torque in the system)
$[C_D]$	damping matrix in the rotordynamic equation (symmetrical or nonsymmetrical depending upon the bearing)
$[C_G]$	gyroscopic matrix in the rotordynamic equation (skew-symmetric, causes forward and backward whirl in the shaft)
C	blade chord length at the tip
$[C_{i,j}]$	a typical ith row and jth column term in the velocity-dependent matrix
D_L	axial length of the rigid disk
D_{XX}^b, D_{YY}^b	damping in the bearing support location of $Z = Z_i$ at respective bearings
e	nondimensional factor used to linearize the effect of blade rotation
$(EI)_s$	flexural rigidity of the rotor shaft
$(EI)_b$	flexural rigidity of the blade
F_a	axial load on the blade along the longitudinal axis (global radial direction)
F_G	gas load perpendicular to the blade pressure surface on each blade
F_U	unbalance load $F_U = \left(m_r\Omega^2\right)$
$\{F(t)\}$	column vector containing external forces on the dynamical system
$\{f\}$	column vector of generalized co-ordinates of the dynamical system (dimension: length unit)
H	critical damping parameter in the shaft
I_d	shaft diametral mass moment of inertia (per unit length) $= (\rho A)_s \left(r_o^2 + r_i^2\right)/4$
I_p	shaft polar mass moment of inertia (per unit length), $I_p = 2I_d$
J_D	diametral mass moment of inertia of fan disk, $= \left(M_D r^2\right)/4$ with r = outer radius of the disk or, bladed disk)
K_{rub}	radial stiffness of the outer case filler material during rub
K_{shaft}	lateral stiffness of the shaft at the disk center
$[\mathbf{K}_S]$	general stiffness matrix of the rotor shaft (usually symmetrical, but may be nonsymmetrical due to nonsymmetrical bearing in the horizontal and vertical directions) and also due to torque in the system)
$[K_{i,j}]$	a typical ith row and jth column term in the stiffness matrix
L	span length of the cantilever blade
ℓ	span length of the shaft
$[M]$	mass matrix (usually symmetrical, but may become unsymmetrical and time-dependent due to missing release blade in FBO analysis)
$[M_{i,j}]$	a typical ith row and jth column term in the mass matrix
M_D	mass of the disk or bladed disk
M_b	mass of each blade $= (\rho A)_b L$
m_r	mass moment of unbalance (unit: mass times radial distance "r")
N_b	number of discrete bearings supporting the shaft
N	number of blades on the fan disk

P_0	steady-state axial load on the shaft
P_t	time-dependent axial load on the shaft
R	blade-tip radius
r	blade root radius = disk outer radius
s	blade local co-ordinates in the global direction
S^b_{XX}, S^b_{YY}	stiffness in the bearing support
T_0	steady-state torque on the shaft
T_t	time-dependent torque on the shaft
$u(Z,t)$	transient deflection of the shaft in the global *X* direction
$U_m(t)$	time-dependent generalized co-ordinates for deflection of the shaft (*X* direction)
$v(Z,t)$	transient deflection of the shaft in the global *Y* direction
$V_m(t)$	time-dependent generalized co-ordinates for deflection of the shaft (*Y* direction)
$W_m(Z)$	deformation shape functions or mode shapes of the rotor shaft
$\eta(s,t), \zeta(s,t)$	transverse and longitudinal deformation functions of the cantilever blade
Z	axial co-ordinate of a point on the shaft

Greek letters

β	blade stagger angle (radian), i.e., angle between the blade chord and the engine axis (axis of rotation) at the blade tip
δ	effective logarithmic decrement as a measurement of damping
$\delta(Z-Z_i)$	Dirac Delta function for $Z = Z_i$
ε	radial incursion of the blade tip in the case filler material during rub (interference)
θ	angle of the rigid body rotation of the shaft about the spin axis at time "t" from Initial condition of time "$t = 0$" ($\theta = \Omega t$ for constant angular velocity "Ω")
λ	eigenvalues
λ_r, λ_i	real and imaginary parts of the eigenvalue
μ	dynamic coefficient of friction between the blade tip and the outer case
ξ	nondimensional beam frequency parameter
$(\rho)_s$	mass density of the shaft material
τ	ultimate shear strength of the case filler material
ω_n	natural frequency (radians/s)
Ω	rotor spin velocity in radians/s

REFERENCES

1. S. K. Sinha (1989) "Stability of a Viscoelastic Rotor-Disk System under Dynamic Axial Loads", *AIAA Journal*, Vol. 27, No. 11, pp. 1653–1655. https://doi.org/10.2514/3.10317
2. S. K. Sinha (1992) "On General Conditions of Rotor Dynamic Stability under Combined Axial Force and Torque", *ASME Journal of Applied Mechanics,* Vol. 59, pp. 225–228. https://doi.org/10.1115/1.2899439

3. S. K. Sinha (2004) "Dynamic Characteristics of a Flexible Bladed-Rotor with Coulomb damping due to Tip-Rub", *Journal of Sound and Vibration*, Vol. 273, No. 4–5, pp. 875–919. https://doi.org/10.1016/S0022-460X(03)00647-3
4. S. K. Sinha (2005) "Non-linear Dynamic Response of a Rotating Radial Timoshenko Beam with Periodic Pulse Loading at the Free-End", *International Journal of Non-linear Mechanics*, Vol. 40, No. 1, pp. 113–149. https://doi.org/10.1016/j.ijnonlinmec.2004.05.019
5. S. K. Sinha (2013) "Rotordynamic Analysis of Asymmetric Turbofan Rotor due to Fan Blade-Loss Event with Contact-Impact Rub Loads", *Journal of Sound and Vibration*, Vol. 332, No. 9, pp. 2253–2283. https://doi.org/10.1016/j.jsv.2012.11.033
6. S. K. Sinha (2021) "Transient Vibratory Response of Turbofan Engine Rotor Impacted by Bird-Strike", *ASCE Journal of Aerospace Engineering*, Vol. 34, No. 4, p. 04021027. https://doi.org/10.1061/(ASCE)AS.1943-5525.0001250
7. K. G. VanDuyn (2006), Bearing Support, United States Patent No. US7097413B2, United Technologies Corporation, Hartford, CT, August 29, 2006.
8. F. A. Favara (2003), "Turbine Engine Continued Rotation and Rotor Locking", Advisory Circular No. 33.74/92-1A, Initiated by ANE-110, US Department of Transportation, Federal Aviation Administration, July 23, 2003. https://www.govinfo.gov/content/pkg/GOVPUB-TD4-PURL-LPS106427/pdf/GOVPUB-TD4-PURL-LPS106427.pdf

6 Composite Impact

6.1 INTRODUCTION TO LAMINATED COMPOSITE RESPONSE UNDER IMPACT LOADING

Due to the advantage of significant reduction in weight, the multilayered composites made up of carbon fiber-reinforced epoxy-based resin are gaining popularity in several aerospace applications including fan blades. The composite fan blades exhibit highly anisotropic elastic behavior, especially in their dynamic response and damage characteristics such as delamination. In most of the analytical modeling or numerical simulations, airfoil blades are usually considered orthotropic plates and shells of homogeneous materials with bidirectional equivalent in-plane elastic properties along their span and chord axes. Currently, a subcell approach is used to determine the single-ply and multi-ply response of braided composite materials. In the subcell approach, the equivalent orthotropic material properties are developed using a volume-fraction ratio methodology. This chapter discusses in-depth analytical methodology and the numerical approach to determine the dynamic response and impact resistance capability of a composite fan blade in a closed form by numerically solving the highly complex eigenvalue problem using a stiffness and mass matrices formulation. The solution of the partial differential equation allows us to get a better physics-based insight into the complex dynamics of aerodynamically driven flutter on a curved composite airfoil and also to carry out parametric studies, which may be difficult to perform using a finite-element-based approach. In this chapter, we discuss the analytical formulation as to how the natural frequencies of a composite airfoil in a rotating condition can be used for optimizing the dynamic response of any generic pretwisted fan blade during the preliminary design phase of an aeroengine.

The technical contents of this chapter are completely based upon the author's own lecture notes as well as his other original publications and presentations [see References 1–4] in archival journals and other conferences.

6.1.1 Volume Fraction Unit Cell Methodology

In a laminated shell formulation, each fiber-reinforced layer or ply is considered a 2-D orthotropic element for which we must develop an equivalent set of material properties in shell coordinates. To this end, we start with a set of unidirectional material properties in the fiber longitudinal directions and the transverse directions of the ply layup of a lamina (see Figure 6.1). At this point, we will assume that there are no voids or air gaps inside the matrix, and as such porosity is zero. We also assume that the strength in the longitudinal direction of the fibers is provided by only the load-carrying capability of the lamina, as such we ignore any additional contribution of the epoxy-based resin in the matrix. Similarly, in the transverse direction of the plane lamina, its mechanical load-carrying capability, which is very low, is affected

DOI: 10.1201/9781003749752-6

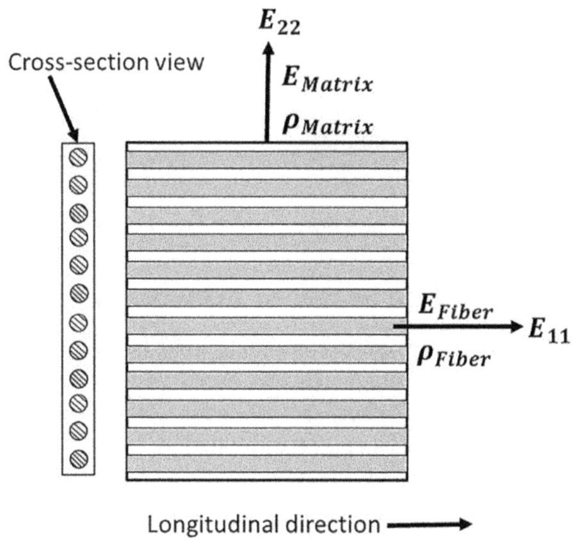

FIGURE 6.1 Schematic diagram of unidirectional fibers as reinforcement in an epoxy resin-based matrix.

by the elastic properties of the matrix. After introducing these assumptions, we determine the homogenized orthotropic mechanical properties of a 2-D thin laminate using a volume ratio fraction method. To that end, we start with using properties with usual notations like E_{Fiber}, E_{Matrix}, ρ_{Fiber}, and ρ_{Matrix} (see Figure 6.1).

Then using a volume fraction ratio method formulation, we can estimate the equivalent material data to represent the orthotropic material characteristics of a homogenized ply for any dynamic analysis. These equivalent mechanical properties of a laminate are treated as a continuum. For any detailed dynamic analysis or high-velocity impact simulations involving composites, these steps are necessary in order to solve any realistic problem by available numerical techniques. To that objective in consideration, the term "Fiber Volume Fraction Ratio V_{fiber}" is defined as follows:

$$V_{\text{fiber}} = \frac{\dfrac{W_{\text{fiber}}}{\rho_{\text{fiber}}}}{\left(\dfrac{W_{\text{fiber}}}{\rho_{\text{fiber}}}\right) + \left(\dfrac{W_{\text{matrix}}}{\rho_{\text{matrix}}}\right)} \tag{6.1}$$

where

V_{fiber} – Volume fraction of fiber used for reinforcement
W_{fiber} – Weight of reinforcing fiber
W_{matrix} – Weight of resin matrix
ρ_{fiber} – Density of reinforcing fiber
ρ_{matrix} – Density of resin matrix

V_{matrix} – Volume fraction of matrix = $(1 - V_{\text{fiber}})$
V_c – Volume fraction of composite = 1.0

Here, the equivalent smeared-up density of the single composite ply, henceforth, called $\rho_{\text{Composite}}$ is determined as follows:

$$\rho_{\text{Composite}} = V_{\text{fiber}}\left(\rho_{\text{fiber}}\right) + \left(1 - V_{\text{fiber}}\right)\left(\rho_{\text{matrix}}\right) \tag{6.2}$$

6.1.2 Bulk Material Characteristics of a Homogeneous Laminate or Ply

The following relationships are most commonly used for composite plies made up of reinforced carbon fibers impregnated and cured with epoxy-based resin as the matrix. In these very thin 2-D plies, the following orthotropic elastic parameters to be used in any impact analysis can be estimated as follows. Here, the term "V_f" abbreviated for the "Fiber Volume Fraction Ratio V_{fiber}" is an important parameter as shown earlier. It is shown here as how to determine it in plies with unidirectional fiber orientation, also called "Unidirectional Tape" data. On using subscripts "f" for fiber and "m" for matrix, we can write the following equations for the equivalent effective orthotropic material characteristics data for the lamina or the ply for a full 3-D LS-DYNA modeling.

$$E_{11} = V_f\ E_F + V_m\ E_m = V_f\ E_F + \left(1 - V_f\right) E_m \tag{6.3}$$

$$E_{22} = \frac{E_m}{1 - V_f\left(1 - \dfrac{E_m}{E_{22\text{-fiber}}}\right)} \tag{6.4}$$

$$G_{12} = \frac{G_m}{1 - V_f\left(1 - \dfrac{G_m}{G_{12\text{-fiber}}}\right)} \tag{6.5}$$

$$G_{23} = \frac{G_m}{1 - V_f\left(1 - \dfrac{G_m}{G_{23\text{-fiber}}}\right)} \tag{6.6}$$

$$G_{13} = G_{12} \tag{6.7}$$

$$\nu_{12} = V_f\left(\nu_{12\text{-fiber}}\right) + \left(1 - V_f\right)\nu_{\text{matrix}} \tag{6.8}$$

$$\nu_{23} = \frac{E_{22}}{2G_{23}} - 1 \tag{6.9}$$

$$\nu_{13} = \nu_{12} \tag{6.10}$$

$$\nu_{21} = \nu_{12}\frac{E_{22}}{E_{11}} \tag{6.11}$$

It should be noted that for most of the epoxy-based carbon fiber composite applications $(1+\nu_{12}\nu_{21}) \approx 1.1$. For some known strain values in the lamina or the ply, the corresponding stress field will have three components of stresses, viz., σ_{11}, σ_{22}, and τ_{12} in the thin 2-D ply, which are computed as follows:

$$\sigma_{11} = \frac{E_{11}}{(1-\nu_{12}\nu_{21})}\,\varepsilon_{11} + \frac{\nu_{12}E_{22}}{(1-\nu_{12}\nu_{21})}\,\varepsilon_{22} \tag{6.12}$$

$$\sigma_{22} = \frac{\nu_{21}E_{11}}{(1-\nu_{12}\nu_{21})}\,\varepsilon_{11} + \frac{E_{22}}{(1-\nu_{12}\nu_{21})}\,\varepsilon_{22} \tag{6.13}$$

$$\tau_{12} = G_{12}\ \gamma_{12} \tag{6.14}$$

Most of the time, all these material characteristics data shown above are not very precise nor are they readily available either. In those situations, they can be approximated as shown below:

$$E_{11} = V_{\text{fiber}} \times E_{\text{fiber}} \tag{6.15}$$

$$E_{22} = (1 - V_{\text{fiber}}) \times E_{\text{matrix}} \tag{6.16}$$

$$\nu_{21} = \nu_{\text{fiber}} \tag{6.17}$$

$$\nu_{12} = \frac{E_{11}}{E_{22}}\,\nu_{21} \tag{6.18}$$

$$G_{12} = \frac{1}{(1+2\nu_{12})}\,E_{\text{matrix}} \tag{6.19}$$

$$\text{Density (mass units)} = \rho_{\text{Composite}} \tag{6.20}$$

These orthotropic material data are useful in a typical simple LS-DYNA modeling of an impact event involving composite plate types of simulations.

6.1.3 Orthotropic Lamina Ply Oriented at Some Arbitrary Rotation Angle

From Mohr's circle transformation equations and referring to Figure 6.2, we know from the 2-D linear elasticity equations that the respective terms of stress tensor σ_{ij} and strain tensor ε_{ij} in the rotated coordinate system follow the following relationships:

$$\begin{Bmatrix} \sigma_{XX} \\ \sigma_{YY} \\ \tau_{XY} \end{Bmatrix} = \begin{bmatrix} \cos^2\theta & \sin^2\theta & -2\sin\theta\cos\theta \\ \sin^2\theta & \cos^2\theta & 2\sin\theta\cos\theta \\ \sin\theta\cos\theta & -\sin\theta\cos\theta & (\cos^2\theta - \sin^2\theta) \end{bmatrix} \begin{Bmatrix} \sigma_{11} \\ \sigma_{22} \\ \tau_{12} \end{Bmatrix} \tag{6.21}$$

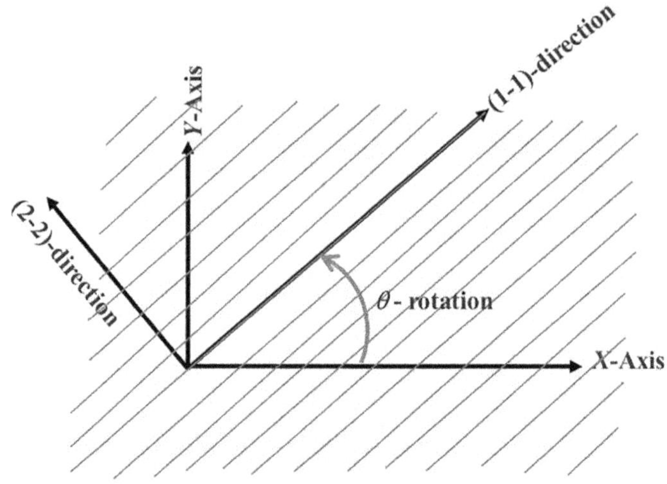

FIGURE 6.2 θ-angular rotation of principal material axes (1-1 and 2-2 directions) with respect to arbitrary X and Y axes.

$$\left\{\begin{array}{c}\varepsilon_{XX}\\ \varepsilon_{YY}\\ \dfrac{\gamma_{XY}}{2}\end{array}\right\}=\left\{\begin{array}{c}\varepsilon_{XX}\\ \varepsilon_{YY}\\ \varepsilon_{XY}\end{array}\right\}=\begin{bmatrix}\cos^2\theta & \sin^2\theta & -2\sin\theta\cos\theta\\ \sin^2\theta & \cos^2\theta & 2\sin\theta\cos\theta\\ \sin\theta\cos\theta & -\sin\theta\cos\theta & (\cos^2\theta-\sin^2\theta)\end{bmatrix}\left\{\begin{array}{c}\varepsilon_{11}\\ \varepsilon_{22}\\ \dfrac{\gamma_{12}}{2}\end{array}\right\} \tag{6.22}$$

In the general design of multilayered composite plates, different plies can be oriented at various angles to enable the structural designer to obtain the desired values of effective orthotropic elastic properties such as $(E_{11})_{\text{Effective}}$ and $(E_{22})_{\text{Effective}}$ with that combination of the stacked set of individual plies. To that goal in mind, let us consider a thin 2-D laminate ply with typical orthotropic properties such as E_{11}, E_{22}, G_{12}, and ν_{21} with its fiber orientation angle θ shown in Figure 6.2. Here, we must determine the respective E_{XX}, E_{YY} of the rotated ply in some other reference directions.

Then, per Reference [1] derivations, its equivalent elastic properties in a new arbitrary global $(X-Y)$ direction are shown as follows:

$$\frac{1}{E_{XX}}=\frac{1}{E_{11}}\cos^4\theta+\left(\frac{1}{G_{12}}-\frac{2\nu_{12}}{E_{11}}\right)\sin^2\theta\cos^2\theta+\frac{1}{E_{22}}\sin^4\theta \tag{6.23}$$

$$\nu_{XY}=E_{XX}\left[\frac{\nu_{12}}{E_{11}}\left(\sin^4\theta+\cos^4\theta\right)-\left(\frac{1}{E_{11}}+\frac{1}{E_{22}}-\frac{1}{G_{12}}\right)\sin^2\theta\cos^2\theta\right] \tag{6.24}$$

$$\frac{1}{E_{YY}}=\frac{1}{E_{11}}\sin^4\theta+\left(\frac{1}{G_{12}}-\frac{2\nu_{12}}{E_{11}}\right)\sin^2\theta\cos^2\theta+\frac{1}{E_{22}}\cos^4\theta \tag{6.25}$$

$$\frac{1}{G_{XY}} = 2\left(\frac{2}{E_{11}} + \frac{2}{E_{22}} + \frac{4\nu_{12}}{E_{11}} - \frac{1}{G_{12}}\right)\sin^2\theta\cos^2\theta + \frac{1}{G_{12}}\left(\sin^4\theta + \cos^4\theta\right) \quad (6.26)$$

These orthotropic material data can directly be used in any LS-DYNA modeling of an impact event involving composite plate types of simulations in a more general case. Here, the reinforcing fibers may be oriented in any arbitrary direction not necessarily coinciding with the geometry of the structure. The equations shown in Eqs. (6.23–6.26) can easily be extended to a composite plate made up of multiple plies each oriented in a different direction.

To this end, in order to model a thick composite plate made up of multiple plies, each with different angular orientations of the fibers such as +90°/–90°/+45°/–45°/0 °/–45°/+45°/–90°/+90°, we will present the 3-D elasticity equations applicable in the orthotropic material model in the LS-DYNA® notations. The respective stress–strain relationships in 3-D elasticity formulation for orthotropic material model, commonly used for modeling composite, are as follows:

$$\varepsilon_{11} = \frac{\sigma_{11}}{E_{11}} - \nu_{21}\frac{\sigma_{22}}{E_{22}} - \nu_{31}\frac{\sigma_{33}}{E_{33}} \quad (6.27)$$

$$\varepsilon_{22} = -\nu_{12}\frac{\sigma_{11}}{E_{11}} + \frac{\sigma_{22}}{E_{22}} - \nu_{32}\frac{\sigma_{33}}{E_{33}} \quad (6.28)$$

$$\varepsilon_{33} = -\nu_{13}\frac{\sigma_{11}}{E_{11}} - \nu_{23}\frac{\sigma_{22}}{E_{22}} + \frac{\sigma_{33}}{E_{33}} \quad (6.29)$$

$$\gamma_{12} = 2\varepsilon_{12} = \frac{\tau_{12}}{G_{12}} \quad (6.30)$$

$$\gamma_{23} = 2\varepsilon_{23} = \frac{\tau_{23}}{G_{23}} \quad (6.31)$$

$$\gamma_{31} = 2\varepsilon_{31} = \frac{\tau_{31}}{G_{31}} \quad (6.32)$$

$$\nu_{12} = \frac{E_{11}}{E_{22}}\nu_{21}, \nu_{13} = \frac{E_{11}}{E_{33}}\nu_{31}, \text{ and } \nu_{23} = \frac{E_{22}}{E_{33}}\nu_{32} \quad (6.33)$$

In the above equation, it should be noted that the six dynamic strain components ε_{11}, ε_{22}, ε_{33}, γ_{12}, γ_{23}, γ_{31} are computed in this manner in any brick element are the element centroidal values. In the composite failure model in LS-DYNA, the progressive failure of the composite element is invoked by comparing the transient dynamic strain components with respect to the limiting values of the respective strain components.

6.2 DYNAMIC BEHAVIOR UNDER IMPACT AND DELAMINATION

In a laminated shell formulation, the delamination poses an important challenge, because each ply is considered a separate 2-D shell element. The dynamic loads get transferred from one ply to the next adjacent ply through interlaminar shear stresses such as τ_{31} and τ_{32}. When the shear stress exceeds the matrix resin material's bonding strength capability, it starts delaminating the plies. In the dynamic loading scenarios, these delamination failures travel through the structural components very rapidly. The delamination failure of adjacent plies is more prevalent under low-velocity impact than with a projectile moving with very high velocity. Delamination along the free edge of composite plies under an in-plane uniaxial load has been observed for a long time. The free-edge delamination is attributed to the existence of interlaminar stresses, which are highly localized in the vicinity of free edge under an in-plane loading. The stacking sequence of plies plays a very important role in predicting the delamination failure.

In classical lamination theory based upon considering only 2-D plane stress formulations with terms like σ_{XX}, σ_{YY}, and τ_{XY}, delamination is not possible to be analyzed. The interlaminar stresses such as σ_{ZZ}, τ_{ZX}, and τ_{ZY} are not even considered, hence the delamination failure is not even possible to account for. The delamination failure essentially observed at the free edge is directly affected by the normal component of the stress σ_{ZZ}. When the sign of σ_{ZZ} component changes from compression to tension, the peeling of the plies takes place. It happens due to Poisson's ratios like ν_{31} and ν_{32} mismatch between adjacent layers in the angled ply stacking of the layers. Designing the ply stacking sequence for angled plies like +15°/–15°/+45°/–45°/0°/–45°/+45°/–15°/+15° for a symmetric configuration is better than an antisymmetric ply stacking sequence. Additionally, another stacking sequence for angled plies like +15°/+45°/–45°/0°/–45°/+45°/+15° may have a less tendency to delaminate under impact.

The existence of all angled plies such as +15°/–15° exhibit shear coupling and no Poisson ratio mismatch effect. On the other hand, the existence of all cross plies such as 0°/90° exhibit the Poisson ratio mismatch effect and no shear coupling. One can conclude that a combination of angled plies and normal plies in the same stacking sequence exhibits both shear coupling and Poisson's ratio mismatch effect. As such, in any impact loading of such a composite plate will yield nonzero values of all three thickness Z-direction components such as σ_{ZZ}, τ_{ZX}, and τ_{ZY}. The existence of angled plies helps in reducing the possibility of delamination during impact.

6.2.1 Normal Impact on a Flat Composite Panel

A typical engine component may have hundreds of plies with each layer of ply oriented in various fiber directions. These types of ply layups could be symmetric or asymmetric in nature. Due to the limitations on the size of any computer model (in terms of total nodes and elements and its running time in LS-DYNA®), the common practice is not to model each individual ply. Instead, a typical composite element is represented with an orthotropic elastic material model considered an equivalent bulk material representation of many individual plies. The method to convert individual ply mechanical characteristics into bulk orthotropic properties is based upon in-plane two-dimensional Mohr's circle transformation of stresses and strains in two orthogonal directions (usually, 0° and 90° with respect to a reference direction).

Since most of the finite-element meshing patterns follow some natural directions of the component being analyzed, it is very important to be careful (in the meshing of composite structures) that the nodal connectivity for all elements on a particular surface maintains a certain order. In the majority of commercially available automatic mesh generator schemes, such an order of elemental nodal connectivity is not usually maintained. In addition, the accuracy of the LS-DYNA® results for the orthotropic material model can be further improved by taking into consideration that the stress wave velocities V_{11}, V_{22}, and V_{33} are different in different directions, i.e.,

$$V_{11} = \sqrt{\frac{E_{11}}{\rho}},\ V_{22} = \sqrt{\frac{E_{22}}{\rho}},\ \text{and}\ V_{33} = \sqrt{\frac{E_{33}}{\rho}} \tag{6.34}$$

It should be noted that the stress wave velocities shown above are typical membrane wave velocities, which involve stretching and shrinking of the fibers. The corresponding flexural bending wave effects traveling through the plate have a relatively slower velocity, and the resulting deformation is visible to human eyes. In an ideal LS-DYNA® mesh for a composite plate, the element size should be chosen in such a way that, for a given time-step size Δt, the stress wave travels one element long in each direction. For example, in an orthotropic plate with (1–1) and (2 – 2) as the two natural directions, the in-plane best aspect ratio of a typical element should be determined as follows:

$$\text{In-plane aspect ratio}\ \left(1\text{–}2\ \text{material axes directions}\right)\ \text{of elements} = \sqrt{\frac{E_{11}}{E_{22}}} \tag{6.35}$$

In an attempt to validate the above-mentioned approach of using equivalent bulk material properties in the LS-DYNA® model instead of ply-by-ply properties, a composite panel was tested. The details of this test have been discussed earlier in an ASME journal paper published in December 2011 [see Reference 2]. With the permission from the ASME, the results of that test are being reproduced here for the benefit of the readers. The controlled test condition simulating a typical normal impact of the composite panel was represented in the analytical LS-DYNA® modeling scheme. The rectangular panel was made up of multiple symmetrical plies, each ply being 0.1448 mm thick. In the test setup, a rectangular plate (dimension: 43.18 cm long × 30.48 cm wide × 1.12 cm thick) was impacted in the middle by a gelatin bird weighing 161.7 g moving at 92.66 m/s (see Figure 6.3). The gelatin-bird cylindrical impactor had a length-to-diameter ratio of 2. The impactor was a simulated gelatin bird (10% air + 90% water) hitting it in the middle at 92.66 m/sec. The numerical values of equivalent bulk orthotropic elastic properties of the test panel are shown below:

$$E_{11} = 22.815\ \text{GPa}$$

$$E_{22} = 80.120\ \text{GPa}$$

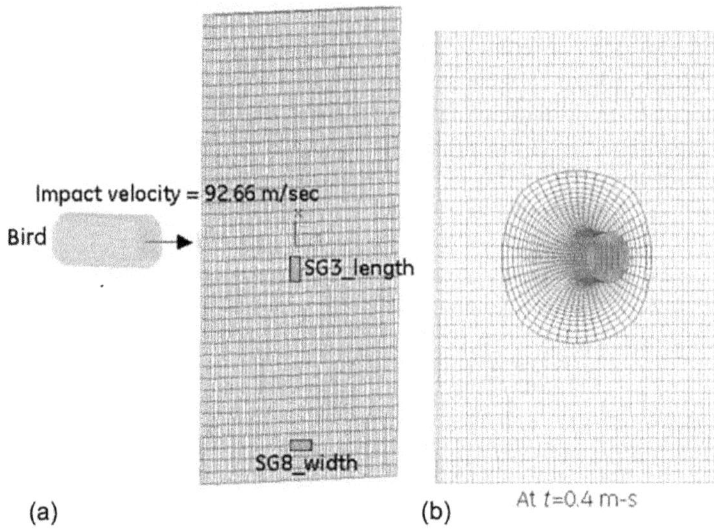

FIGURE 6.3 (a) Rectangular composite panel with strain gauges (SG3 and SG8) being impacted by a gelatin cylinder moving at 92.66 m/s. (b) LS-DYNA® results showing the deformation of the gelatin cylinder after 0.4 ms of initial contact. (Reproduced with permission from Reference [2].)

$$E_{33} = 7.440 \text{ GPa}$$

$$\nu_{21} = 0.716$$

$$\nu_{31} = 0.107$$

$$\nu_{32} = 0.0172$$

$$G_{12} = 21.071 \text{ GPa}$$

It had several strain gauges on the nonimpact surface at different locations of the panel, out of which two typical gauges called SG_3 and SG_8 are shown in Figure 6.3a.

The dynamic strain history in the composite panel due to the normal impact of a gelatin cylindrical soft-body impactor was measured in two directions shown as SG_3 lengthwise and SG_8 widthwise on the back surface of the target plate.

Figures 6.4 and 6.5 show the comparison of measured versus analytically predicted strain results for two typical gauges using the equivalent bulk material properties, instead of ply-by-ply orthotropic data to represent the dynamic response characteristics of a multilayered composite plate. Numerically computed strain gauge response results based upon the equivalent bulk properties compare very well with the measured data from the actual test both in magnitude and in frequency contents

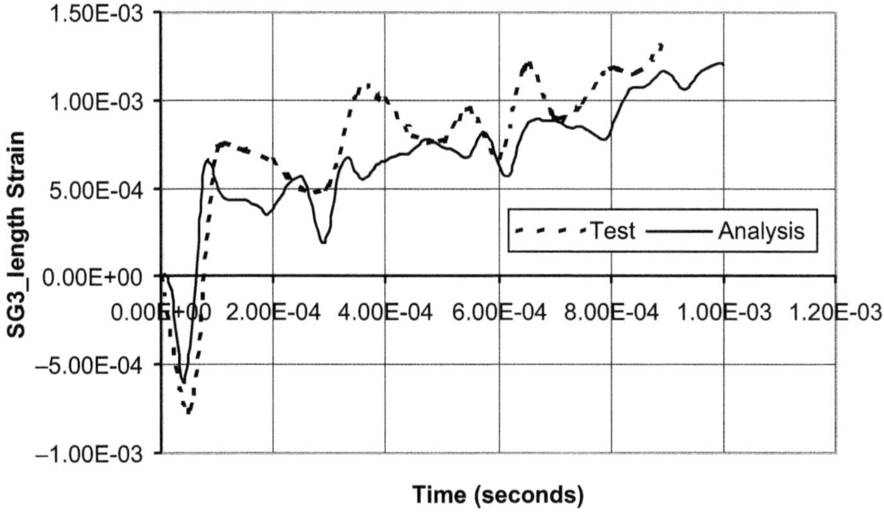

FIGURE 6.4 Comparison of strain gauge dynamic response from LS-DYNA® vs test data (lengthwise strain gauge – SG3). (Reproduced with permission from Reference [2].)

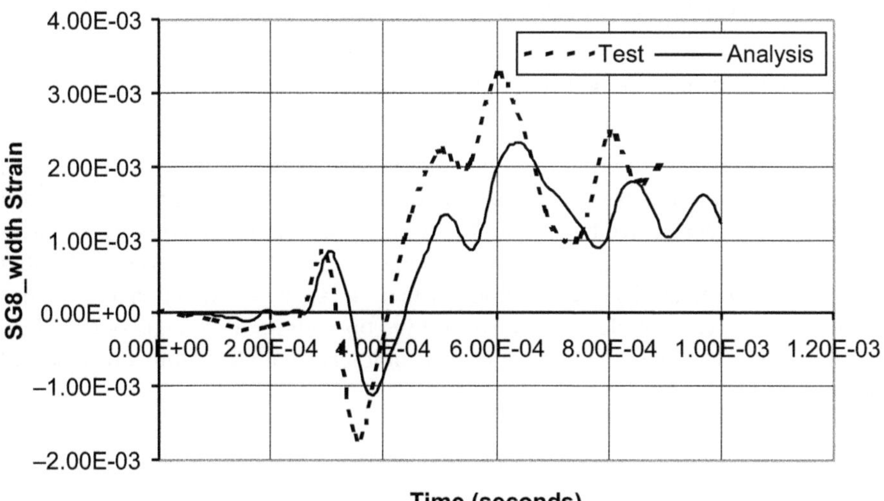

FIGURE 6.5 Comparison of strain gauge dynamic response from LS-DYNA® vs test data (widthwise strain gauge – SG8). (Reproduced with permission from Reference [2].)

of the dynamic response. From Figure 6.5, the time-history plot of SG-8, it should be noted that the LS-DYNA® composite model captures the dynamic response in the width direction of the plate very well in terms of its frequency contents, but misses the peak magnitude of strain values in both the positive (+) and negative (–) directions almost by the same amount. This observation is attributed to the fact that the strain-gauge measures the surface of the corresponding eight-noded brick element in that location, whereas the LS-DYNA® model computes the corresponding element strain parameter ε_{22} at the element centroid of the brick element.

Overall, the comparison of the predicted versus measured surface strain results shows that the material model developed in the current context of composite impact simulation produces reasonably good correlation with the test data and can be used reliably for analyzing various types of complex both hard and soft bodies impact loading scenarios including containment structure and rotating airfoil fan blades.

6.3 DYNAMIC RESPONSE OF COMPOSITE AIRFOILS

Under the dynamic response of composite airfoils, we will focus our attention on both the free vibrational modes of the composite blade and its dynamic response due to the impact of foreign objects and Fan Blade Out (FBO) such as birds or hailstones ingested into the engine during normal operations. The technical contents of this section are based upon the author's previous publications in archival journals such as *AIAA Journal* [see Reference 5] and the *International Journal of Mechanical Sciences* [see Reference 3]. Some of the results presented in these academic papers are being reproduced here with permission from the respective publishers. To that end, first we discuss as how to determine the important vibrational mode frequencies of an airfoil turbofan blade made up of laminated multilayered composite.

6.3.1 Vibrational Characteristics of Composite Airfoil Fan Blades

Using the thin shell theory in the curvilinear coordinate system, we start with the laminate equations with orthotropic material properties (see Figure 6.6) in the span and chordwise directions of an airfoil. Here, we assume that the pretwisted airfoil mid-surface is represented by an orthogonal curvilinear thin shell coordinate system called (ξ, η, ζ). This Lagrangian coordinate system is such that the origin is located at the airfoil root, and its cross-section twists in the spanwise direction as we move from the root's twist angle being ϕ_0 toward the tip with its twist angle of ϕ_L. Thus, the rate of pretwist ϕ' in the units of radians/unit length in the airfoil is represented as follows:

$$\phi' = \frac{(\phi_L - \phi_0)}{L} \tag{6.36}$$

It should be noted that in the curved airfoil terminology, the concave curved surface is also called "Pressure surface". Similarly, the convex surface of the airfoil is referred to as "Suction surface". Additionally, the forward side of the airfoil is called "Leading Edge (LE)" and the aft side is known as "Trailing Edge (TE)". Then, the strain equations in 2-D curvilinear formulation are as follows. Assuming that in the Lagrangian frame of the curvilinear shell coordinate reference (ξ, η, ζ) system, the corresponding deflection of any typical point is (u, v, w), which enables us to write the following strain deflection equations. In our composite material notation, we will use the term ε_{11} in the spanwise ξ-direction and term ε_{22} in the chordwise η-direction,

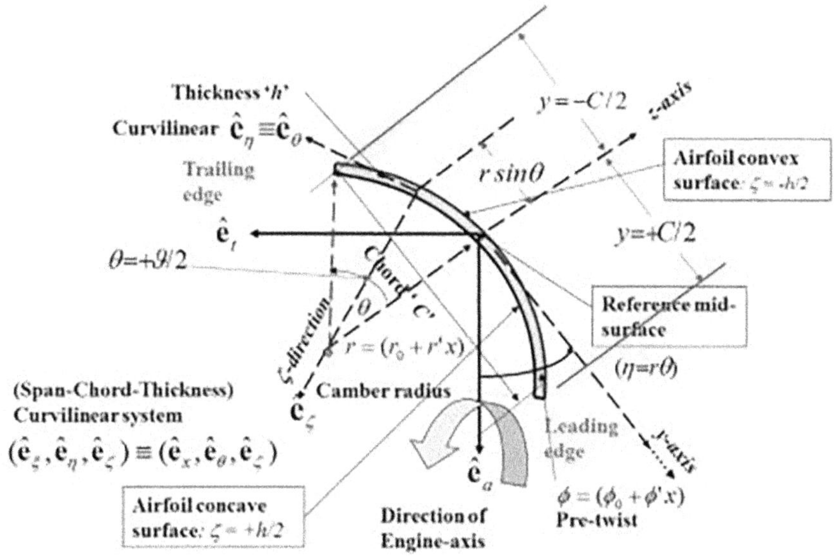

(a)

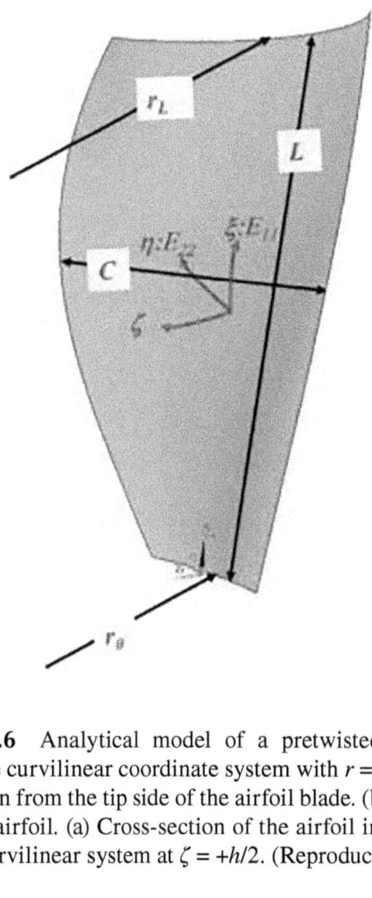

(b)

FIGURE 6.6 Analytical model of a pretwisted composite airfoil and the respective mid-surface curvilinear coordinate system with r = camber radius: (a) View of (η, ζ) or (y, z) cross-section from the tip side of the airfoil blade. (b) View of (ξ, η) surface from the concave side of the airfoil. (a) Cross-section of the airfoil in the (η, ζ) plane. (b) Concave surface in the (ξ, η) curvilinear system at $\zeta = +h/2$. (Reproduced with permission from Reference [3].)

respectively. In this particular case, the span direction x-axis coincides with the ξ -direction.

$$\varepsilon_{11} = \bar{\varepsilon}_{\xi\xi} + \zeta\kappa_{\xi} = \frac{\partial u}{\partial x} + (r\theta\phi')\left[\frac{\partial w}{\partial x} + \phi' v\right] - \zeta\left[\frac{\partial^2 w}{\partial x^2} - (\phi')^2\left(w + x\frac{\partial w}{\partial x}\right) + \phi'\frac{\partial v}{\partial x}\right] \tag{6.37}$$

$$\varepsilon_{22} = \bar{\varepsilon}_{\eta\eta} + \zeta\kappa_{\eta} = \left[\frac{1}{r}\frac{\partial v}{\partial \theta} - \frac{w}{r} + \frac{\theta v}{r} - \frac{x}{r}\frac{\partial w}{\partial x}\right] - \zeta\begin{bmatrix}\frac{1}{r^2}\frac{\partial^2 w}{\partial \theta^2} - \left[1 - (\phi' x)^2\right] \\ \frac{w}{r^2} - \frac{x}{r^2}\frac{\partial w}{\partial x} + \frac{1}{r^2}\frac{\partial v}{\partial \theta}\end{bmatrix} \tag{6.38}$$

$$\gamma_{12} = \bar{\gamma}_{12} + \zeta\kappa_{12} = \frac{1}{r}\frac{\partial u}{\partial \theta} + \frac{\partial v}{\partial x} + (1+\theta)\frac{\partial w}{\partial x} + x\frac{\partial^2 w}{\partial x^2} - (\phi' w) + 2\theta\phi'\left[v + \frac{\partial w}{\partial \theta}\right]$$
$$-\zeta\left[\left(\frac{w}{r}\right)\right]\frac{\partial^2 w}{\partial r\,\partial \theta} + (\phi')^2\left[\frac{x}{2r}\frac{\partial w}{\partial \theta} - \frac{2xw}{r} - \frac{2x^2}{r}\frac{\partial w}{\partial x}\right] - \phi'\left[\frac{w}{r} + \frac{x}{r}\frac{\partial w}{\partial x}\right] \tag{6.39}$$

$$\gamma_{13} = x\frac{\partial^2 w}{\partial x^2} \tag{6.40}$$

$$\gamma_{23} = \frac{\theta}{r}\frac{\partial^2 w}{\partial \theta^2} \tag{6.41}$$

Using the shell approximation for a 2-D surface in a curvilinear coordinate system, the corresponding stress–strain relationships for any typical point can be written as follows:

$$\varepsilon_{11} = \frac{\sigma_{11}}{E_{11}} - \nu_{21}\frac{\sigma_{22}}{E_{22}} \tag{6.42}$$

$$\varepsilon_{22} = -\nu_{12}\frac{\sigma_{11}}{E_{11}} + \frac{\sigma_{22}}{E_{22}} \tag{6.43}$$

Equations (6.42) and (6.43) can be simplified as follows:

$$\sigma_{11} = \frac{E_{11}}{(1 - \nu_{12}\nu_{21})}\left[\varepsilon_{11} + \nu_{21}\varepsilon_{22}\right] \tag{6.44}$$

$$\sigma_{22} = \frac{E_{22}}{(1 - \nu_{12}\nu_{21})}\left[\varepsilon_{22} + \nu_{12}\varepsilon_{11}\right] \tag{6.45}$$

$$\sigma_{33} = 3K(\varepsilon_{11} + \varepsilon_{22} + \varepsilon_{33}) - (\sigma_{11} + \sigma_{22}) \tag{6.46}$$

$$\gamma_{12} = \frac{\tau_{12}}{G_{12}} \tag{6.47}$$

$$\frac{\nu_{21}}{E_{22}} = \frac{\nu_{12}}{E_{11}} \tag{6.48}$$

$$D_{11} = \frac{E_{11}\ h^3}{12(1-\nu_{12}\nu_{21})} \tag{6.49}$$

$$D_{22} = \frac{E_{22}\ h^3}{12(1-\nu_{12}\nu_{21})} \tag{6.50}$$

$$D_{12} = \frac{\nu_{21}E_{11}\ h^3}{12(1-\nu_{12}\nu_{21})} = \frac{\nu_{12}E_{22}\ h^3}{12(1-\nu_{12}\nu_{21})} \tag{6.51}$$

$$D_{66} = \frac{G_{12}\ h^3}{12} \tag{6.52}$$

$$D_{16} = \frac{G_{13}\ h^3}{12} \tag{6.53}$$

$$D_{26} = \frac{G_{23}\ h^3}{12} \tag{6.54}$$

The full bending moment terms in a curved composite thin shell element (see Figure 6.7) such as (M_x, M_θ, $M_{\theta x}$), which also include the second-order negligible effects of small membrane-type stretch loading ($\bar{\sigma}_{xx}$, $\bar{\sigma}_{\theta\theta}$, $\bar{\tau}_{\theta x}$) of the mid-surface of the thin shell, are as follows:

$$M_\xi = M_x = \int_{-h/2}^{+h/2} \zeta\left[\sigma_{xx} - \zeta\phi'\bar{\sigma}_{xx}\right]d\zeta \approx D_{11}\kappa_x + D_{12}\kappa_\theta$$

$$= -D_{11}\left[\frac{\partial^2 w}{\partial x^2} - (\phi')^2\left[w + x\frac{\partial w}{\partial x}\right]\right] - D_{12}\left[\frac{1}{r^2}\frac{\partial^2 w}{\partial \theta^2} - \left[1-(\phi' x)^2\right]\frac{w}{r^2} - \left(\frac{x}{r^2}\right)\frac{\partial w}{\partial x}\right] \tag{6.55}$$

$$M_\eta = M_\theta = \int_{-h/2}^{+h/2} \zeta\left[\sigma_{\theta\theta} - \frac{\zeta}{r}\ \bar{\sigma}_{\theta\theta}\right]d\zeta \approx D_{12}\kappa_x + D_{22}\kappa_\theta$$

$$= -D_{22}\left[\frac{1}{r^2}\frac{\partial^2 w}{\partial \theta^2} - \left[1-(\phi' x)^2\right]\frac{w}{r^2} - \left(\frac{x}{r^2}\right)\frac{\partial w}{\partial x}\right] - D_{12}\left[\frac{\partial^2 w}{\partial x^2} - (\phi')^2\left[w + x\frac{\partial w}{\partial x}\right]\right] \tag{6.56}$$

Using the shear balance equation of $\tau_{x\theta} = \tau_{\theta x}$, we obtain the following equation:

$$M_{\eta\xi} = M_{\theta x} = \int_{-h/2}^{+h/2} \zeta \left[\tau_{\theta x} - \zeta \sqrt{\frac{\phi'}{r}} \bar{\tau}_{\theta x} \right] d\zeta \approx -2D_{66}\kappa_{x\theta}$$

$$= -2D_{66} \left[\frac{\partial^2 w}{r\,\partial x\,\partial\theta} - (\phi')^2 \left[\frac{2xw}{r} + \frac{2x^2}{r}\frac{\partial w}{\partial x} - \frac{x}{2r}\frac{\partial w}{\partial\theta} - \phi'\frac{w}{r} - \frac{x\phi'}{r}\frac{\partial w}{\partial x} \right] \right] \tag{6.57}$$

The corresponding shear force terms like Q_x and Q_θ for force per unit length are as follows:

$$Q_\xi \equiv Q_x = \int_{-h/2}^{+h/2} \tau_{\theta x} d\zeta = \frac{\partial M_x}{\partial x} + \frac{\partial M_{\theta x}}{r\,\partial\theta} = \frac{\partial}{\partial x}\left[D_{11}\kappa_x + D_{12}\kappa_\theta\right] + \frac{2D_{66}}{r}\frac{\partial\kappa_{\theta x}}{\partial\theta} \tag{6.58}$$

$$Q_\eta \equiv Q_\theta = \int_{-h/2}^{+h/2} \tau_{\theta\zeta} d\zeta = \frac{\partial M_\theta}{r\,\partial\theta} - \frac{\partial M_{x\theta}}{\partial x} = \frac{\partial}{r\,\partial\theta}\left[D_{22}\kappa_\theta + D_{12}\kappa_x\right] - 2D_{66}\frac{\partial\kappa_{x\theta}}{\partial x} \tag{6.59}$$

In a laminated shell formulation, the dynamic equilibrium of the shell element is shown in Figure 6.7 in the form of an airfoil blade by considering its inertia terms as

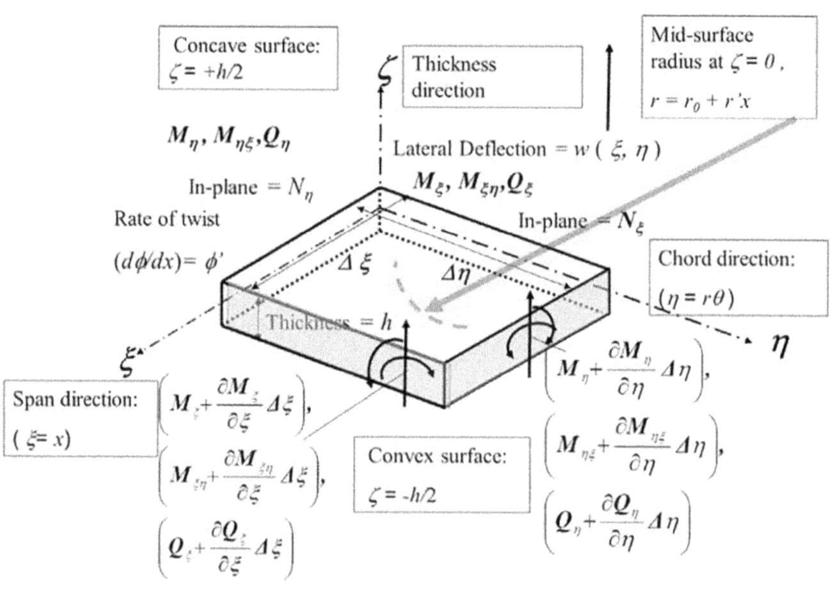

FIGURE 6.7 Force and moment balance in a typical infinitesimal plate element (radius of curvature or camber $r = r_0 + r'x$, spanwise length $= \Delta\xi$, chordwise length $= \Delta\eta$). (Reproduced with permission from Reference [3].)

well as the centrifugal terms, and the corresponding equation is written as follows. To that end, if we ignore the second-order effect due to a change in the camber radius of the airfoil cross-section, then for shallow thin shells using the shear force (dimension: force/unit length) terms like Q_ξ and Q_η, we can write the equation for dynamic equilibrium of a typical infinitesimal shell element of thickness "h" and in-plane dimensions of $(\Delta\xi \times \eta)$ as follows:

$$\frac{\partial Q_\xi}{\partial \xi} + \frac{\partial Q_\eta}{\partial \eta} + \frac{\partial}{\partial \xi}\left(N_\xi \frac{\partial w}{\partial \xi}\right) + \frac{\partial}{\partial \eta}\left(N_\eta \frac{\partial w}{\partial \eta}\right) \cong -q(\xi,\eta,t) + \rho h \ddot{w} - \frac{\rho h^3}{12}\left[\frac{\partial^2 \ddot{w}}{\partial \xi^2} + \frac{\partial^2 \ddot{w}}{\partial \eta^2}\right] \tag{6.60}$$

The above equation is for the dynamic equilibrium of a pretwisted airfoil for the out-of-plane deflection $w(\xi,\eta,t)$ in the thickness direction, normal to its mid-surface, made up of multilayered laminated composite plies with its equivalent bulk orthotropic properties. It should be noted that on the right-hand side of Eq. (6.60), there are two $\ddot{w}$ acceleration-related terms are being included. The first term with (ρh) is due to translational inertia of the infinitesimal element, and the second term with $(\rho h^3/12)$ is due to rotational inertia of the cross-section, especially important for thick shells. After some algebraic manipulations, Eq. (6.60) becomes

$$\begin{aligned}
&D_{11}\frac{\partial^4 w}{\partial x^4} + \frac{2(2D_{66}+D_{12})}{r^2}\frac{\partial^4 w}{\partial x^2 \partial \theta^2} + \frac{D_{22}}{r^4}\left[\frac{\partial^4 w}{\partial \theta^4} - x^2\frac{\partial^2 w}{\partial x^2} - 2x\frac{\partial w}{\partial x} - \frac{\partial^2 w}{\partial \theta^2}\right] - \frac{3D_{12}}{r^2}\frac{\partial^2 w}{\partial x^2} \\
&- (\phi')^4\left[D_{11}\,x^2\frac{\partial^2 w}{\partial x^2} + 2D_{11}x\frac{\partial w}{\partial x} + D_{66}\frac{x^2}{r^2}\frac{\partial^2 w}{\partial \theta^2}\right] + (\phi')^2\left[\begin{array}{l} 3D_{11}\dfrac{\partial^2 w}{\partial x^2} + \dfrac{(2D_{66}+D_{12})}{r^2}\dfrac{\partial^2 w}{\partial \theta^2} \\ -2D_{66}\left[\dfrac{3}{r^2}\dfrac{\partial^2 w}{\partial \theta^2} + \dfrac{x}{r^2}\dfrac{\partial^3 w}{\partial x \partial \theta^2}\right] \end{array}\right] \\
&-h\frac{\partial(\sigma_{cf})_x}{\partial x}\frac{\partial w}{\partial x} - h(\sigma_{cf})_x\frac{\partial^2 w}{\partial x^2} - h\frac{\partial(\sigma_{cf})_\theta}{r\partial \theta}\frac{\partial w}{r\partial \theta} - \rho h\Omega^2\cos^2(\phi_0 + \phi' x)\,w \\
&= -\rho h\ddot{w} + \frac{\rho h^3}{12}\left[\frac{\partial^2 \ddot{w}}{\partial x^2} + \frac{1}{r^2}\frac{\partial^2 \ddot{w}}{\partial \theta^2}\right] - \rho h(\phi' x)^2\left[\ddot{w} - \nu_{12}\left(\frac{r\vartheta\theta}{L}\right)^2\frac{\partial^2 \ddot{w}}{\partial \theta^2}\right]
\end{aligned} \tag{6.61}$$

It should be noted that in the above equation, the two additional terms on the right-hand side of the equation containing acceleration effect $\ddot{w}$ are due to contributions of rotary inertia and warping inertia of an airfoil cross-section in the thin shell formulation. Readers should recall that warping deformation in a thin shell is the characteristic of a noncircular cross-section being subjected to some kind of torsional loading. Here, the numerical values of the free vibrational mode frequencies of the

TABLE 6.1
Composite Airfoil Parameters Considered for a Sample Example

Longitudinal Span Length of the Airfoil	$L = 100$ cm
Blade outer radius on the rotor	$R = (1.5)L$
Blade inner radius at its base (clamped airfoil root)	$b = (0.5)L$
Average thickness of the airfoil	$h = (0.02)L$
Blade average aspect ratio	$L/C = 2.0$
Average chord width of the airfoil	$C = (0.5)L$
a. Chord width at the airfoil root	$C_0 = (0.4)L$
b. Chord width at the airfoil tip	$C_L = (0.6)L$
Blade average camber radius	$r = (3.5)L$
a. Airfoil camber radius at the root	$r_0 = 3L$
b. Airfoil camber radius at the tip	$r_L = 4L$
Total twist in the blade	$(\phi_L - \phi_0) = -60.0°$
a. Stagger angle at the airfoil root	$\phi_0 = 0.0°$
b. Stagger angle at the airfoil tip	$\phi_L = -60.0°$
Rotational speed	rpm $= 3069$
Angular velocity	$\Omega = 321.4159$ radian/s
Orthotropic material properties of the airfoil:	
$E_{11} = 100$ GPa	
$E_{22} = 0.4E_{11}$	
$G_{12} = 0.5E_{11}$	
$\nu_{12} = 0.55$	
$\nu_{21} = (E_{22} / E_{11})\nu_{12}$	
$\rho = 1.6$ gm/cc	

composite airfoil can be determined using Galerkin's method to solve the Eigenvalue problem, as described in Reference [4].

The free vibration analysis for this hypothetical composite blade has been carried out for all the dynamical modes of interest in a typical compressor or fan blade. In the following sample problem, blade dimensions are arbitrary, and all the relevant key geometrical features have been nondimensionalized with respect to the assumed span length of the airfoil being $L = 100$ cm. In this example, the numerical values are typical and are representative of a realistic blade design, which are shown in Table 6.1.

The first 20 fundamental modal frequencies determined for a stationary airfoil blade called "Bench Frequencies" along with rotating frequencies are shown in Table 6.2. It should be noted that after the first five modes, the majority of blade vibrational mode shapes are extremely complex to classify as a pure torsional or flexural bending.

These numerically computed frequencies are consistent with typical blade frequencies, and the results of the current thin shell model are in close agreement

TABLE 6.2
Comparison of Composite Blade Natural Frequencies for the First 20 Modes Between Stationary (Bench Frequencies) and Rotating (3069 RPM) Conditions

	Stationary Blade	Blade Rotating at 3069 RPM
Composite Blade Airfoil Vibrational Mode	**Nondimensional Frequency Parameter $\bar{\omega}_N$**	**Nondimensional Frequency Parameter $\bar{\omega}_N$**
1. First flexural bending (1F)	3.46704	8.20094
2. First torsion (1T)	7.99381	8.04074
3. Second torsion (2T)	12.84426	12.97331
4. Third torsion (3T)	17.52426	17.77457
5. Second flexural bending (2F)	19.48870	28.19237
6. Fourth torsion (4T)	22.93593	23.31909
7. Fifth torsion (5T)	29.43606	29.95096
8. Sixth torsion (6T)	37.25264	37.89617
9. Seventh torsion (7T)	46.80615	47.58909
10. Third flexural bending (3F)	51.41827	62.18006
11. First edge bending (1EB)	54.29583	54.29629
12. Eighth torsion (8T)	58.50909	59.43396
13. Second edge bending (2EB)	62.69783	62.69926
14. Third edge bending (3EB)	70.89889	70.90166
15. Ninth torsion (9T)	74.82443	75.92477
16. Fourth edge bending (4EB)	79.72839	79.79629
17. Fifth edge bending (5EB)	89.48008	89.48499
18. Fourth flexural bending (4F)	99.15452	111.28983
19. Sixth edge bending (6EB)	100.51140	100.51694
20. Seventh edge bending (7EB)	113.14414	113.15013

with the finite-element model results using ANSYS®. As an example, the comparison of the first three vibrational mode frequencies of the airfoil blade is shown in Figure 6.3a–c. It can be seen that the first three modes, which are usually of the most interest in the preliminary blade design stage, are predicted with less than 2.5% numerical error (Figure 6.8).

The mathematical formulation presented in this section enables us, the reader, to determine the natural frequencies of all the fundamental modes of vibration in a physics-based analytical model of a composite fan blade under rotating conditions. This modeling method can also be used for determining the optimum ply layup, which is an important design consideration for the separation of any potential excitation frequencies from the natural frequency of the blade in the rotating condition.

6.3.2 Composite Airfoil Blade Under Slicing Impact Loading

As an example of carbon fiber-reinforced multilayered plies with a combination of angled plies along with a ply layup of +90°/–90°/0°/–90°/+90° plies, a composite

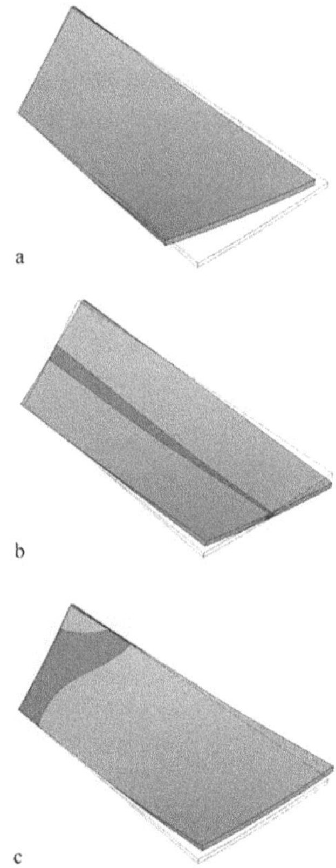

FIGURE 6.8 (a) First mode: ANSYS® 25.26 Hz vs thin shell bulk orthotropic analytical model: 25.90 Hz (difference = 2.5%). (b) Second mode: ANSYS® 69.83 Hz vs thin shell bulk orthotropic analytical model: 70.16 Hz (difference = 0.4%). (c) Third mode: ANSYS® 107.37 Hz vs thin shell bulk orthotropic analytical model: 108.51 Hz (difference = 1.06%). (Reproduced with permission from Reference [3].)

airfoil blade response is shown under the slicing action of an incoming bird by its LE. The contents of this section are based upon the recently published paper by the author in the *AIAA Journal*. The summarized results shown here are based upon the findings in that paper and are being reproduced here with permission from the AIAA. The impact conditions and bulk equivalent orthotropic properties of the composite blade are similar to those listed in Table 6.1. The blade is impacted by a simulated gelatin bird cylinder of 2 kg mass on the blade spinning about the engine axis at 3000 rpm. The analytical model considers both the spanwise radial growth of the blade under centrifugal force field and out-of-plane flexural bending due to the impact of the bird slice. The in-plane surface will be represented by $(x-y)$ plane, with x being radially outward in the spanwise direction and y being in the chordwise direction of the fan blade. The thickness direction of the pretwisted plate, the out-of-plane deformation $w(x,y,t)$ will be in the local z-axis direction normal to the mid-surface.

For accurate bird-strike simulation, we must consider the full coupling of the spanwise radial growth of the blade represented by $u(x, y, t)$ with the lateral out-of-plane movement $w(x, y, t)$ of the blade mid-surface due to bird impact.

For simplification purposes, one fundamental assumption has been made that, in a multilayered fiber-reinforced composite component, it is possible to capture its impact response by considering it as a monolithic anisotropic plate with its equivalent bulk elastic properties in two orthogonal directions such as E_{xx}, E_{yy}, ν_{xy}, ν_{yx}, and G_{xy}. Also, we will ignore the effect of camber radius change, and effectively we will assume that the camber radius is so large that the airfoil can be considered a flat plate in simplifying the partial differential equation (Eq. 6.61). Hence, any typical point located at $(x - y)$ on the mid-surface of the pretwisted plate represents the composite airfoil (see Figure 6.9) in the local shell coordinate system.

As such, the equation of longitudinal motion due to spanwise deformation $u(x, y, t)$ in the spanwise x direction of the blade is expressed as follows:

$$-\frac{hE_{xx}}{\left[1-\nu_{xy}\nu_{yx}\right]}\frac{\partial^2 u}{\partial x^2} - \rho h\Omega^2 u = -\rho h\ddot{u} + \rho h\Omega^2\,(R-L+x) - 2\rho h\Omega\cos(\phi_0+\phi' x)\,\dot{w} \tag{6.62}$$

$$D_{11}\frac{\partial^4 w}{\partial x^4} + 2(2D_{66}+D_{12})\frac{\partial^4 w}{\partial x^2\partial y^2} + D_{22}\frac{\partial^4 w}{\partial y^4} + (\phi')^2\begin{bmatrix} 3D_{11}\dfrac{\partial^2 w}{\partial x^2} + (2D_{66}+D_{12})\dfrac{\partial^2 w}{\partial y^2} \\ -2D_{66}\left[3\dfrac{\partial^2 w}{\partial y^2} + x\dfrac{\partial^3 w}{\partial x\,\partial y^2}\right]\end{bmatrix}$$

$$-(\phi')^4\left[D_{11}\,x^2\frac{\partial^2 w}{\partial x^2} + 2D_{11}x\frac{\partial w}{\partial x} + D_{66}\,x^2\frac{\partial^2 w}{\partial y^2}\right] - \frac{\partial}{\partial x}\left[N_x\frac{\partial w}{\partial x}\right] - \frac{\partial}{\partial y}\left[N_y\frac{\partial w}{\partial y}\right]$$

$$-\rho h\Omega^2\cos^2(\phi_0+\phi' x)\,w - 2\rho h\Omega\cos(\phi_0+\phi' x)\,\dot{u} = -\rho h\ddot{w} + \frac{\rho h^3}{12}\left[\frac{\partial^2\ddot{w}}{\partial x^2} + \frac{\partial^2\ddot{w}}{\partial y^2}\right]$$

$$-\rho h(\phi' x)^2\left[\ddot{w} - \nu_{12}\left(\frac{Cy}{L}\right)^2\frac{\partial^2\ddot{w}}{\partial y^2}\right] + \left[-B_m\ddot{w} + F_{\text{Bird-slice}}\right]\delta(x-s)\,\delta(y-c) \tag{6.63}$$

We also have a rotating cantilever blade with its clamped end being at a radial distance of b from the engine axis of rotation, which enables us to write the components of centrifugal membrane loading in the local span and chordwise directions $(\sigma_{cf})_x$ and $(\sigma_{cf})_y$, respectively, as follows:

$$\frac{N_x}{h} = (\sigma_{cf})_x = \frac{\rho\Omega^2}{2}\left[R^2 - b^2 - 2bx - x^2\right] \tag{6.64}$$

$$\frac{N_y}{h} = (\sigma_{cf})_y = \frac{\rho C^2\Omega^2\sin^2(\phi_0+\phi' x)}{8} \tag{6.65}$$

$$\frac{\partial\left(\sigma_{cf}\right)_x}{\partial x}=-\rho\Omega^2(b+x) \tag{6.66}$$

$$\frac{\partial\left(\sigma_{cf}\right)_y}{\partial y}=0 \tag{6.67}$$

Based upon the above expressions of Eqs. (6.64–6.67), we can write the effect of membrane components of centrifugal loading on the blade bending equation in Eq. (6.63) as follows:

$$\frac{\partial}{\partial x}\left[N_x\frac{\partial w}{\partial x}\right]+\frac{\partial}{\partial y}\left[N_y\frac{\partial w}{\partial y}\right]=h\frac{\partial\left(\sigma_{cf}\right)_x}{\partial x}\frac{\partial w}{\partial x}+h\left(\sigma_{cf}\right)_x\frac{\partial^2 w}{\partial x^2}+h\frac{\partial\left(\sigma_{cf}\right)_y}{\partial y}\frac{\partial w}{\partial y}$$

$$+h\left(\sigma_{cf}\right)_y\frac{\partial^2 w}{\partial y^2}=-\rho h\Omega^2\begin{bmatrix}(b+x)\dfrac{\partial w}{\partial x}-\dfrac{\left[R^2-b^2-2bx-x^2\right]}{2} \\ \dfrac{\partial^2 w}{\partial x^2}-\dfrac{C^2\sin^2\left(\phi_0+\phi' x\right)}{8}\dfrac{\partial^2 w}{\partial y^2}\end{bmatrix} \tag{6.68}$$

The previous set of Eqs. (6.62) and (6.63) exhibits the dynamical coupling characteristics of the plate vibratory motion in an angular velocity field Ω. It can be easily seen that the longitudinal (extensional wave) and the transverse (flexural wave) direction motions are cross-coupled through the skew-symmetric velocity-dependent gyroscopic terms like $-2\rho h\Omega\cos\left(\phi_0+\phi' x\right)\dot{w}$ and $+2\rho h\Omega\cos\left(\phi_0+\phi' x\right)\dot{u}$, respectively. In a fan-rotor composite airfoil, the coupling effect is further complicated due to the presence of Coriolis forces in the dynamical system. The Coriolis forces caused by the rotation of the engine make the phase velocity of the forward-traveling stress wave faster than the backward-traveling stress wave from the point of initial contact of the airfoil by the bird impactor.

Using the classical Galerkin's method discussed in earlier chapters, the above set of coupled partial differential equations (Eqs. 6.62 and 6.63) are converted into a set of ordinary second-order differential equations in a matrix form. The resulting equations of motion are solved numerically by using a time-marching forward integration approach. The results and the key findings of the transient solution, especially during the slicing action of the bird cylinder by the LE of the blade, are solved numerically. To apply Galerkin's method, we start the solution of the previous set of coupled partial differential equations shown in Eqs. (6.62) and (6.63) by assuming the following set of deformation shape functions $u(x,t)$ and $w(x,y,t)$ for the composite blade. It should be noted that the assumed deformation shape functions satisfy the geometric boundary conditions at the clamped end of the blade root arbitrarily, located at the top of the fan disk, which are as follows:

$$u(x,t)=\sum_{i=1}^{n_1}X_i(t)\left[U(x)\right]_i=\sum_{i=1}^{n_1}X_i(t)\sin\left(\varphi_i x\right) \tag{6.65}$$

$$w(x,y,t)=\sum_{m=1}^{3}\left[\sum_{i=1}^{n_2} Z_i(t)\left[W(x)\right]_i\right]_m V_m(y)=\sum_{m=1}^{3}\left[\sum_{i=1}^{n_2} Z_i(t)\left[W(x)\right]_i\right]_m V_m(y)$$

$$=\sum_{m=1}^{3}\left[\sum_{i=1}^{n_2} Z_i(t)\left[W(x)\right]_i\right]_m \cos\alpha_m\left(y+\frac{C}{2}\right)$$

$$=\sum_{m=1}^{3}\left[\sum_{i=1}^{n_2} Z_i(t)\left[1-\cos(\varphi_i x)\right]\right]_m \cos\alpha_m\left(y+\frac{C}{2}\right) \tag{6.66}$$

Here, it should be noted that in the above equations, the terms $X_i(t)$ and $Z_i(t)$ have the dimension of length units. As such, it should be obvious that the shape function terms like $\left[U(x)\right]_i$ and $\left[W(x)\right]_i$ are dimensionless periodic functions and should be chosen arbitrarily.

In the series form of the blade deformation shape function, the coefficients such as α_m, and φ_i of the sinusoidal functions for sequential terms in the series (with $m=1,2,3.....$ and $i=1,2,3,4.......$) are defined as follows:

$$\alpha_m=\frac{(m-1)\pi}{C} \tag{6.67}$$

$$\varphi_i=\frac{(2i-1)\pi}{2L} \tag{6.68}$$

Here, the individual nonzero typical ith row and jth column term of the equivalent mass [M], damping [C], and stiffness [K] submatrices are as shown next. In the following two submatrices shown in Eqs. (6.69) and (6.70), the range of indices is $[i=1,2......\ n_1]$; $[j=1,2......\ n_1]$; as shown below:

$$[\mathrm{K}]_{(i,j)}=[\boldsymbol{K}_{uu}]_{(i,j)}=-\frac{hE_{xx}}{\left[1-\nu_{xy}\nu_{yx}\right]}\left[\int_0^L CU_iU_j''dx-CU_i(L)U_j'(L)\right]-\rho h\Omega^2\int_0^L CU_iU_jdx \tag{6.69}$$

$$[\mathrm{M}]_{(i,j)}=[\boldsymbol{M}_{uu}]_{(i,j)}=\rho h\int_0^L CU_iU_jdx \tag{6.70}$$

In the following two submatrices shown in Eqs. (6.71) and (6.72), for varying values of $[m=1,2,3]$ the size of stiffness and mass matrices in the range of $[i=1,2......m(n_2),\ j=1,2......m(n_2)\ ]$, the corresponding terms are given by the following equations:

$$
\begin{aligned}
[\mathrm{K}]_{(n_1+i,n_1+j)} = [\boldsymbol{K_{ww}}]_{(i,j)} = \int_0^L C\Big[& D_{11}W_iW_j'''' - 2(2D_{66}+D_{12})(\alpha_m)^2 W_iW_j'' + D_{22}(\alpha_m)^4 W_iW_j \\
& + (\phi')^2 \begin{bmatrix} 3D_{11}W_iW_j'' - (2D_{66}+D_{12})(\alpha_m)^2 W_iW_j + 2D_{66} \\ \left[3(\alpha_m)^2 W_iW_j + x(\alpha_m)^2 W_iW_j'\right] \end{bmatrix} \\
& - (\phi')^4\left[D_{11}\ x^2W_iW_j'' - D_{66}\ x^2(\alpha_m)^2 W_iW_j + 2D_{11}xW_i\,W_j'\right]\Big]dx + \rho hC\ \Omega^2 \\
& \Bigg[\int_0^L x\left(W_iW_j'\right)dx + b\int_0^L\left(W_iW_j'\right)dx - \frac{\left(R^2-b^2\right)}{2}\int_0^L\left(W_iW_j''\right)dx + b\int_0^L x\left(W_iW_j''\right)dx \\
& + \frac{1}{2}\int_0^L x^2\left(W_iW_j''\right)dx - \left(\left(\alpha_m^2\right)/8\right)\int_0^L C^2\sin^2(\phi_0+\phi'x)\left(W_iW_j\right)dx - \int_0^L \cos^2(\phi_0+\phi'x)\left(W_iW_j\right)dx\Bigg] \qquad (6.71) \\
& + \left[(4D_{66}+D_{12})C(\alpha_m)^2 W_i(L)W_j'(L) - (\phi')^2\begin{bmatrix} 2D_{11}CW_i(L)W_j'(L) - 2D_{66} \\ C(\alpha_m)^2\ L\ W_i(L)\ W_j(L) \end{bmatrix}\right] \\
& - \left[D_{12}\ C(\alpha_m)^2\ W_i'(L)\ W_j(L) - D_{11}\ C(\phi')^2\left[W_i'(L)\ W_j(L) + L\ W_i'(L)\ W_j'(L)\right]\right]
\end{aligned}
$$

$$[\mathrm{M}]_{(n_1+i,n_1+j)} = [\boldsymbol{M}_{ww}]_{(i,j)} = \rho hC\left[\int_0^L (W_iW_j)\,dx\right]$$
$$-\frac{\rho h^3C}{12}\left[\int_0^L \left(W_iW_j''\right)dx - (\alpha_m)^2\int_0^L (W_iW_j)\,dx\right] \tag{6.72}$$
$$+\ \rho hC(\phi')^2\left[1+\frac{\nu_{xy}}{12}\left(\frac{C(m-1)\pi}{L}\right)^2\right]\int_0^L x^2(W_iW_j)\,dx$$

In the following submatrix shown in Eq. (6.73), the range of indices $\left[i = 1,2......n_1,\ j = 1,2......m(n_2)\right]$ are as shown below:

$$[C]_{(i,n_1+j)} = [\boldsymbol{C}_{uw}]_{(i,j)} = 2\rho hC\Omega\int_0^L (U_iW_j)\cos(\phi_0+\phi' x)\,dx \tag{6.73}$$

In the following submatrix shown in Eq. (6.74), the range of indices $\left[i = 1,2......m(n_2),\ j = 1,2......\ n_1\right]$ are as shown below:

$$[C]_{(n_1+i,j)} = [\boldsymbol{C}_{wu}]_{(i,j)} = -2\rho hC\Omega\int_0^L (W_iU_j)\cos(\phi_0+\phi' x)\,dx \tag{6.74}$$

In the current example, we assume that the blade chord C is constant for the entire span of the blade. In the analytical simulation, the individual terms of the respective matrices $[M]$, $[C]$, and $[K]$ are determined using the numerical integration with the displacement shape function outlined in Eqs. (6.65) and (6.66). For the sample example problem illustrated in Figure 6.9, we have used $m = 3$, $n_1 = 10$, and $n_2 = 10$, which results in the global matrices being 40×40.

Figures 6.10–6.13 illustrate the time-history plots of various dynamic response parameters of interest in any typical bird-strike analysis. As an example, Figure 6.10 shows the computed span and chordwise strains in the vicinity of impact using the present analytical model. Figure 6.9 highlights the oscillatory behavior of the span and chordwise strains in the vicinity of impact at the LE of the blade airfoil for the first 20 ms from the time of initial contact.

Similarly, Figure 6.11 shows the time-history plot of the lateral deflection of the blade airfoil in the vicinity of impact at the LE for the first 20 ms of the time duration. The corresponding peak local out-of-plane deformation in the blade (see Figure 6.11) has been determined as 6 cm at the impact location. It is interesting and counterintuitive that peak deformations in a composite airfoil blade may be larger at a distant location than at the point of initial impact. For example, at the blade tip, the peak lateral deformations (see Figure 6.11) are more at the free end of the TE, with 10.8 cm, than at the blade LE, with a smaller value of 9.5 cm.

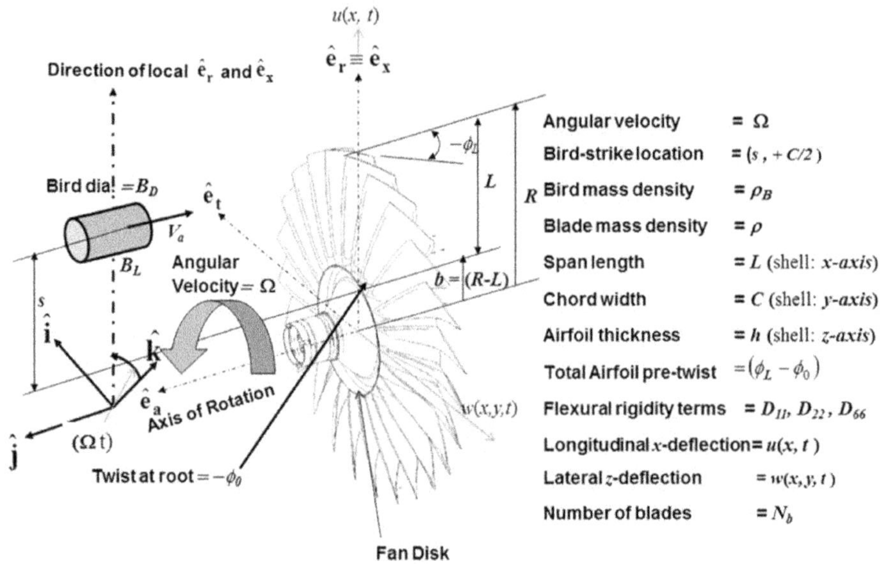

FIGURE 6.9 Schematic diagram of a bird cylinder of diameter B_D coming in contact with the blade airfoil rotating at angular velocity Ω. (Reproduced with permission from Reference [5].)

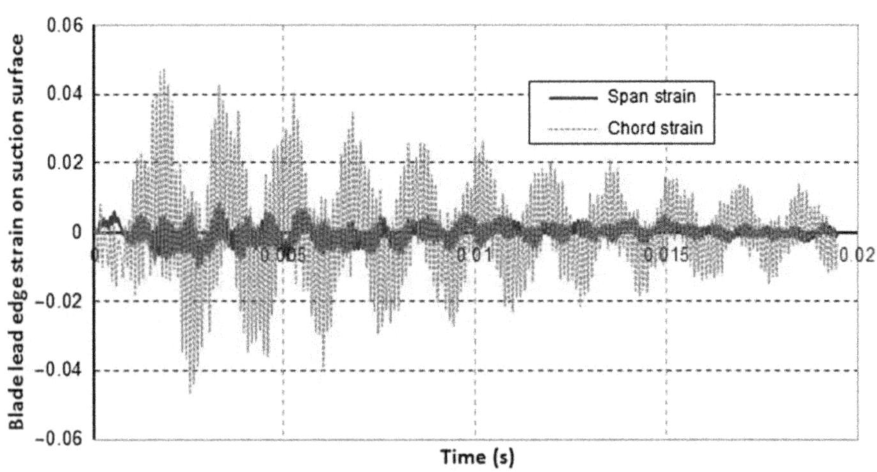

FIGURE 6.10 Time history of surface strain at the leading edge of the blade-impact location. (Reproduced with permission from Reference [5].)

Figure 6.11 highlights the transient behavior of the radial and tangential components of the blade response at the root of the composite airfoil. It should be noted that the oscillating nature of the radial load with its peak of 40 kN is caused by the dynamic coupling of the longitudinal and transverse motions of the blade. For the first time, it is possible to analytically predict the effective imbalance load on a

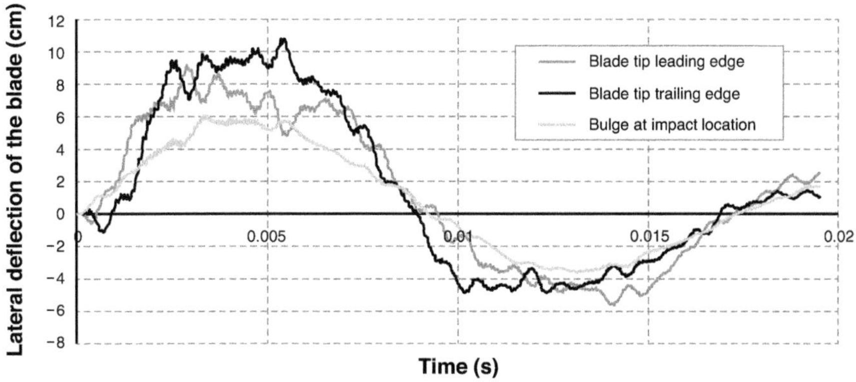

FIGURE 6.11 Time history of blade tip deflection in the transverse direction at the bird-impact location and the blade tip. (Reproduced with permission from Reference [5].)

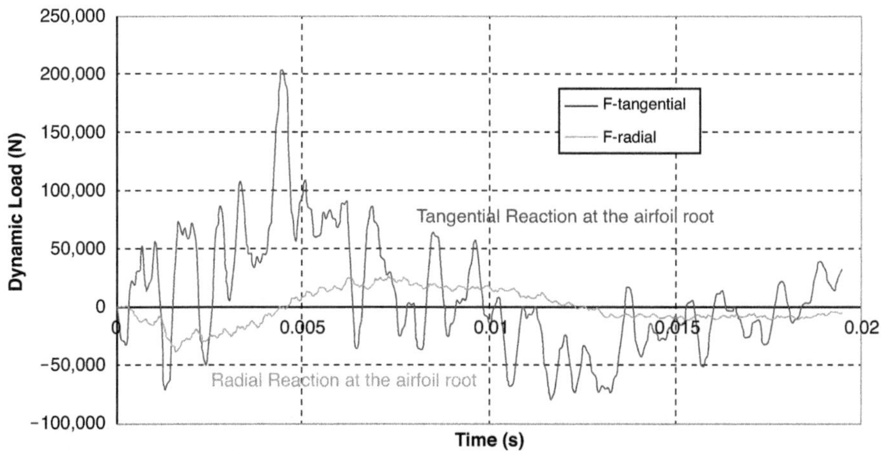

FIGURE 6.12 Time-history dynamic response at the airfoil root in the tangential and radial direction of the rotor shaft. (Reproduced with permission from Reference [5].)

rotating shaft caused by the bird impact. Because of the bird's impulsive loading on its pressure surface, the clamped end of the airfoil root also experiences a significantly large bending moment, which in the present condition reaches up to 18,000 N · m (see Figure 6.13). This dynamic moment produces tension in the fibers on the pressure side of the blade.

6.4 DYNAMICS OF UNIDIRECTIONAL YARN AND WOVEN FABRIC COMPOSITES

Woven fabrics have various applications in the field of impact dynamics and to enhance the impact damage resistance capability of textile fibers such as the design of airbags in the automative crash simulations and protection from serious injury to

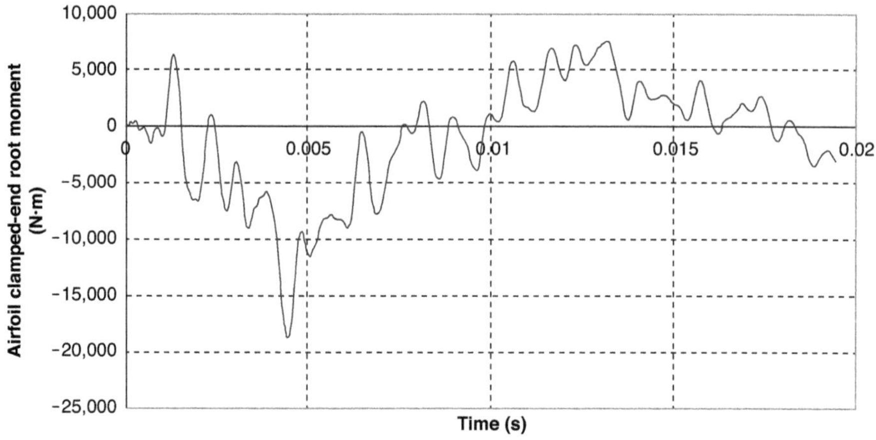

FIGURE 6.13 Time history of bending moment at the clamped-end root of the airfoil. (Reproduced with permission from Reference [5].)

the driver and the passengers. The fabric weaving pattern plays an important role in determining the equivalent bulk directional material properties such as E_{11} and E_{22}. There are two main types of weaving patterns, which are as follows:

a. Plain weave
b. Braided weave

Weaving patterns also introduce small bending in the fibers, which is called "Crimp". In analytical modeling of Crimp, it is very critical, because during the deformation of the yarn inside the fabric it tries to straighten up first. Up to that point, the deformation can be considered elastic, and the remainder of the time until the "Failure" point of fiber it deforms and forms a pocket to nest the release blade. This phase of deformation of the fabric never recovers, even after the dynamic loads are gone. Due to this reason, this phase of deformation can be considered pseudo-plastic deformation of the fabric, until it fails.

6.4.1 Aramid and Kevlar Fabric Discussion

The best-known application of Kevlar fabric is in the design of bulletproof vests for human body protection, usually among the law enforcement personnel. In the aerospace applications, they are one of the preferred materials to design the fan blade containment structure discussed earlier in Chapter 4. Aramid fibers can absorb a significant amount of kinetic energy of the impacting projectile due to relatively large deflection. Aramid fibers are synthetic man-made materials with high-performance mechanical properties in terms of their strength-to-weight ratio, especially in comparison to metals such as high-strength steel and titanium. Aramid fabrics are a lightweight solution for ballistic protection used in the defense sector for a large variety of applications including soft containment structures and body armors. Their melting temperature is in excess of 950°F and is known for their extremely flame and fire-resistant characteristics. Due to their exceptional impact-resistant toughness capability, they have wide applications in aerospace, automotive, and various types of body armors and in bulletproof jackets. Chemically speaking, they are polymers with small molecules and long chains, which make these fibers ideal for spinning into yarns.

Aramid yarns are of bright yellow color, straight, insoluble, and made up of aramid molecules, which are well-suited for spinning into high-performance synthetic fibers and fabrics. They derive their strength from strong bonding between relatively short molecules, transferring mechanical stress rather efficiently. When it comes to protecting the human body, these durable, high-performance fibers are excellent for use in fire-resistant clothing, cut-resistant gloves, and aprons. Aramids were first developed in the 1960s by the DuPont chemical company. Kevlar is the commercial name of aramid fibers developed and marketed by DuPont. There are two basic types of Kevlar fibers known as Kevlar 29 and Kevlar 49.

Depending on their architecture from yarn level to the fabric level usage, the aramid fabrics can be classified into two categories: woven and nonwovens. In woven fabrics, fibers are bundled in yarns woven into a regular pattern, whereas nonwovens consist of a random fiber network connected through local bonds consolidated by thermal fusion, chemical binding, or resin-based bonding with their adhesive behavior. These bonds influence the interaction between fibers and have a primary role in the deformation and the local strain characteristics of the nonwoven.

So far, the numerical simulations are concerned, and they are modeled like a composite with orthotropic material characteristics. In developing their equivalent material properties, one can consider the fabric made up of Kevlar fiber as yarn being treated as unidirectional fibers with air space being the matrix. In the woven form, Kevlar yarn modeling is based upon the model suggested by Abood in 1980. To use the mathematical formulation of "Volume fraction ratio effect", at this point, we will introduce another term called "Solidity Factor". Solidity factor is defined as follows. Thus, neglecting any Poisson's ratio effect in the Kevlar fiber direction, one can write equivalent elastic properties in the 1-1 direction as follows:

$$E_{11\text{-Kevlar Fabric-at-ply-level}} = (\text{Solidity factor}) \times E_{\text{Yarn}} \tag{6.75}$$

$$\text{Effective density of Kevlar Blanket} = (\text{Solidity factor}) \times \rho_{\text{Yarn}} \tag{6.76}$$

In the balanced plain weave configuration, as shown in Figure 6.13, the corresponding equivalent elastic properties in the 2-2 direction will be similar. It should be noted that their Poisson's ratio numbers ν_{12} and ν_{21} will be different.

$$E_{22\text{-Kevlar Fabric-at-ply-level}} = (\text{Solidity factor}) \times E_{\text{Yarn}} \times (\text{Poisson's ratio}) \tag{6.77}$$

TABLE 6.3
Characteristic Material Property Data of Kevlar "Yarn"

Kevlar 49 Yarn Properties	Metric Units	British Units (Lbs-in-Seconds)
1. Material density	1.440 g/cc	0.052 Lbs/in^3
2. Young's modulus	100 GPa	14.46×10^6 psi
3. Extensional wave propagation speed	8334 m/s	32.81×10^4 in/s
4. Fiber breaking failure strain	4%	4%

The groundbreaking work in the field of Kevlar Fabric modeling was first done by Leech and Abood, which was presented in at the December 1989 ASME conference in San Francisco [see Reference 6]. The material presented here is based upon their findings and recommendations. With the permission from ASME, some of their illustrations are being reproduced here for the benefit of the readers. However, for a detailed description and in-depth discussion on this topic, the reader should refer to the original research paper. The basic building block of woven Kevlar fabric, sometimes also called "Blanket" or "Wraps", is known as "yarns", as supplied and marketed by its manufacturer DuPont. Yarns are made up of Kevlar filaments or fibers.

The Kevlar's yarn level material characteristics data, needed for any dynamic analysis or numerical simulations, as reported by Leech and Abood in their classic 1989 paper are as shown in Table 6.3.

It should be noted that above all stress and effective density-related material data are for the yarn level configuration only, and as such for any fabric level conversion, they must be derated by the respective "Solidity Factor", based upon the "Volume Fraction Ratio" formula. It should also be emphasized that based upon a prudent and conservative design best practice, the total failure strain $\varepsilon_{\text{Failure}}$ remains the same 4% and neither it should be modified nor corrected in any form from the "yarn" level data. However, a typical value of 1% "Crimp strain correction" is introduced in such a way that the remainder strain capability of the woven fabric is left with 3% capability. The details of the Crimp correction were first proposed and its mathematical formulation based upon the crimp geometry was derived by Leech and Abood [see Reference 6].

Now, we will consider the geometrical effect of weaving pattern and its effect on the strength capability of fabric versus yarn characteristics. Referring to Figure 6.14, the weaving process introduces "Crimp stress" due to local bending of the yarn.

In a Kevlar wrap or blanket numerical simulations, the yarn properties listed in Table 6.3 need to be corrected per its "Solidity Factor" and also by introducing "Crimp Stress" as Pseudo-elastic stress limit, which may appear like a very low value of "Pseudo-Yield stress σ_Y". This "Pseudo-Yield stress σ_Y" is also sometimes called "Crimp stress σ_{Crimp}". Usually, this limit for most of the Kevlar woven fabric blanket is about 1% strain out of the "total failure limit being 4%", as shown in Table 6.3. This typical artificial elastic–plastic representation of a Kevlar fabric blanket is schematically shown in Figure 6.15.

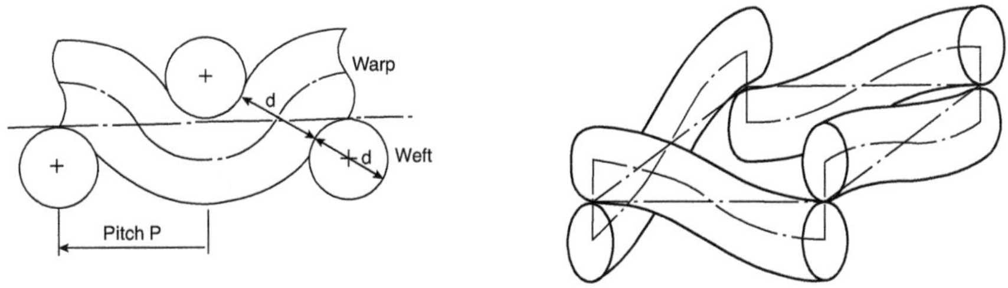

FIGURE 6.14 Schematic representation of two different views of the Crimp geometry with yarn fiber's cross-overs. (Reproduced with permission from Reference [6].)

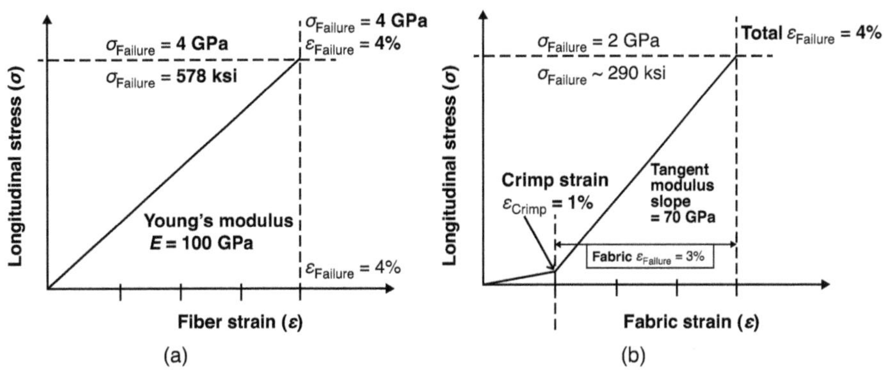

FIGURE 6.15 Schematic representation of Kevlar stress–strain characteristics (a) for unidirectional fibers packed with no air gaps and (b) for woven yarn with "Crimp" correction.

Finite-element modeling of Kevlar fabric wrap or blanket in any numerical simulations is a challenge, but here are some recommendations as the best practices specifically useful in explicit method solving schemes.

1. Each Kevlar wrap must be modeled independently using one layer of an eight-noded constant strain brick element in LS-DYNA®.
2. All dynamic parameters such as equivalent material density, elastic Young's modulus, tangent modulus, pseudo-yield stress, and failure stress must be derated from the basic value of respective parameters shown in Figure 6.14b by the required "Solidity factor" parameter. The actual "Solidity Factor" is due to air gaps between the Kevlar layers and other voids introduced during the installation process.

6.5 COMPLEX NATURE OF DYNAMIC FAILURE MODES IN COMPOSITES

In a laminated fiber-reinforced layered composite structure, the failure modes under dynamic loading scenarios are extremely complex. The failures can be of various variations, such as fiber failure, matrix failure, and a combination thereof as well.

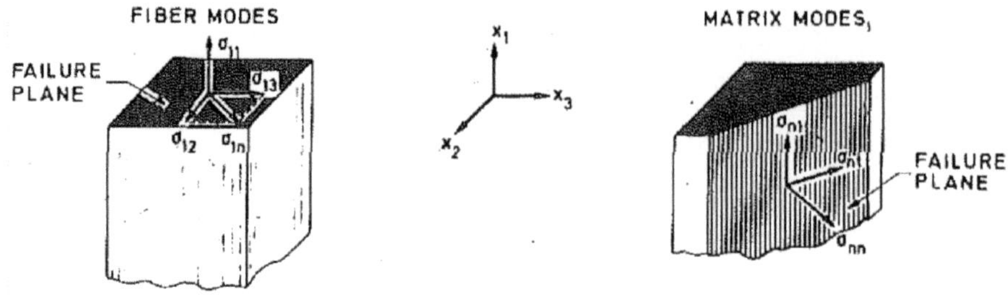

FIGURE 6.16 Schematic representation of failure modes and failure planes in composites. (Reproduced with permission from Reference [8].)

These complex failure modes also involve interlaminar shear as well as intralaminar crack propagation (see Figure 6.16). Usually, these failure modes under dynamic conditions depend upon the type of impact such as normal, oblique, and glancing impact as well as on the magnitude of the relative velocity vector of contact. In that context in the following subsection, we discuss two different composite failure analytical methodologies as the current state of the art in the field of numerical simulation and failure evaluation of composite plate and shell types of structures and mechanical components.

6.5.1 Tsai–Wu Criteria

In 1965, Tsai originally assumed that the failure criterion of a unidirectional fiber composite has the same mathematical form as the yield criterion of an orthotropic ideally plastic material as given by Hill's Theory of the Yielding and Plastic Flow of Anisotropic Materials in the form of a three-dimensional yield surface. Later on, Tsai and Wu [see Reference 7] represented the failure criterion as a general quadratic in the stresses, thereby eliminating the special dependence on the normal stress components. It is more like a weighted failure criteria based on components of normal stresses such as σ_{11} and σ_{22} in a 2-D composite orthotropic plate type of structures.

Figure 6.17 shows schematically how different combinations of σ_{22} will be affected depending upon the choice of constant, linear, or quadratic combinations

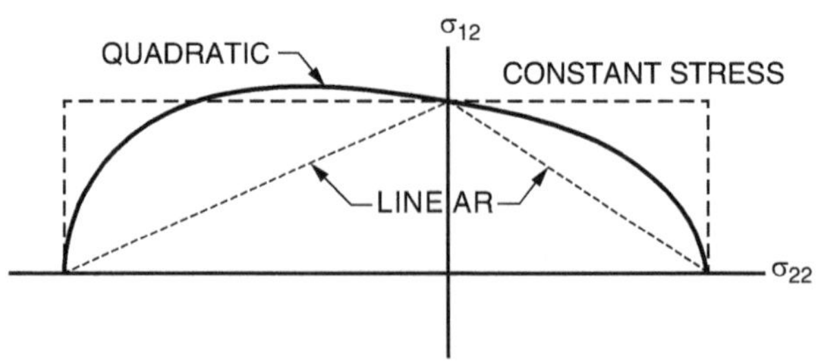

FIGURE 6.17 Schematic representation of failure surfaces as (a) constant stress, (b) linear, or (c) quadratic with the combined effect of σ_{22} and σ_{12}. (Reproduced with permission from Reference [8].)

of these two stress components. Tsai and Wu introduced something called "Strength tensors terms" represented as F_i as the second-order tensor and F_{ij} as the fourth-order tensor, such that the general failure surface in the stress space exists, which can be expressed in the following form:

$$F_i\sigma_i + F_{ij}\ \sigma_i\sigma_j = 1 \tag{6.78}$$

It should be noted that the dimensions of "Strength tensor terms" as F_i and F_{ij} are such that they are inverse of respective failure stress limits in tension and compression like $(\sigma_{\text{Failure}})_T$ and $(\sigma_{\text{Failure}})_C$, respectively, as shown below. For simplicity, and a better understanding of the limiting (τ_{Failure}) term, we will derive the F_1, F_2, F_{11}, F_{22}, F_6, and F_{66} terms only in 2-D lamina applications only, which are as follows:

a. In the E_{11}-direction linear strength tensor: $F_1 = \dfrac{1}{(\sigma_{\text{Failure}})_T} + \dfrac{1}{(\sigma_{\text{Failure}})_C}$
b. In the E_{11}-direction quadratic strength tensor: $F_{11} = -\dfrac{1}{(\sigma_{\text{Failure}})_T} \times \dfrac{1}{(\sigma_{\text{Failure}})_C}$
c. In the E_{22}-direction linear strength tensor: $F_2 = \dfrac{1}{(\sigma_{\text{Failure}})_T} + \dfrac{1}{(\sigma_{\text{Failure}})_C}$
d. In the E_{22}-direction quadratic strength tensor: $F_{22} = \dfrac{1}{(\sigma_{\text{Failure}})_T} \times \dfrac{1}{(\sigma_{\text{Failure}})_C}$
e. In the $(1-2)$ shear plane, linear strength tensor: $F_6 = 0$
f. In the $(1-2)$ shear plane, quadratic strength tensor: $F_{66} = \dfrac{1}{(\tau_{\text{Failure}})^2_{\text{Shear}}}$

In an orthotropic lamina under plane stress conditions, the very complex form of 3-D general Tsai–Wu failure criteria (6.65) reduces to the following equation:

$$F_1\sigma_{11} + F_2\sigma_{22} + F_6\tau_{12} + F_{11}\ (\sigma_{11})^2 + F_{22}\ (\sigma_{22})^2 + F_{66}\ (\tau_{12})^2 + 2\ = 1 \tag{6.79}$$

Here, in the most general 3-D space using the Tsai–Wu formulation, with the normal stress notation terms, all six different components of a stress tensor, we can write as follows:

$$\sigma_1 \equiv \sigma_{11}, \sigma_2 \equiv \sigma_{22}, \sigma_3 \equiv \sigma_{33}, \sigma_4 \equiv \tau_{23}, \sigma_5 \equiv \tau_{31} \text{ and } \sigma_6 \equiv \tau_{12} \tag{6.80}$$

It should be obvious that these equations are extremely complicated and are very difficult to implement, and as such are never used in solving any real-life problems. Thus, in a 3-D composite structure, the failure surface is a six-dimensional ellipsoid. And, anywhere in the stress field, if point lies inside the ellipsoid, then the equivalent combination of six parameters stress field will not make it to fail. However, if the stress field is such that the point is outside the "Failure ellipsoid" surface, then it is likely to fail.

It also takes into account varying strength and failure limits in compression and tension. According to Tsai–Wu criteria, the shear strength of composites such as liming values of τ_{12}, τ_{13} and τ_{23} must be determined by actual test, which are very expensive and cumbersome. This difficult method of testing was simplified by Hashin's failure model, which is relatively much simpler to implement in any typical finite-element schemes, such as the LS-DYNA® material model like MAT_161 and MAT_162 invoked.

6.5.2 Hashin's Failure Model

In his ASME paper published in 1980, Hashin proposed a three-dimensional failure criterion of unidirectional fiber composites, which is established in terms of quadratic stress polynomials. The failure parameters are expressed in terms of the transversely isotropic invariants of the applied average stress state. Four distinct failure modes – tensile and compressive fiber and matrix modes – are modeled separately, resulting in a piecewise smooth failure surface. In the present discussion, the fiber failure criteria of Hashin for a unidirectional layer are generalized to characterize the fiber damage in terms of strain components for a plain weave layer under the plane stress failure criteria. It should be emphasized that the case of plane stress is of considerable practical significance since such is the state of stress in the laminate ply at a sufficient distance from the free edges. Here, the original three-dimensional failure criteria are reduced for the relevant four different modes of the current plane stress situation of composite laminate ply with unidirectional fiber orientation, as follows. In the following derivations, it is assumed that the strength in the transverse direction to the fiber longitudinal direction is completely governed by the strength of the matrix only, and fibers do not contribute toward the transverse direction strength at all.

a. Tensile fiber failure mode for $\sigma_{11} > 0$.

$$\left(\frac{\sigma_{11}}{(\sigma_{\text{Failure}})_T}\right)^2 + \left(\frac{\tau_{12}}{(\tau_{\text{Failure}})_{\text{Shear}}}\right)^2 = 1 \tag{6.81}$$

b. Compressive fiber failure mode for $\sigma_{11} < 0$.

$$\sigma_{11} = -(\sigma_{\text{Failure}})_C \tag{6.82}$$

c. Tensile matrix failure mode for $\sigma_{22} > 0$.

$$\left(\frac{\sigma_{22}}{(\sigma_{\text{Failure-Matrix}})_T}\right)^2 + \left(\frac{\tau_{12}}{(\tau_{\text{Failure}})_{\text{Shear}}}\right)^2 = 1 \tag{6.83}$$

d. Compressive matrix failure mode for $\sigma_{22} < 0$.

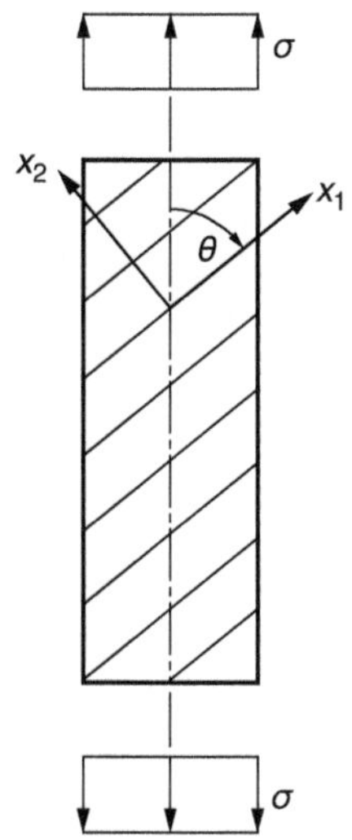

FIGURE 6.18 Schematic representation of an off-axis specimen. [Reproduced with permission from Reference [8].)

$$\left(\frac{\sigma_{22}}{2(\tau_{\text{Failure-Matrix}})_{\text{Shear}}}\right)^2+\left[\left(\frac{(\sigma_{\text{Failure-Matrix}})_C}{2(\tau_{\text{Failure-Matrix}})_{\text{Shear}}}\right)^2-1\right]\left(\frac{\sigma_{22}}{(\sigma_{\text{Failure-Matrix}})_C}\right)$$
$$+\left(\frac{\tau_{12}}{(\tau_{\text{Failure}})_{\text{Shear}}}\right)^2=1 \quad (6.84)$$

The case of tensile modes failure outlined in Eqs. (6.81) and (6.83) can be easily tested and verified by means of an off-axis specimen (see Figure 6.18) under uniaxial tension σ. This state of stress with respect to the material axes can be written as follows:

$$\sigma_{11}=\sigma\cos^2\theta \quad (6.85)$$

$$\sigma_{22}=\sigma\sin^2\theta \quad (6.86)$$

$$\tau_{12}=\sigma\sin\theta\cos\theta \quad (6.87)$$

On substitution of Eqs. (6.85–6.87) in Eqs. (6.81–6.84) yields the off-axis (direction normal to the direction of external load application in the test specimen, which happens to be in the horizontal direction in this figure) conditions for failure in the fiber and matrix, respectively, as follows:

$$\sigma_{\text{failure-fiber}}^{\text{Transverse}} = \frac{1}{\cos^2\theta\left[\frac{\cos^2\theta}{(\sigma_{\text{Failure}})_T} + \frac{\sin^2\theta}{(\tau_{\text{Failure}})_{\text{Shear}}}\right]} \tag{6.88}$$

$$\sigma_{\text{failure-matrix}}^{\text{Transverse}} = \frac{1}{\sin^2\theta\left[\frac{\sin^2\theta}{(\sigma_{\text{Failure}})_T} + \frac{\cos^2\theta}{(\tau_{\text{Failure}})_{\text{Shear}}}\right]} \tag{6.89}$$

Thus, we obtain the failure limits in the transverse direction from a unidirectional angled ply test data as shown in Figure 6.18. In summary, the composite failure analysis especially under the transient dynamic loading situations is extremely challenging, because the composite failure modes are still very difficult to predict with sufficient confidence and numerical accuracy.

In LS-DYNA implementation of progressive failure of composites during dynamic impact simulation, the most recommended approach is always to use the strain-based rather than a stress-based failure criterion.

6.6 CMC BEHAVIOR UNDER IMPACT AND HIGH-TEMPERATURE APPLICATIONS

Composites made up of epoxy-based matrix with carbon fibers as the reinforcement have extensive applications in many structural components; however, they cannot be used in high-temperature applications. As such, a high-temperature environment needs either metal-matrix composites or ceramic matrix composites "CMCs". The limiting temperature for epoxy-based matrix is about 800–900°F.

The most common CMC is the Si–C composites. They use SiC fibers up to a temperature of about 2200°F. However, they are extremely brittle, and their impact tolerance capability is very poor. Most of the composite material's mechanical properties such as E_{11}, E_{22}, G_{12}, and ν_{21} degrade at higher temperatures, and as such all the elastic properties needed for numerical simulation must be corrected. It should be noted that most of these properties are of a proprietary nature and are not readily available in the public domain published literature. For a typical reference, the reader may refer to Reference [9] listed at the end of this chapter.

REFERENCES

1. R. M. Jones (1975) *Mechanics of Composite Materials*, McGraw-Hill Book Company, New York, pp. 1–84.
2. S. K. Sinha, K. E. Turner and N. Jain (2011) "Dynamic Loading on Turbofan Blades Due to Bird-strike", *ASME Journal of Engineering for Gas Turbines and Power*, Vol. 133, No. 12, pp. 122504-1–122504-13. https://doi.org/10.1115/1.4004126
3. S. K. Sinha and R.P. Zylka (2017) "Vibration Analysis of Composite Airfoil Blade Using Orthotropic Thin Shell Bending Theory", *International Journal of Mechanical Sciences*, Vol. 121, No. 2, pp. 90–105. https://doi.org/10.1016/j.ijmecsci.2016.12.012

4. S. K. Sinha and K. E. Turner (2011) "Natural Frequencies of Pre-twisted Blade in a Centrifugal Force Field", *Journal of Sound and Vibration*, Vol. 330, No. 11, pp. 2655–2681. https://doi.org/10.1016/j.jsv.2010.12.017
5. S. K. Sinha (2014) "Transient Response of a Multi-Layered Composite Rotating Airfoil Under Slicing-Impact Loading", *AIAA Journal*, Vol. 52, No. 12, pp. 2701–2711. https://doi.org/10.2514/1.J052888
6. C. M. Leech and S. M. Abood (1989) "Modelling of the Dynamics of Woven Constructions", *Computational Techniques for Contact, Impact, Penetration, and Perforation of Solids: Presented at the Winter Annual Meeting of the American Society of Mechanical Engineers*, San Francisco, CA, December 10–15, 1989, pp. 191–214.
7. S. W. Tsai and E. M. Wu (1971) "A General Theory of Strength for Anisotropic Materials", *Journal of Composite Materials*, Vol. 5, p. 58.
8. Z. Hashin (1980) "Failure Criteria for Unidirectional Fiber Composites", *ASME Journal of Applied Mechanics*, Vol. 47, No. 2, pp. 329–334. https://doi.org/10.1115/1.3153664
9. J. A. DiCarlo, H.-M. Yun, G. N. Morscher and R. T. Bhatt (2005) "SiC/SiC Composites for 1200 C and Above". In *Handbook of Ceramic Composites*, pp. 77–98. Boston, MA: Springer US. https://ntrs.nasa.gov/api/citations/20040191405/downloads/20040191405.pdf

7 Vibro-Impact Problems of Blade Tip-Rub Dynamics

7.1 INTRODUCTION TO SLIDING IMPACT AND BLADE TIP-RUB IN ROTATING MACHINERY

In Chapter 1, we introduced the topic of oblique impact with special emphasis on aerospace applications. In the aero-engine applications, to analyze the dynamic characteristics of turbine and compressor blades, it is a common practice to consider them as a rotating radial cantilever beam. At the same time, turbine and compressor blade designers have long felt that this characterization ignores some vital geometrical details of a real blade such as lean and twist in the blade, which limits the applicability of such simplified analytical models, especially in the area of aerodynamic flutter and rub-induced dynamic instabilities in the blade. One such aspect with direct applications to turbine and compressor blades is the vibration of pre-twisted beams, which is commonly referred to as "twist-bend coupling characteristics of airfoils." During the last 30 years, there has been a lot of interest in applying the free vibration characteristics of pretwisted beams simulating the airfoil, and there is a considerable current interest to extend it to analyze the nonlinear vibration to the dynamic stability of asymmetric airfoils cross sections, especially rotating blades with aerodynamic excitations. The dynamic stability of cantilever beams with varying levels of complexity and different types of loading conditions has been investigated by the current author of this book. The technical contents of this chapter are based upon the author's groundbreaking work published earlier in the *International Journal of Nonlinear Mechanics* and the *ASME Journal of Applied Mechanics* during the 2005–2007 time frame [see References 1,2]. For the benefit of current readers, some of the illustrations and the research findings published in those publications are being reproduced here with the permission of the respective publishers.

In any high-performance turbo-machinery, rub between the stator structure and the rotor blade is a commonly occurring vibro-impact problem. For example, in a jet engine high-pressure compressor and fan, usually, the outer casing ovalization results in a blade or a sector of blades rubbing against the case. Specifically, in aero-engines, the case distortion may be caused either by a temperature gradient or due to various maneuvering loads usually during takeoff and landing of the aircraft. Depending upon the stabilizing or destabilizing nature of the rub, the engine may correct itself and will keep on running or it may cause extensive damage to the rotor blades to a point that the engine may have to be shutdown. Higher energy-efficiency of the engine is achieved by reducing the tip clearance between the rotor and the

DOI: 10.1201/9781003749752-7

stator components, but as the clearance is reduced, the probability of rub taking place during small changes in the operating conditions also increases. For example, in a large jet engine, the typical tangential velocity of the blade tip will vary in the range of 400–600 m/s. During the initiation of local tip rub, the typical contact duration would last anywhere between 0.1 and 0.8 ms, and usually it will repeat for several revolutions with one pulse per revolution. The periodic nature of the radial impact pulse along the beam-column axis may initiate parametric instability in the blade, which is further complicated due to the nonlinear effect of the Coulomb friction at the contact location. It is known that the rub forces can be extremely high causing severe damage, but due to the short duration involved, determining their true magnitude has been a real challenge. During the rub, the peak amplitude of the compressive radial impact pulse can be many times more than Euler's static critical buckling load F_{cr} for the similar size beam-column. In short, rubbing of blade tips against the stationary outer case is a highly nonlinear contact-impact problem of a rotating Timoshenko beam, and until now no one has attempted to solve it analytically, which covers its full dynamic characteristics including the Coriolis forces due to the coupling of the flexural and axial motions of the beam. After the importance of transverse shear and rotary inertia in the beam formulation was shown by Timoshenko [3], various aspects of his beam theory have been studied by several authors over the past 40 years.

In this chapter, our main focus is on developing the governing dynamical equations to analyze the effect of rub-induced contact-impact forces at the free end of the rotating blade modeled as a cantilever Timoshenko beam with a pretwist (see Figure 7.1). Starting from the basic deformation and velocity equations along with the rotary inertia and gyroscopic effect terms, a complete set of coupled dynamic

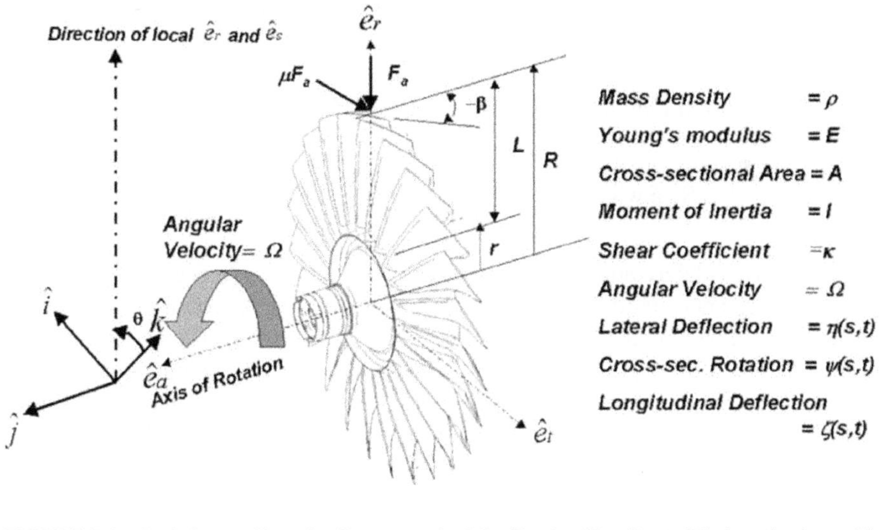

FIGURE 7.1 Rub forces F_a and μF_a on a typical Jet Engine Fan Rotor Blade spinning with constant angular velocity vector Ω with respect to a stationary global frame of reference and the resulting transverse deformations η, cross-sectional rotation ψ and longitudinal membrane extension ζ. (Reproduced with permission from Reference [1].)

equations has been derived for this problem. The eigenvalue problem of these equations in a matrix form is solved to determine the fundamental natural frequencies for a combination of varying geometrical parameters to characterize the Timoshenko beam. We have also solved the corresponding transient dynamics problem due to time-dependent contact-impact loading at the free end for twisted and untwisted blades.

7.2 CANTILEVER TIMOSHENKO BEAM WITH AXIAL IMPACT AT FREE END

In this analysis, we consider that elastically deformable radial blades of outer radius "R" with the stagger angle "β", which the blade chord makes with the engine axis, are mounted on a rigid disk of hub radius "r" (see Figure 7.1). The individual blades behave like a cantilever beam column of span length "L" and are subjected to a centripetal force field generated due to the rotor spin velocity "Ω" with the blade-tip rub load F_a (+ sign: tension and – sign: compression) acting along the beam axis.

The tip-rub force F_a is a dynamic contact load, which is non-zero only during the tip travel through the rub zone on the stator and will always have a negative (–) sign due to the contact load being compressive in nature. In addition, the blade is deformed in bending by applying bending moment about its chord in such a way that a typical cross section of the deformed blade produces the cross-sectional rotation "ψ", the lateral deflection "η", and axial deflection "ζ." It is assumed that all three components of deformation are functions of spatial coordinate "s", measured along the beam axis and the temporal parameter time "t". It is also assumed that the minor principal moment of inertia of the blade cross section "I" coincides with the chord direction so that under pure bending moment the blade lateral deflection $\eta(s,t)$ takes place in the direction normal to the chord with the neutral surface passing through the radial-chord plane. For large transient motions of the beam, the geometrically nonlinear finite strain of the mid-surface should be considered, which per Green–Saint Venant's definition in terms of the current notation is written as follows:

$$\text{Spanwise strain} = \varepsilon_s = \frac{\partial \zeta}{\partial s} + \frac{1}{2}\left[\left(\frac{\partial \eta}{\partial s}\right)^2 + \left(\frac{\partial \zeta}{\partial s}\right)^2\right] \tag{7.1}$$

For the analytical derivation, we will use the usual notations such as elastic Young's modulus "E", shear modulus "G", material mass density "ρ", Poisson's ratio "ν", and cross-sectional area "A". It should be noted that the stagger angle β is defined such that in a turbine β and Ω have the same sign and in a compressor β and Ω have the opposite sign. We define two different local coordinate systems called "Axial-Tangential-Radial" with unit vectors as $(\hat{e}_a, \hat{e}_t, \hat{e}_r)$ and "Chord-Normal-Span" with unit vectors as $(\hat{e}_c, \hat{e}_n, \hat{e}_s)$ (see Figure 7.2), respectively.

With respect to the stationary global Cartesian unit vectors $(\hat{i}, \hat{j}, \hat{k})$, the local unit vectors attached to the beam $(\hat{e}_c, \hat{e}_n, \hat{e}_s)$, rotating at a constant angular velocity Ω such that $\theta = \Omega t$, are related to each other as shown below:

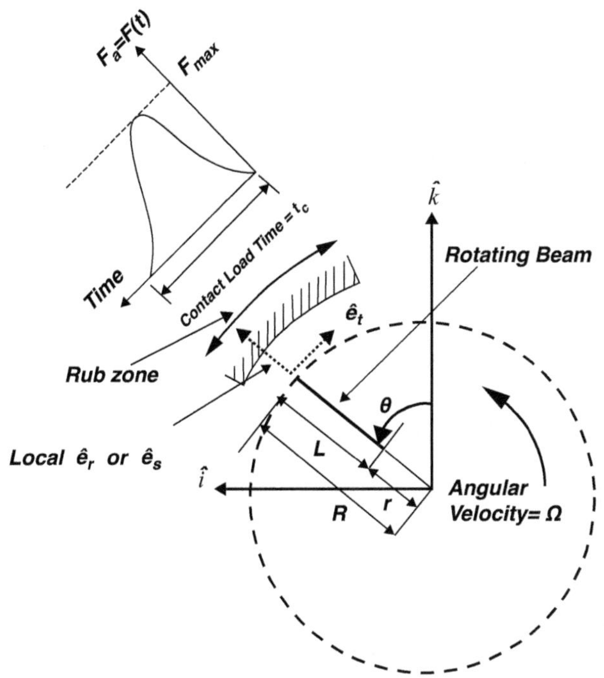

FIGURE 7.2 Schematic view of the rub zone on the stator with respect to the stationary global frame of reference, and the relative orientation of the local coordinate system attached to the beam spinning with angular velocity vector Ω during a typical rub (time period of one revolution = 2π/Ω second). Rub zone contact load of magnitude $F_a = F(t)$ active for the contact duration time = t_c second. (Reproduced with permission from Reference [1].)

$$\begin{Bmatrix} \hat{e}_c \\ \hat{e}_n \\ \hat{e}_s \end{Bmatrix} = \begin{bmatrix} -\sin\beta\cos\theta & \cos\beta & \sin\beta\sin\theta \\ -\cos\beta\cos\theta & -\sin\beta & \cos\beta\sin\theta \\ \sin\theta & 0 & \cos\theta \end{bmatrix} \begin{Bmatrix} \hat{i} \\ \hat{j} \\ \hat{k} \end{Bmatrix} \tag{7.2}$$

The deformation vector of a typical point on the beam with large rotation at a distance "z" from the mid-surface can be written (see Figure 7.3) as shown below:

$$\begin{aligned} &\text{Deformation vector} = [0]\hat{e}_c + \eta(s,t)\hat{e}_n + \\ &\left[s\left(\cos\psi(s,t) - 1\right) + \zeta(s,t) - z\sin\psi(s,t)\right]\hat{e}_s \end{aligned} \tag{7.3}$$

In the present analysis, it is considered that during the transient motion of the beam, the cross-sectional rotation term $\psi(s,t)$ is sufficiently small such that $\cos\psi \approx 1$ and $\sin\psi \approx \psi$. Thus, under the small rotation assumption, the deformation vector of a typical point on the beam simplifies as shown below:

$$\text{Deformation vector} = [0]\hat{e}_c + \eta(s,t)\hat{e}_n + \left[\zeta(s,t) - z\sin\psi(s,t)\right]\hat{e}_s \tag{7.4}$$

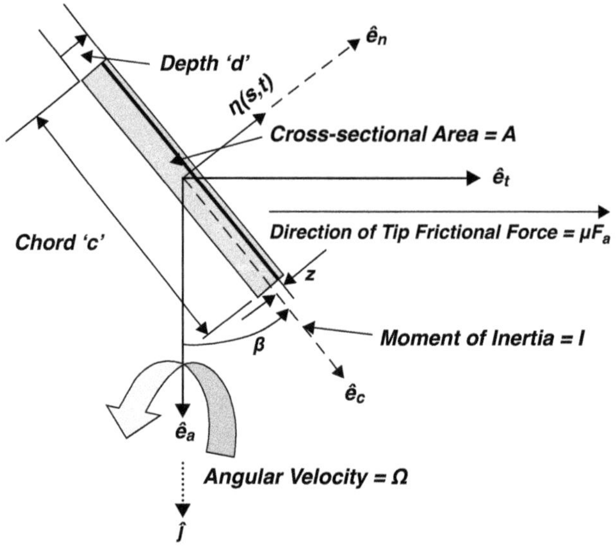

FIGURE 7.3 Cross section of the Timoshenko beam as viewed from the free end of the cantilever beam along the local radial direction (z=distance of a typical bending-induced deformed layer from the mid-surface). Beam cross section rotated by an angle with respect to the axis of spin. (Reproduced with permission from Reference [1].)

Here, the total slope $\eta'(s,t)$ of the lateral deformation of the beam is the sum of the rotation $\psi(s,t)$ and the rotation of the cross section due to shear force $Q(s,t)$, which is expressed as $-(Q/\kappa AG)$, i.e.,

$$\eta'(s,t) = \psi(s,t) - \frac{Q(s,t)}{\kappa AG} \tag{7.5}$$

The angular velocity vector of the beam column due to spin velocity Ω is given by the following relationship:

$$\vec{\Omega} = \Omega\hat{j} = [\Omega \cos\beta]\,\hat{e}_c + [-\Omega \sin\beta]\,\hat{e}_n + [\,0\,]\,\hat{e}_s \tag{7.6}$$

Thus, the velocity vector of any point in the deformed configuration of the beam is the cross-product of the angular velocity vector "$\vec{\Omega}$"and its position vector $\vec{r}$ as shown below:

$$\text{Velocity} = \vec{V}_0 + \vec{\Omega}\times\vec{r} = \frac{\partial\eta}{\partial t}\hat{e}_n + \left[\frac{\partial\zeta}{\partial t} - z\frac{\partial\psi}{\partial t}\right]\hat{e}_s + \begin{vmatrix} \hat{e}_s & \hat{e}_c & \hat{e}_n \\ 0 & \Omega\cos\beta & -\Omega\sin\beta \\ (s+r+\zeta-z\psi) & 0 & (z+\eta) \end{vmatrix} \tag{7.7}$$

The velocity vector relationship outlined in Eq. (7.7) is simplified as follows:

$$\text{Velocity} = \left[\dot{\zeta} - z\,\dot{\psi} + (z+\eta)\Omega\cos\beta\right]\hat{e}_s - \left[(s+r+\zeta-z\psi)\Omega\sin\beta\right]\hat{e}_c + \left[\dot{\eta} - (s+r+\zeta-z\psi)\Omega\cos\beta\right]\hat{e}_n$$

$$= \left[\left[\dot{\zeta} - z\,\dot{\psi} + (z+\eta)\Omega\cos\beta\right]\sin\theta + \left[(s+r+\zeta-z\psi)\Omega - \dot{\eta}\cos\beta\right]\cos\theta\right]\hat{i} - \left[\dot{\eta}\sin\beta\right]\hat{j}$$

$$+ \left[\left[\dot{\zeta} - z\,\dot{\psi} + (z+\eta)\Omega\cos\beta\right]\cos\theta - \left[(s+r+\zeta-z\psi)\Omega - \dot{\eta}\cos\beta\right]\sin\theta\right]\hat{k} \quad (7.8)$$

7.3 DYNAMICAL EQUATIONS FOR THE TIMOSHENKO BEAM

To analyze the dynamic response due to the coupling effect of longitudinal stress wave with flexural bending wave in a Timoshenko beam-column problem, we will start with the Lagrange equation of motion. To that end, in the typical Lagrangian equation in terms of Lagrangian function "Λ" is derived as follows. The simplified form of Lagrangian term "Λ" in terms of kinetic energy term "*T*" and the strain energy term "*U*" is expressed as shown below:

$$\Lambda = T - U \quad (7.9)$$

Thus, using the Lagrangian term "Λ" in the general Lagrange's equation for three deformation variables ψ, η, and ζ, one obtains the following set of coupled partial differential equations to describe the entire dynamical system:

$$\frac{d}{dt}\left(\frac{\partial \Lambda}{\partial \dot{\psi}}\right) - \frac{\partial \Lambda}{\partial \psi} + \frac{d}{ds}\left(\frac{\partial \Lambda}{\partial \psi'}\right) = q_\psi \quad (7.10a)$$

$$\frac{d}{dt}\left(\frac{\partial \Lambda}{\partial \dot{\eta}}\right) - \frac{\partial \Lambda}{\partial \eta} + \frac{d}{ds}\left(\frac{\partial \Lambda}{\partial \eta'}\right) = q_\eta \quad (7.10b)$$

$$\frac{d}{dt}\left(\frac{\partial \Lambda}{\partial \dot{\zeta}}\right) - \frac{\partial \Lambda}{\partial \zeta} + \frac{d}{ds}\left(\frac{\partial \Lambda}{\partial \zeta'}\right) = q_\zeta \quad (7.10c)$$

Using the deformation-related variables ψ, η, and ζ and the corresponding velocity terms like $\dot{\psi}$, $\dot{\eta}$, and $\dot{\zeta}$, in the velocity vector equation shown in Eq. (7.8), we obtain the equations for the kinetic energy term "*T*" and the strain energy term "*U*" in the Timoshenko beam as shown below. Thus, based upon Eq. (7.8) for the velocity vector with respect to the global fixed Cartesian frame of reference, the kinetic energy "*T*" of the beam is expressed as follows:

$$T = \frac{\rho A}{2}\int_0^L \left[\begin{array}{l} \left[\left[\dot{\zeta} - z\,\dot{\psi} + (z+\eta)\Omega\cos\beta\right]\sin\theta + \left[\begin{array}{l}(s+r+\zeta - z\psi)\Omega\ - \\ \dot{\eta}\cos\beta\end{array}\right]\cos\theta\right]^2 + \\ \left[\dot{\eta}\sin\beta\right]^2 + \left[\begin{array}{l}\left[\dot{\zeta} - z\,\dot{\psi} + (z+\eta)\Omega\cos\beta\right]\cos\theta - \\ \left[(s+r+\zeta - z\psi)\Omega\ - \dot{\eta}\cos\beta\right]\sin\theta\end{array}\right]^2 \end{array}\right] ds \tag{7.11}$$

On using the minor principal moment of inertia "I" of the cross section (see Figure 7.3) about the neutral axis (axis along the local $\hat{e}_c$ direction) and also the property of the CG (Center of Gravity) of the cross section inside the integral as shown below, we have

$$I = \iint_{\text{Area}} z^2\, dA = c\int_{-d/2}^{+d/2} z^2 dz \text{ and, } \iint_{\text{Area}} z dA = c\int_{-d/2}^{+d/2} z dz \tag{7.12}$$

We have the potential energy "U" due to bending, transverse shear, and centrifugal stresses "σ_{cf}" as well as the axial force acting in the longitudinal direction of the beam column "F_a" are given as follows. It should be noted that the F_a, being a contact load at the tip of the rotating beam-column, is always going to be negative and will be acting in the direction opposite to the direction of the centrifugal forces,

$$\begin{aligned} U = &\frac{1}{2}\int_0^L \left[EI(\psi')^2 + \kappa AG\,(\eta' - \psi)^2 + AE(\varepsilon_s)^2\right] ds \\ &- \frac{F_a}{2}\int_0^L (\eta')^2\, ds - \frac{A}{2}\int_0^L \left[\sigma_{cf}(\eta')^2\right] ds \end{aligned} \tag{7.13}$$

The simplified form of Lagrangian "Λ" for this rotating Timoshenko's beam-column motion and deformation is as follows:

$$\Lambda = T - U = \frac{1}{2}\int_0^L \begin{array}{l} [\rho A \left[\left[\begin{array}{l}\left[\dot{\zeta} - z\,\dot{\psi} + (z+\eta)\Omega\cos\beta\right]\sin\theta + \\ \left[(s+r+\zeta - z\psi)\Omega - \dot{\eta}\cos\beta\right]\cos\theta\end{array}\right]^2 + \\ \left[\dot{\eta}\sin\beta\right]^2 + \left[\begin{array}{l}\left[\dot{\zeta} - z\,\dot{\psi} + (z+\eta)\Omega\cos\beta\right]\cos\theta - \\ \left[(s+r+\zeta - z\psi)\Omega - \dot{\eta}\cos\beta\right]\sin\theta\end{array}\right]^2\right] \\ -\left[EI(\psi')^2 + \kappa AG(\eta' - \psi)^2 + AE(\varepsilon_s)^2\right] + \left[A\sigma_{cf} + F_a\right](\eta')^2\,] ds \end{array} \tag{7.14}$$

Thus, using the Lagrangian term "Λ" obtained in Eq. (7.14) and substituting in the Lagrange's equations described in Eqs. (7.10a)–(7.10c) yield the following three coupled equations of motion for dependent variables ψ, η, and ζ, respectively, in the local frame of reference attached to the rotating beam-column with angular velocity Ω as follows:

$$\frac{\partial}{\partial s}\left(EI\frac{\partial\psi}{\partial s}\right)+\kappa AG\left(\frac{\partial\eta}{\partial s}-\psi\right)-\rho I\left(\frac{\partial^2\psi}{\partial t^2}-\cos\beta\frac{d\Omega}{dt}-\Omega^2\psi\right)=0 \quad (7.15a)$$

$$-\frac{\partial}{\partial s}\left((F_a+A\sigma_{cf})\frac{\partial\eta}{\partial s}+\kappa AG\left[\frac{\partial\eta}{\partial s}-\psi\right]\right)+$$
$$\rho A\left(\frac{\partial^2\eta}{\partial t^2}-(s+r+\zeta)\cos\beta\frac{d\Omega}{dt}-\Omega^2(\cos^2\beta)\eta-2\frac{\partial\zeta}{\partial t}\Omega\cos\beta\right)=q_\eta(s,t)$$
$$(7.15b)$$

$$\frac{\partial^2\zeta}{\partial s^2}=\rho\left[\frac{\partial^2\zeta}{\partial t^2}-\Omega^2(s+r+\zeta)+\eta\cos\beta\frac{d\Omega}{dt}+2\frac{\partial\eta}{\partial t}\Omega\cos\beta\right] \quad (7.15c)$$

In Eq. (7.15b), $q_\eta(s,t)$ accounts for the distributed external transverse load on the beam-column, if there is any. In the above equations, it should be noted that in a rotating beam; its lateral motion $\eta(s,t)$ is coupled with the longitudinal motion $\zeta(s,t)$ due to the Coriolis effect, which introduces velocity-dependent skew-symmetric terms in the equations of motion. For linearization, we have used infinitesimal definition of longitudinal strain as $\varepsilon_s = \partial\zeta/\partial s$, and on substituting in place of $(R-r)=L$, we can write the centrifugal stress relation in terms of $\rho\Omega^2$ and the longitudinal force F_a, which gives the following equation for membrane in-plane stress $\sigma(s)$:

$$\sigma(s)=\sigma_{cf}+\frac{F_a}{A}=\rho\Omega^2\frac{\left(R^2-r^2-2rs-s^2\right)}{2}+\frac{F_a}{A} \quad (7.16)$$

The above equation also yields another equation as follows:

$$\frac{\partial\sigma(s)}{\partial s}=-\rho\Omega^2(s+r)=-\rho\Omega^2\ (s+R-L) \quad (7.17)$$

The external shear force due to Coulomb friction on the blade tip μF_a is acting opposite to the direction of motion, which in the local coordinate $(\hat{e}_a,\hat{e}_t,\hat{e}_r)$ system would be in the direction of $\hat{e}_t$. Its component along $\hat{e}_n$ would act like a point-load in the beam lateral direction, and as such mathematically with the use of Dirac's Delta function $\delta(s-L)$ can be treated like a continuous external force, which is written as follows:

$$q(s,t)=-\mu F_a\cos\beta\ \delta(s-L) \quad (7.18)$$

The geometric boundary conditions for Timoshenko's beam under consideration are as follows.

$$\psi(0,t)=0,\ \eta(0,t)=0,\text{ and }\zeta(0,t)=0 \tag{7.19}$$

The natural boundary conditions at the free-end of the cantilever beam for $(s=L)$ are as follows:

a. Bending moment : $$(BM)_{\text{at } s=L} = EI\left.\frac{\partial\psi}{\partial s}\right|_{s=L} = 0 \tag{7.20}$$

b. $$\begin{aligned}\text{Shear force : }\quad (SF)_{\text{at } s=L} &= -\kappa AG\left[\frac{\partial\eta}{\partial s} - \psi\right]_{s=L} \\ &= -\mu F_a\cos\beta + F_a\left[\frac{\partial\eta}{\partial s}\right]_{s=L}\end{aligned} \tag{7.21}$$

c. Axial force : $$(F_a)_{\text{at } s=L} = EA\left.\frac{\partial\zeta}{\partial s}\right|_{s=L} = F_a \tag{7.22}$$

Due to the pulse-loading nature of the axial force, we will assume the longitudinal load F_a to be time-dependent and henceforth will be written as $F(t)$. Using the Rayleigh–Ritz method [see Reference 4], the solution of the earlier set of Eqs. (7.10a)–(7.10c) can be assumed with a set of displacement shape functions $U_n(s)$, $V_n(s)$, and $W_n(s)$ such that they satisfy all the geometric boundary conditions arbitrarily. Thus, using corresponding dimensionless time-dependent generalized coordinates $X_n(t)$, $Y_n(t)$, and $Z_n(t)$ respectively, we assume the solution in the form of the following sets of infinite trigonometric series:

$$\psi(s,t) = \sum_{n=0}^{\infty} U_n(s)X_n(t) = \sum_{n=0}^{\infty}\left[\sin\alpha_n s\right]X_n(t) \tag{7.23a}$$

$$\eta(s,t) = \sum_{n=0}^{\infty} V_n(s)Y_n(t) = \sum_{n=0}^{\infty}\left[\frac{[1-\cos\alpha_n s]}{\alpha_n}\right]Y_n(t) \tag{7.23b}$$

$$\zeta(s,t) = \sum_{n=0}^{\infty} W_n(s)Z_n(t) = \sum_{n=0}^{\infty}\left[\frac{\sin\alpha_n s}{\alpha_n}\right]Z_n(t) \tag{7.23c}$$

where

$$\alpha_n = \frac{(2n-1)\pi}{2L} \tag{7.23d}$$

The first term in each of the above sequences of trial displacement functions corresponding to $n=0$ pertains to the rigid body movement of the beam caused by the rotation of the clamped end located at $s = 0$, S usually at the top of the fan disk. Since our interest is mainly in the strain energy-related deformation of the beam and not in the rigid body movement, we will consider the terms in the series starting at $n = 1$. Thus, the substitution of the assumed displacement shape functions expressed as Eqs. (7.23a)–(7.23d) in the equations of motion described by Eqs. (7.15a)–(7.15c) along with the assumption of constant angular velocity Ω yields the following set of equations:

$$\sum_{n=1}^{\infty}\left[-EIX_nU_n'' - \kappa AG\left(Y_nV_n' - X_nU_n\right) + \rho I\left(\ddot{X}_nU_n - \Omega^2 X_nU_n\right)\right] = 0 \quad (7.24a)$$

$$\sum_{n=1}^{\infty}\Bigg[-\rho A\Omega^2\frac{\left(R^2 - r^2 - 2rs - s^2\right)}{2}\left(Y_nV_n''\right) + \rho A\Omega^2(r+s)\left(Y_nV_n'\right) - F(t)\left(Y_nV_n''\right)$$
$$-\kappa AG\left(Y_nV_n'' - X_nU_n'\right) + \rho A\left(\ddot{Y}_nV_n - \Omega^2\cos^2\beta Y_nV_n\right)$$
$$-2\Omega\cos\beta\dot{Z}_nW_n - F(t)\left(Y_nV_n'\right)\delta(s-L)\Bigg] = q_\eta(s,t) \equiv q(s,t)$$

(7.24b)

$$\sum_{n=1}^{\infty}\left[-EZ_nW_n'' + \rho\left(\ddot{Z}_nW_n - \Omega^2 Z_nW_n + 2\Omega\cos\beta\dot{Y}_nV_n\right)\right] = \rho\Omega^2(s+r) \quad (7.24c)$$

In the above sets of equations, the spatial derivative terms can be written as a function of a set of differential operators $\mathfrak{R}$. In addition, Eqs. (7.24a)–(7.24c) are complete only when all the terms in the infinite series for the displacement functions $U_n(s)$, $V_n(s)$, and $W_n(s)$ are considered. However, in our numerical scheme, we will truncate the infinite series into a finite series by considering only first "N" number of terms of the deformation shape functions. In the ideal condition, they may satisfy all the geometric as well as force boundary conditions of the present problem; however, it is not necessary in general. On applying the Rayleigh–Ritz method with relaxed boundary condition requirements [see Reference 4] and the assumed displacement functions, one obtains a set of ordinary differential equations in terms of time-dependent variables of which a typical mth set of equation for each of the trial function terms viz. $U_m(s)$, $V_m(s)$, and $W_m(s)$ can be written as follows. Thus, using Galerkin's weighted residual method, Eqs. (7.24a)–(7.24c) can be written in an integral form for their respective differential operators referred to as $\mathfrak{R}_1(\psi,\eta,\zeta)$, $\mathfrak{R}_2(\psi,\eta,\zeta)$, and $\mathfrak{R}_3(\psi,\eta,\zeta)$ by multiplying with their corresponding deformation shape functions shown in Eqs. (7.23a)–(7.23c), which are as follows:

$$\int_0^L \mathfrak{R}_1(\psi,\eta,\zeta)U_m(s)ds = 0 \quad (7.25a)$$

$$\int_0^L \Re_2(\psi,\eta,\zeta) V_m(s)\, ds = \int_0^L q(s,t) V_m(s)\, ds \tag{7.25b}$$

$$\int_0^L \Re_3(\psi,\eta,\zeta) W_m(s)\, ds = \int_0^L \rho\Omega^2 (s+r) W_m(s)\, ds \tag{7.25c}$$

Since, the assumed displacement functions $U_n(s)$, $V_n(s)$, and $W_n(s)$ do not satisfy all the force boundary conditions necessarily, then with the unbalanced weighted-residual force and moment terms at the free-end boundary (at $s = L$) being applied for the rotating Timoshenko beam with tip-rub, we obtain the following three dynamical equations. Also, instead of having an infinite series of deformation shape function terms, we will consider only "N" number of terms in the finite series such that indices $m = 1,2,\,\ldots\ldots\ldots,(N-1),\ N)$ and $n = 1,2,\,\ldots\ldots\ldots,(N-1),\ N)$.

$$\sum_{n=1}^{N} \begin{bmatrix} -EI \int_0^L U_m U_n'' ds (X_n) + \left(\kappa AG - \rho I \Omega^2\right) \\ \int_0^L U_m U_n ds (X_n) - \kappa AG \int_0^L U_m V_n' ds (Y_n) + \rho I \int_0^L U_m U_n\ ds \left(\ddot{X}_n\right) \end{bmatrix} = 0 \tag{7.26a}$$

$$\sum_{n=1}^{N} \Bigg[-\rho A\Omega^2 \int_0^L \frac{\left(R^2 - r^2 - 2rs - s^2\right)}{2} V_m V_n'' ds (Y_n) + \rho A\Omega^2 \int_0^L (r+s) V_m V_n' ds (Y_n)$$
$$-\rho A\Omega^2 \cos^2\beta \int_0^L V_m V_n ds\ (Y_n) - (\kappa AG + F(t)) \left[\int_0^L V_m V_n'' ds - V_m(L)\ V_n'(L) \right] (Y_n)$$
$$+\kappa AG \left[\int_0^L V_m U_n' ds - V_m(L) U_n(L) \right] (X_n) - 2\rho A\Omega \cos\beta \int_0^L V_m W_n ds \left(\dot{Z}_n\right)$$
$$+\rho A \int_0^L V_m V_n ds \left(\ddot{Y}_n\right) \Bigg] = -\mu F(t) \cos\beta V_m(L) \tag{7.26b}$$

$$\sum_{n=1}^{N} \Bigg[-EA \left(\int_0^L W_m W_n''\ ds\ -\ W_m(L)\ W_n'(L) \right) (Z_n) - \rho A\Omega^2 \int_0^L W_m W_n ds (Z_n)$$
$$+2\rho A\Omega \cos\beta \int_0^L W_m V_n ds\ \left(\dot{Y}_n\right) + \rho\, A \int_0^L W_m W_n\ ds \left(\ddot{Z}_n\right) \Bigg] = \rho A\Omega^2 \int_0^L (s+r) W_m ds \tag{7.26c}$$
$$+F(t) W_m(L)$$

7.3.1 Dynamic Coupling of Longitudinal and Transverse Motion

The general equation of motion for the Timoshenko beam under consideration including all the dynamic effects in a matrix form can be written in the form of a general Mathieu–Hill Equation [see References 5,6], which is as follows:

$$[\boldsymbol{M}]\{\ddot{f}(t)\}+[\boldsymbol{C}]\{\dot{f}(t)\}+[[\boldsymbol{K}]+[\boldsymbol{B}(t)]]\{f(t)\}=\{\boldsymbol{P}(t)\} \tag{7.27}$$

It can be seen that in Eq. (7.27) the terms containing generalized coordinates in the column vector $\{f(t)\}$ are X, Y, and Z's, which are dimensionless. Suppose we consider "N" number of terms for each of the three basic deformation trial functions outlined in Eqs. (7.23a)–(7.23c), then for brevity we can introduce the following notations in lieu of the generalized coordinates $\{f(t)\}$:

Beam cross-sectional rotations "ψ":

$$f_1(t)=X_1(t),\, f_2(t)=X_2(t),\, f_3(t)=X_3(t),\, f_4(t)=X_4(t),\ldots f_N(t)=X_N(t) \tag{7.28a}$$

Beam lateral deflections "η":

$$f_{N+1}(t)=Y_1(t),\, f_{N+2}(t)=Y_2(t),\, f_{N+3}(t)=Y_3(t),\, f_{N+4}(t)=Y_4(t),\ldots f_{2N}(t)=Y_N(t) \tag{7.28b}$$

Beam axial deflections "ζ":

$$\begin{aligned} f_{2N+1}(t)&=Z_1(t),\, f_{2N+2}(t)=Z_2(t),\, f_{2N+3}(t)=Z_3(t),\, f_{2N+4}(t)\\ &=Z_4(t),\ldots f_{3N}(t)=Z_N(t) \end{aligned} \tag{7.28c}$$

In the above expressions, the subscripts on $X(t)$, $Y(t)$, and $Z(t)$'s refer to the term sequence number from the admissible sets of displacement functions for the beam. Also, the accent signs like $(\dot{\ })$ and $(\ddot{\ })$ denote their respective first and second derivatives with respect to time "t". Thus, the transpose of the column vector represented as $\{f(t)\}^T$ is shown below:

$$\{f(t)\}^T=\{f_1\ \ f_2\ \ f_3\ \ f_4\ \ f_5 \ldots f_{3N-1}\ \ f_{3N}\} \tag{7.29}$$

It should be obvious that with the "N" number of terms being used to represent each of the three displacement functions viz. (ψ,η,ζ), the total number of degrees of freedom in the numerical scheme will be "$3N$". Thus, terms for the right-hand side force vector $\{\boldsymbol{P}(t)\}$ turn out to be as follows:

$$\left\{\begin{array}{c} \{0\}_N \\ \{-\mu F(t)\cos\beta\, V_m(L)\}_N \\ \left\{\rho A\Omega^2 \int_0^L (s+r)W_m\, ds + F(t)W_m(L)\right\}_N \end{array}\right\}_{3N} \tag{7.30}$$

The individual nonzero terms in the equivalent inertia matrix $[\boldsymbol{M}]$, velocity-dependent coefficient matrix $[\boldsymbol{C}]$, elastic stiffness matrix $[\boldsymbol{K}]$, and the time-varying force-dependent matrix $[\boldsymbol{B}(t)]$ are as follows.

$$[\mathrm{K}]_{(m,n)} = -EI\int_0^L U_m U_n''\, ds + \left(\kappa AG - \rho I\Omega^2\right)\int_0^L U_m U_n ds \tag{7.31}$$

$$[\mathrm{M}]_{(m,n)} = \rho I \int_0^L U_m U_n\, ds \tag{7.32}$$

$$[\mathrm{K}]_{(m,n+N)} = -\kappa AG \int_0^L U_m V_n' ds \tag{7.33}$$

$$[\mathrm{K}]_{(m+N,n)} = \kappa AG\left[\int_0^L V_m U_n' ds - V_m(L)\, U_n(L)\right] \tag{7.34}$$

$$[\mathrm{C}]_{(m+N,n+2N)} = -2\rho A\Omega\cos\beta \int_0^L V_m W_n ds \tag{7.35}$$

$$\begin{aligned} [\mathrm{K}]_{(m+N,n+N)} = {} & -\kappa AG\left[\int_0^L V_m V_n'' ds - V_m(L)\, V_n'(L)\right] \\ & -\rho A\Omega^2 \int_0^L \frac{\left(R^2 - r^2 - 2rs - s^2\right)}{2} V_m V_n''\, ds + \\ & \rho A\Omega^2 \int_0^L (r+s)\, V_m V_n' ds - \rho A\Omega^2 \cos^2\beta \int_0^L V_m V_n ds \end{aligned} \tag{7.36}$$

$$[\mathrm{B}]_{(m+N,n+N)} = -F(t)\left[\int_0^L V_m V_n'' ds - V_m(L)\, V_n'(L)\right] \tag{7.37}$$

$$[\mathrm{M}]_{(m+N,n+N)} = \rho A \int_0^L V_m V_n ds \tag{7.38}$$

$$[\mathrm{K}]_{(m+2N,n+2N)} = -EA\left(\int_0^L W_m W_n'' ds - W_m(L)\, W_n'(L)\right) - \rho A \Omega^2 \int_0^L W_m W_n \; ds \tag{7.39}$$

$$[\mathrm{M}]_{(m+2N,n+2N)} = \rho A \int_0^L W_m W_n \; ds \tag{7.40}$$

$$[\mathrm{C}]_{(m+2N,n+N)} = 2\rho A \Omega \cos\beta \int_0^L W_m V_n ds \tag{7.41}$$

The above nonzero terms in the velocity-dependent coefficient matrix $[\mathrm{C}]_{(m,n)}$ are all due to rotational velocity Ω, which generates a Coriolis force in the system by being skew-symmetric in nature, which in the governing equations are usually shown as $[\mathrm{C}]_{\text{Coriolis}}$. The time-dependent nature of the longitudinal load $F(t)$ creates an extensional wave in the Timoshenko beam, which travels up-and-down its longitudinal axis with very high velocity, giving rise to large Coriolis forces, which are oscillatory in nature. It should also be noted that when the extensional wave returns from a free-edge boundary, it changes its sign from a compressive wave to tensile wave and vice versa. Similarly, when the stress wave returns from the clamped end of the cantilever beam, it reflects back with the same sign, and in that process its magnitude doubles immediately. The longitudinal wave's vibratory nature combined with the Corolis effect of rotating beam produces the coupling of extensional wave with the flexural bending wave of a such large magnitude that the dynamical system can become unstable. Mathematically, this Coriolis force is a distributed load, which is represented by the term such as $\pm 2\rho A L \Omega \cos\beta \dot{\zeta}$ shown in Eqs. (7.35) and (7.41), and its value at the blade root is computed as the integrated sum over the span length of the beam column can be determined by the following equation:

$$\text{Coriolis force at the clamped end} = -2\rho A \Omega \int_0^L \cos\beta \dot{\zeta} ds \tag{7.42}$$

The material internal damping can be taken into account as a function of the nondimensional factor φ of the critical damping of the beam material and the spin angular velocity Ω. In this situation, the typical terms in the damping matrix $[\mathrm{C}]_{\text{Damping}}$ due to the material internal damping are computed as a function of stiffness matrix terms $[\mathrm{K}]_{\text{Stiffness}}$ containing material parameters Young's modulus E and shear modulus G as well as shear coefficient κ of the beam, which are written as follows:

$$[\mathrm{C}]_{\text{Damping}} = \frac{2\varphi}{\Omega} [\mathrm{K}]_{\text{Stiffness}} \tag{7.43}$$

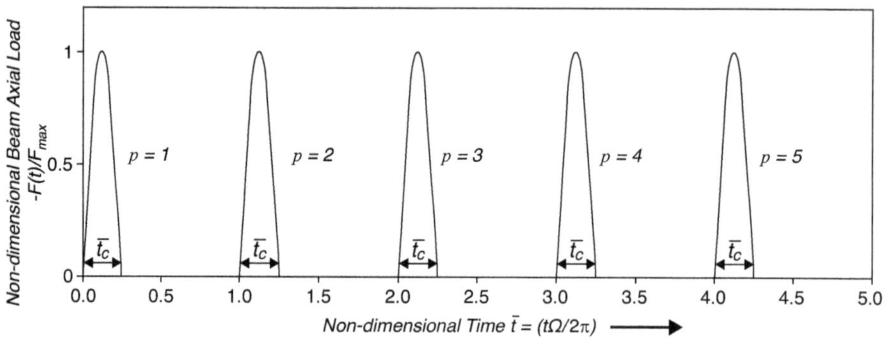

FIGURE 7.4 Periodic nature of the axial pulse $F(t)$ on the free end of the beam with sinusoidal temporal distribution given by Eqs. (7.44) and (7.45) during rub (where pulse sequence number $p = 1,2,3,4,5,\ldots$) and the frequency of pulse $f_p = 1$ pulse per revolution (nondimensional contact load time $= \bar{t}_c$). (Reproduced with permission from Reference [1].)

In general, the waveform of the periodic impact force $F(t)$ along the axis of the Timoshenko beam can have various time-dependent distributions, such as half-sine wave, rectangular pulse, or saw-tooth profile. In Section 5.2, the mathematical representations of all commonly used waveforms to simulate the impact force distribution in the time domain with a typical pulse frequency of f_p-Hz was discussed. In the current investigation, the following periodic load with half-sine wave distribution function of frequency f_p-Hz, a peak magnitude of $F_{\max}$, and a load-application time of "t_c" has been used (see Figure 7.4).

$$F(t) = 0 \text{ for} \left[\left(\frac{(p-1)}{f_p} + t_c \right) \leq t \leq \frac{p}{f_p} \right] \tag{7.44}$$

and

$$F(t) = -F_{\max} \sin\left(\frac{\pi}{t_c} \left(t - \frac{(p-1)}{f_p} \right) \right) \text{for} \left[\left(\frac{(p-1)}{f_p} \right) \leq t \leq \left(\frac{(p-1)}{f_p} + t_c \right) \right] \tag{7.45}$$

In the above relationship for the force-history of the radial impact force, "p" represents the sequence number of the pulse starting from time $t = 0$ and as such $p = 1$. For example, in Eq. (7.45), $p = 1$ is the first pulse, $p = 2$ is the second pulse, $p = 3$ is the third pulse, and so on. The general sinusoidal pulse equation shown in Eq. (7.45) can easily be reduced to a special case of one pulse per revolution in a rotating system by setting f_p = RPM/60 = $\Omega/2\pi$. Figure 7.4 shows the time history of such a special case of half-sine wave pulse with the pulse frequency "f_p" of one pulse per revolution and a load-application time of "t_c". In general, the contact load time "t_c" plays an important role in determining the dynamic response of an impacted beam, and it can be seen that in the limit case as $t_c \to 0$, the force-history becomes perfect impulse function. Some very important wave propagation types of investigations in Timoshenko beam are carried out by sending a single pulse through the beam and analyzing its dispersion at different locations [see Reference 6]. The above multiple-pulse force function shown in Eq. (7.45) can easily be simplified for simulating such a single pulse impact by setting $p = 1$, which yields $F(t) = 0$ for $p > 1$, i.e.,

$$F(t) = -\ F_{\max} \sin\left(\frac{\pi t}{t_c}\right) \text{ for } 0 \le t \le t_c$$

and

$$F(t) = 0 \text{ for } t \ge t_c \tag{7.46}$$

In case of large deformation of the beam during the impact, the radial force along the beam axis should be corrected for the changing directions of the slope-dependent follower forces. This correction may be also required in a displacement-controlled external tip load (see Figure 7.5), when large radial incursion from the outer case makes it necessary that the right-hand-side column vector $\{\boldsymbol{P}(t)\}$ representing the external forces on the dynamical system be updated according to the changing slope at the tip $\eta'(L,t)$ as

$$F_{\text{Longitudinal}} = F(t) = F_a \cos\left(\eta'(L,t)\right) \text{ and } F_{\text{Transverse}} = -F_a \sin\left(\eta'(L,t)\right) \tag{7.47}$$

The current formulation is also directly applicable for studying the characteristics of coupled nonlinear axial and flexural wave propagation generated by sending a single pulse through the Timoshenko beam. For discussion purposes, we will consider a Timoshenko beam of rectangular cross section of chord length "*c*" and depth "*d*", the structural properties of which are shown in Table 7.1.

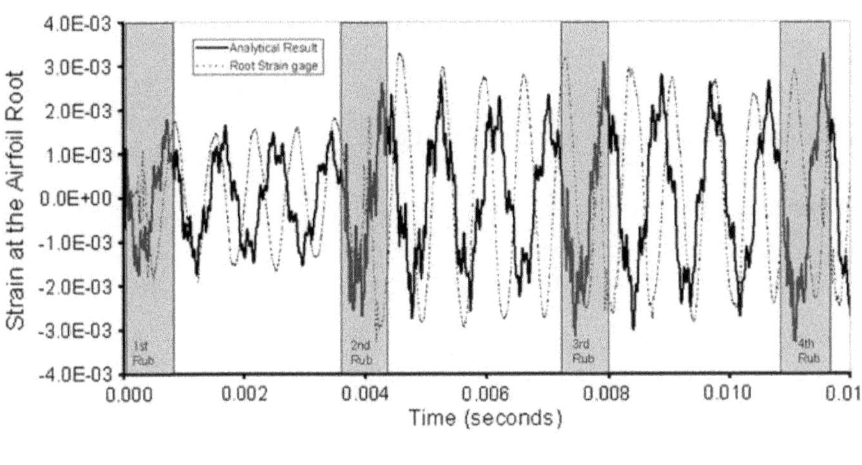

FIGURE 7.5 Comparison of measured airfoil root (***L*** = 4 cm, ***L/c*** = 1.51745, ***h*** = 0.125 cm, β = −45°, ***N*** = 16,500 rpm) strain gage data (………) versus results from the present analytical model (———-) during repeated radial incursion of 0.1 mm at the blade tip with 72° circumferential rub zone for the first four rubs. (Reproduced with permission from Reference [2].)

TABLE 7.1
Nomenclatures of Structural and Loading Parameters Used in the Example

Minor principal moment of inertia of the cross section	$I = \frac{1}{12} c d^3$
Cross-sectional area	$A = cd$
Mass of the beam	$\rho A L$
Shear modulus	$G = \frac{E}{2(1+\nu)}$
Shear coefficient shape factor for rectangular cross section:	$\kappa = 0.833$
First flexural mode frequency for the Euler–Bernoulli beam	$\omega_1 = \frac{\xi}{L^2}\sqrt{\frac{EI}{\rho A}}$ $(\xi = 3.51602)$
Pulse frequency (Hz)	f_p
Euler's buckling load	$F_{cr} = \left[\frac{\pi^2 EI}{4L^2}\right]$
Spin velocity (rpm)	N
Rotational velocity (radians/s)	$\Omega = \frac{2\pi N}{60}$
Special case of 1 pulse per revolution	$f_p = \frac{N}{60}$ Hz

At this point, we will also introduce a set of nondimensional parameters with a new sign $(\bar{\ })$ on the top as follows, which make it convenient to discuss the significance of the numerical results:

Nondimensional peak axial force $\bar{F}_{max}$: $\bar{F}_{max} = \frac{F_{max}}{F_{cr}}$

Nondimensional angular velocity of rotation $\bar{\Omega} = \Omega\sqrt{\frac{(\rho R^2)}{E}}$

Nondimensional time $\bar{t} = t\left(\frac{\Omega}{2\pi}\right)$

Nondimensional contact load time $\bar{t}_c = t_c\left(\frac{\Omega}{2\pi}\right)$

Slenderness ratio $\bar{d}$ indicating the relative depth of the beam cross section:

$$\bar{d} = \sqrt{I/(AL^2)}$$

Nondimensional lateral deflection $\bar{\eta} = \eta/L$

Nondimensional axial deflection $\bar{\zeta} = \zeta/L$

Hence, the matrix terms used in Eq. (7.27) can be broken into the following separate matrices:

$$[\boldsymbol{M}]\{\ddot{f}(t)\}+\left[[\mathrm{C}]_D+[\mathrm{C}]_G\right]\{\dot{f}(t)\}+\left[\left[[\mathrm{K}]_S+[\mathrm{K}]_C\right]+[\boldsymbol{B}]F(t)\right]\{f(t)\}=\{\boldsymbol{P}(t)\} \tag{7.48}$$

where $[\boldsymbol{M}]$ is the mass matrix, a function of density (symmetric); $[\mathrm{C}]_D$ is the damping matrix due to the material internal damping $= (2\varphi / \Omega)[\mathrm{K}]_S$; $[\mathrm{C}]_G$ is the gyroscopic matrix (skew-symmetric, causes coupling of axial and lateral motions in the beam); $[\mathrm{K}]_S$ is the structure stiffness matrix, a function of E, G , I, etc. (symmetric); $[\mathrm{K}]_C$ is the stress stiffness matrix due to centrifugal stress field, a function of Ω (symmetric $[\boldsymbol{B}]F(t)$ the time-dependent part of circulatory matrix due to impulsive forces along the longitudinal axis of the beam-column (nonsymmetric)); and $\{\boldsymbol{P}(t)\}$ is the column vector containing external forces on the dynamical system.

In all the above relations, the inertia matrix $[\boldsymbol{M}]$ is always symmetric, but the coefficient matrix $[\boldsymbol{C}]$ for the velocity-dependent forces is a combination of symmetric and skew-symmetric parts. As mentioned earlier, the skew-symmetric terms are due to the Coriolis effect as a result of the coupled lateral and axial movements of the beam. The coefficient matrix $[\boldsymbol{K}]$ for the displacement-dependent-forces can have time-dependent terms due to contact-impact situation in the longitudinal direction making it nonlinear. In general, in a simple linear structure, it will be a symmetric matrix. However, the time-dependence part of the stiffness matrix $[\boldsymbol{B}]F(t)$ exhibits the interesting nonlinear characteristics of the equation of motion.

On the right-hand side of the equation, the column vector $\{\boldsymbol{P}(t)\}$ contains all the terms due to the presence of the external forces on the beam. For example, corresponding to the generalized coordinates $\{Y\}$, it contains the terms related to lateral forces caused by the Coulomb friction at the tip. Similarly, corresponding to the generalized coordinates $\{Z\}$, it contains the terms related to axial forces $F(t)$ caused by the contact with the stator structure at the blade tip. If the rotational speed of the rotor Ω is not constant, then the time duration between each pulse is also changing, which makes $\Omega(t)$ as a function of time as well, and the corresponding impulse force history with one pulse per revolution shown in Eq. (7.45) becomes as follows:

$$F(t)=-\ F_{\max}\sin\left(\frac{\pi}{t_c}\left(t-\frac{2\pi(p-1)}{\Omega(t)}\right)\right)\text{for}\left[\frac{2\pi(p-1)}{\Omega(t)}\leq t\leq\left(\frac{2\pi(p-1)}{\Omega(t)}+t_c\right)\right] \tag{7.49}$$

In addition, the terms containing the dynamic coefficient of friction are non-zero only for the duration the contact-pulse force is active. As a result, all tip rubs are not necessarily unstable. The dynamic stability of the rotating beam can be determined either in a frequency domain or in a time domain. The frequency-domain stability criteria can be used only for linear or quasilinear type of dynamic problems, where it is possible to write the $[\boldsymbol{C}]$, $[\boldsymbol{K}]$, and $[\boldsymbol{B}]$ matrices explicitly without any reference to the displacements such as the effect of gaps or clearances in the system. Still, the frequency-domain stability conditions provide a good insight in a complex dynamic situation, where instability can be initiated due to the excitation of certain mode shapes in the rotor caused by infrequent rubbing of blades. These instabilities that may occur at certain speed ranges of the rotor can be either a simple or parametric resonance type of instability in the flexible radial rotating blade.

Mathematically speaking, a rigorous definition for the stability of a dynamical equation in a multiple degrees-of-freedom system is very difficult to establish. One such approach is to use the direct method of Liapunov, also called the second method of Liapunov, to prove the stability or instability of an equilibrium point of a nonlinear system of differential equations. However, in a practical sense, if the magnitude of the parameter describing the motion grows asymptotically to some unacceptably high level, then it should be considered unstable. For example, if the lateral movement of a Timoshenko beam under some kind of excitation or due to the application of a dynamic force grows to an extent that the outer fiber elastic strain in the beam would become in excess of 5%, it may be taken as the threshold of failure based upon the material stress capability considerations.

In the time-domain stability determination, one uses the direct-time marching-forward-integration technique, which is a preferred approach in a true nonlinear dynamical problem with clearances and varying coefficients of friction as a function of speed. In this method, if the lateral motion of the blade at certain locations such as at the tip is increasing monotonically over a period of time, which can be in the range of milliseconds to seconds, then it is considered unstable. The instability is being caused because the energy being fed into the dynamical system is not being dissipated fully by the effective damping present in the dynamical system.

In the frequency-domain approach, one solves the homogeneous form of the second-order differential equations, where the goal is to establish the boundaries of the stable and unstable response of the dynamical system as certain parameters such as the magnitude of the dynamic load or the frequency of excitation are varied. For solving a nonlinear vibration problem in the frequency domain, first one must simplify the contribution of nonlinear terms by using techniques such as the perturbation method and asymptotic expansion. Bolotin [see Reference 7] describes a method of formal expansion into trigonometric series for solving a general system of nonlinear differential equations for the dynamic stability of a multiple degrees-of-freedom system. The common methods of establishing the stability criteria in the frequency domain are impossible to implement in the current problem due to two main reasons. First, the discontinuous nature of the dynamic axial load $F(t)$, which is active only during the blade tip travel through the rub zone on the stator. Second, the nonconservative nature of Coulomb friction at the blade-tip during the rub cannot be included in the homogeneous form of the equation. As a result of these limitations, to obtain a frequency domain solution for the dynamic stability of a vibro-impact system with contact-type of non-linearity, one must make some gross simplifying approximations about the temporal distribution of the dynamic pulse load so that it can be treated as a continuous periodic function. One such approximation about $F(t)$, which has the same peak-to-peak amplitude and frequency of the axial load as the true rub load described in Eq. (7.45), but it is a continuous cosine function of time "t", can be expressed as (see Figure 7.6)

$$F(t) = -\frac{F_{max}}{2} + \frac{F_{max}}{2}\cos\left(2\pi f_p t\right) \tag{7.50}$$

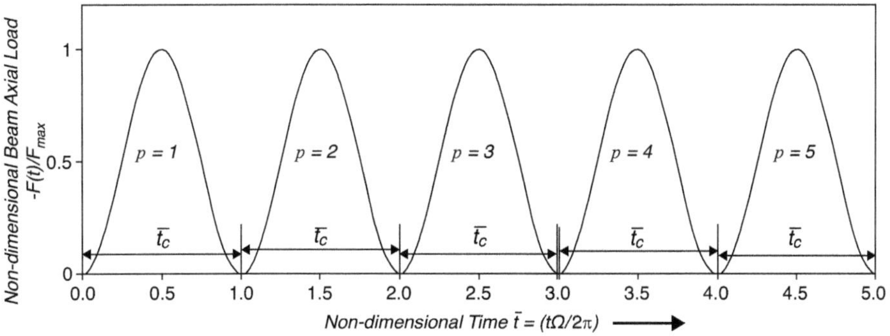

FIGURE 7.6 Periodic nature of the axial pulse $F(t)$ on the free end of the beam with cosine function temporal distribution given by Eq. (7.50) during rub (where pulse sequence number $p = 1, 2, 3, 4, 5$, etc.) and the frequency of pulse $f_p = 1$ pulse per revolution (contact load time $t_c = (2\pi / \Omega)$ second). (Reproduced with permission from Reference [1].)

It should be noted that Eqs. (7.45) and (7.50) have only two parameters in common, that is, the magnitude of the axial load F_{max} and the pulse frequency f_p. They have a completely different temporal distribution and the function shown in Eq. (7.50) disregards the effect of contact-load time "t_c" altogether, by inherently assuming that $t_c = (2\pi/\Omega)$. The net result due to the differences in the two forcing functions is that the amount of energy per revolution being fed to the same dynamical system is different. Thus, such simplifications, which are necessary to make the nonlinear contact-impact problem amenable to solution in the frequency domain, may not capture the full dynamics by neglecting the effects of some important parameters. The time-dependence of the force-history described by Eq. (7.50) is shown in Figure 7.6, which can easily be compared with the one shown in Figure 7.4. These approximations may be valid in some special situations, such as if the spin velocity Ω is very high or the rub-zone is circumferentially so large that it is fair to assume almost a 360° rub on the stator. The underlying assumptions are that the pulse load is being repeated so fast that the inertia of the beam does not allow it to recover into the undeformed state before the next pulse. On introducing the simplifying approximation for $F(t)$ shown in Eq. (7.50) in the relationship for $[\boldsymbol{B}]F(t)$ matrix of Eq. (7.48), it can be broken into two parts called $[\boldsymbol{B}]_1$ and $[\boldsymbol{B}]_2$, such that the first part accounts for the constant force of magnitude $F_{\max}$ and the second part accounts for the time-varying component of the axial force on the beam as follows:

$$[\boldsymbol{B}]F(t) = [\boldsymbol{B}]_1 (F_{\max}) - [\boldsymbol{B}]_2 (F_{\max}) \cos(2\pi f_p t) \tag{7.51}$$

Thus, the equation of motion shown in Eq. (7.48) in a homogeneous form can be written as

$$[\boldsymbol{M}]\{\ddot{\boldsymbol{f}}(t)\} + [\boldsymbol{C}]\{\dot{\boldsymbol{f}}(t)\} + ([\boldsymbol{K}] + [\boldsymbol{B}]_1 - [\boldsymbol{B}]_2 \cos(2\pi f_p t))\{\boldsymbol{f}(t)\} = \{0\} \tag{7.52}$$

where

$$[\boldsymbol{C}] = [\mathrm{C}]_D + [\mathrm{C}]_G \tag{7.53}$$

and

$$[\boldsymbol{K}] = [\mathrm{K}]_S + [\mathrm{K}]_C \tag{7.54}$$

The above equation can easily be identified as a Mathieu equation for a multiple degrees-of-freedom system. The following section describes Bolotin's general method of solving the dynamic stability characteristic in a gyroscopic system for a class of problems with Coriolis effects, which can be reduced in the form of a Mathieu equation. In this approach, we can assume the general solution of the displacement/deformation function $\{f(t)\}$ in Eq. (7.52) in the form of a product of two functions: the first is an exponential function with characteristic exponent λ and the second in an infinite Fourier series form of periodic sine and cosine functions with time-period equal to that of integral multiples of the pulse frequency $(1/f_p)$ as follows:

$$\{f(t)\} = e^{\lambda t}\left[\frac{\{b_0\}}{2} + \sum_{q=1}^{\infty}\left(\{a_q\}\sin\left(2\pi q f_p t\right) + \{b_q\}\cos\left(2\pi q f_p t\right)\right)\right] \tag{7.55}$$

where $\{b_0\}$, $\{a_q\}$, and $\{b_q\}$ are the Fourier coefficients of the sine and cosine terms. By computing the eigenvalues of Eq. (7.52) (roots of λ), the global conditions of the rotating beam stability with a known value of internal damping φ and excitation frequency f_p can easily be determined for any combination of parameters along with $F_{\max}$. The general solution of this set of equations results in eigenvalues λ with complex conjugate roots such that one can write $\lambda = \lambda_r \pm i\lambda_i$, where $i = \sqrt{-1}$; and eigenvalues λ's have dimensions of per unit time (rad/s). The positive and negative signs of the real part of the eigenvalue (λ_r) indicate the unstable and stable behaviors of the dynamical system, respectively. For example, the stability boundaries $(\lambda_r = 0)$ separating the regions of stable and unstable responses can be illustrated by plotting the angular velocity of spin Ω on the x-axis and the peak contact force $F_{\max}$ on the y-axis. On this plot, the unstable regions corresponding to $\lambda_r > 0$ indicate zones of flutter-type instability for a combination of dynamic parameters where the vibratory modal oscillations of the beam associated with the respective eigenfrequency λ_i would grow exponentially with respect to time "t". The real and imaginary parts of the eigenvalues λ_r and λ_i enable us to determine the effective logarithmic decrement as shown below:

$$\delta = -2\pi\left[\frac{\lambda_r}{|\lambda_i|}\right] \tag{7.56}$$

The dynamic system will be stable only when the effective logarithmic decrement for each mode is either zero or positive. In the dynamic stability problems, the usual interest is in identifying the unstable zones, i.e., to determine the boundary excitation frequencies of the "Principal Instability Region".

7.3.1.1 Sample Numerical Example

Any insight about the dynamic characteristics of a vibratory system requires the knowledge of its fundamental modes of vibration and their respective natural frequencies. This becomes a starting point for any dynamic analysis for both types of basic approaches such as

a. *Time-domain analysis for direct-time integration of stress wave equation in a time marching forward approach*: This numerical scheme requires in priori a numerically stable value of time-step size Δt. The numerically stable value of Δt is established by computing the highest frequency of the following homogeneous equation for undamped vibration of the beam after dropping the velocity-dependent coefficient matrix $[\boldsymbol{C}]$ and the force and the time-dependent part of the stiffness matrix $[\boldsymbol{B}]F(t)$ from Eq. (7.52), which yields the following homogeneous equation:

$$[\boldsymbol{M}]\{\ddot{\boldsymbol{f}}(t)\}+[\boldsymbol{K}]\{\boldsymbol{f}(t)\}=\{0\} \tag{7.57}$$

As opposed to the homogeneous form of the original equations of motion given in Eq. (7.52), the above eigenvalue problem is very simple to solve with standard techniques such as the Householder method, Jacobi method, and Matrix-iteration scheme. It yields $3N$ numbers of imaginary roots $\pm i\lambda_i$ giving the undamped frequencies of the simple dynamical system. Using the highest eigenvalue of the set called $|\lambda_i|_{\text{highest}}$ and taking the 1/8 of the time-period of the highest mode, the stable and numerically accurate time-step size Δt is determined as follows:

$$\text{Time-step size } \Delta t \leq \frac{1}{8}\frac{2\pi}{|\lambda_i|_{\text{highest}}} \text{seconds} \tag{7.58}$$

b. *Frequency-domain analysis for modal participation of the first few fundamental modes in determining the dynamic response approach*: This numerical scheme requires the knowledge of natural frequencies of the system in free vibration in the form of eigenvalues and the corresponding mode shapes as the respective eigenvectors. This method is not suitable for stress wave propagation problem and has limited use only in the implicit integration scheme.

To that end, we will develop a complete step-by-step methodology through a numerical example of a typical rotating Timoshenko beam representing a large-size jet engine fan blade. The computer program written for this purpose can simulate any general type of force-history function to represent $F(t)$. For the discussion purposes, the sample numerical results are generated for a typical jet engine fan blade with the parameters shown in Table 7.2. In this chapter for all foregoing discussion, we introduce a "Nondimensional frequency parameter ξ" being defined such that the natural frequency term ω_n (radians/s) is determined as shown below:

TABLE 7.2
Details of the Parameters Considered for the Sample Beam Used in the Discussion

M	Mass of the beam (blade)	10 kg
R	Beam tip radius	150 cm
r	Beam root radius	50 cm
A	Beam cross-sectional area	22.39 cm^2
E	Elastic Young's modulus of the beam material	117 GPa
ρ	Beam mass material density	4.466 g/cm^3
L	Span length of the beam	100 cm
β	Beam stagger angle at the tip	−60°
d	Slenderness ratio $= \sqrt{I/(AL^2)}$	0.010
F_{cr}	Critical Euler buckling load	0.064715 MN
Ω	Angular velocity of rotation (radians/s)	300 radians/s = 47.75 Hz = 2864.8 rpm
v	Beam tip tangential velocity	450 m/s
t_c	Beam-column tip pulse-loading duration (s)	$0.01\left(\frac{2\pi}{\Omega}\right)$
μ	Dynamic coefficient of friction between the blade tip and the inner surface of the outer case	0.0, 0.2
v	Poisson's ratio	0.3
F_{max}	Magnitude of the pulse-load along the beam axis (nondimensional)	0.8, 1.0, 2.0, 5.0, 10.0
t	Time (s)	$\bar{t}\left(\frac{2\pi}{\Omega}\right)$
	Blade-tip running clearance	3 mm

$$\omega_n = \frac{\xi}{L^2}\sqrt{\frac{(EI)}{(\rho A)}} \text{ radians/s} \tag{7.59}$$

The numerical results showing linear displacement and deformations are nondimensionalized with respect to the span length of the blade $L = 100$ cm, and the forces are normalized with respect to the critical Euler's buckling load for the cantilever beam-column. In this example problem, we consider the first 10 terms of the trial functions for each of lateral deflection η, rotation ψ, and the membrane-stretching ζ deformation of the beam, such that $N = 10$. Thus, the sizes of each of the $[\boldsymbol{M}]$, $[\boldsymbol{C}]$, and $[\boldsymbol{K}]$ matrices are 30×30. All the 30 eigenvalues are computed in the current scheme per matrix equation (Eq. 7.57), and the corresponding frequencies expressed in nondimensional parameter ξ and Hz are listed in Table 7.3.

Based upon the time-step size criteria established in Eq. (7.57), Δt for this problem turns out to be 2.32391 μs. It is worth noting that in a Timoshenko beam problem the numerically stable time-step size Δt, to be used in the explicit method, is controlled

TABLE 7.3
Vibrational Mode Frequencies of the Rotating Cantilever Timoshenko Beam in the Numerical Example (Tip Tangential Velocity=450 m/s, $\bar{d}=0.01$)

	Flexural bending mode		Axial stretching mode		Cross-sectional rotation mode	
Mode number	Non-dimensional ξ	(Hz)	Non-dimensional ξ	(Hz)	Non-dimensional ξ	(Hz)
1	8.3177	67.8565	646.8984	5277.4550	5671.3020	46,266.9800
2	28.9632	236.2842	785.3968	6407.3360	5700.8720	46,508.2100
3	68.6869	560.3537	817.2732	6667.3870	5751.1120	46,918.0900
4	127.7992	1042.5970	1099.5530	8970.2530	5818.7410	47,469.8000
5	200.3793	1634.7120	1413.7150	11,533.2100	5907.8370	48,196.6600
6	290.1916	2367.4090	1727.8760	14,096.1600	6010.9200	49,037.6200
7	393.9380	3213.7810	2042.0340	16,659.0900	6135.0360	50,050.1600
8	510.3367	4163.3720	2356.1890	19,222.0000	6270.1780	51,152.6600
9	637.6240	5201.7920	2670.3530	21,784.9800	6424.1020	52,408.3900
10	775.0315	6322.7750	2984.5050	24,347.8600	6593.2950	53,788.6800

by the cross-sectional rotation frequency, which is always much higher than the typical flexural bending mode frequency of the beam. It is obvious from Table 7.3 that for this beam, magnitude wise the typical axial mode frequencies lie in-between the bending mode frequencies on the lower-bound side and the rotational mode frequencies on the upper-bound side.

In the current numerical algorithm, we also compute the (i) internal energy (IE) and (ii) kinetic energy (KE) of the beam at every time step for two main deformation modes, viz., lateral bending and axial movement, which are as follows:

$$(IE)_{\text{Lateral bending}} = \frac{(EI)}{2}\int_0^L (\psi')^2\,ds = \frac{(EI)}{2}\int_0^L \left[\sum_{k=1}^{N} U_k' X_k\right]^2 ds \tag{7.60}$$

$$(IE)_{\text{Axial stretching}} = \frac{(EA)}{2}\int_0^L (\zeta')^2\,ds = \frac{(EA)}{2}\int_0^L \left[\sum_{k=1}^{N} W_k' Z_k\right]^2 ds \tag{7.61}$$

$$(KE)_{\text{Lateral motion}} = \frac{(\rho A)}{2}\int_0^L (\dot{\eta})^2\,ds = \frac{(\rho A)}{2}\int_0^L \left[\sum_{k=1}^{N} V_k \dot{Y}_k\right]^2 ds \tag{7.62}$$

$$(KE)_{\text{Axial motion}} = \frac{(\rho A)}{2}\int_0^L (\dot{\zeta})^2\,ds = \frac{(\rho A)}{2}\int_0^L \left[\sum_{k=1}^{N} W_k \dot{Z}_k\right]^2 ds \tag{7.63}$$

The full tracking of the time history of the transfer of the energy between two deformation modes helps us in interpreting the complex nonlinear dynamic behavior of the Timoshenko beam. In any numerical scheme, this is also a necessary and highly recommended step to ensure that no spurious energy is being introduced in the dynamical system due to some unexpected nonlinearity or other inherent numerical approximations such as round-off or truncation errors.

7.3.1.2 Natural Frequencies of a Timoshenko Beam

Any insight about the dynamic stability of a vibratory system requires the knowledge of its fundamental modes of vibration and its respective natural frequencies. Therefore, it is necessary to systematically compare the trend of results for some limiting cases, such as "slenderness ratio" $\bar{d} \to 0$, of Euler–Bernoulli's beam formulation, which has been analyzed in lot more detail than Timoshenko beams. To that end, the results shown here in this section should serve as a benchmark of further reference for researchers and provide useful information to design engineers. With this aim of obtaining a limiting case solution, Eqs. (7.26a) and (7.26b) are easily reduced to a more familiar form of Euler–Bernoulli's beam-column formulation for span-length L as follows:

$$\sum_{n=1}^{N}\left[\rho A\int_0^L V_m V_n ds\right](\ddot{Y}_n) +$$

$$\sum_{n=1}^{N}\left[EI\left[\int_0^L V_m V_n'''' \, ds - V_m(L)\,V_n'''(L)\right] - F(t)\left[\int_0^L V_m V_n'' \, ds - V_m(L)\,V_n'(L)\right]\right](Y_n)$$

$$= \int_0^L q(s,t)V_m(s)ds \qquad (7.64)$$

The computed nondimensional first flexural mode natural frequency parameters for varying slenderness ratio ($\bar{d}$ = 0.01, 0.05, 0.1, 0.15, 0.2, and 0.3) are plotted for increasing rotational speed up to $\bar{\Omega} = 0.1$ in Figure 7.7.

In the majority of rotating beam applications, the upper limit of the tip tangential velocity is a material stress-capability limiting parameter, which forces the designer not to let the realistic tip velocity to exceed beyond 600 m/s. Corresponding to this limiting value of the maximum tip velocity, the nondimensional angular velocity $\bar{\Omega}$ does not exceed 0.09 in most of the real-life applications. Thus, $\bar{\Omega} = 0.1$ is the practical limit of the highest angular velocity a design engineer will ever need to apply from any of the numerical values presented in this discussion. Here we recall that the nondimensional frequency parameter ξ is defined such that ω_n is determined as shown below:

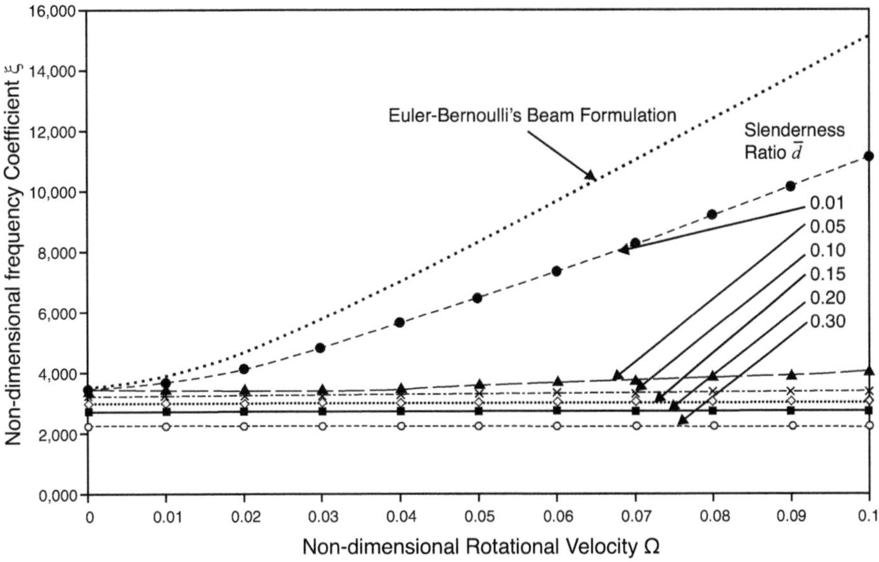

FIGURE 7.7 Increase in the first flexural bending mode frequencies of a rotating cantilever Timoshenko beam (slenderness ratio $\bar{d}$ =0.01, 0.05, 0.1, 0.15, 0.2, 0.3) under increasing rotational speed. (Reproduced with permission from Reference [1].)

$$\omega_n = \frac{\xi}{L^2}\sqrt{\frac{(EI)}{(\rho A)}} \text{ radians/s} \tag{7.65}$$

The computed values of for the static condition match well with the closed-form solution given by all elementary textbooks on beam vibration. The graphical data illustrates the fact that for the lower slenderness ratio, which in the limit case is equivalent to the classical Euler–Bernoulli's beam formulation, moves up rapidly with the increasing rotational speed $\bar{\Omega}$; however, for higher slenderness ratios (short and stubby beams), the change is not significant. The effect of the speed on all the 10 flexural bending mode frequencies of the Timoshenko beam with different slenderness ratios ($\bar{d}$=0.005, 0.01, 0.05, and 0.1) is also shown in Tables 7.4 and 7.5, respectively, for stationary ($\bar{\Omega}$=0.0) and rotating conditions ($\bar{\Omega}$=300 radians/s, i.e., $\bar{\Omega}$=0.0877898) of the beam. The tabulated numerical results are listed here up to four decimal places to demonstrate the accuracy of the eigenvalue extraction. As mentioned in Eq. (7.65), the "Non-dimensional frequency factor ξ" has been defined in such a way that the corresponding frequency in Hz can be determined as follows:

$$\xi = \omega_n L^2\sqrt{\frac{\rho A}{EI}} \text{ and } f_n = \left(\frac{1}{2\pi}\right)\frac{\xi}{L}\bar{d}\sqrt{\frac{E}{\rho}} \text{ Hz.} \tag{7.66}$$

The first flexural mode frequencies of the beam under consideration in the sample example ($\bar{d}$=0.01) for the stationary and rotating conditions are ξ=3.4809 (28.3977 Hz) and 8.3177 (67.8565 Hz), respectively. The decays in the value of

TABLE 7.4
Computed Nondimensional Frequency Parameter ξ for the First 10 Flexural Bending Modes of a Stationary ($\bar{\Omega}=0.0$) Cantilever Timoshenko Beam as a Function of Slenderness Ratio $\bar{d}$

	Slenderness Ratio $\bar{d}$			
	0.005	0.01	0.05	0.10
Mode No.	Nondimensional ξ	Nondimensional ξ	Nondimensional ξ	Non-imensional ξ
1	3.5160	3.4809	3.4395	3.2308
2	22.0971	21.9210	19.1603	14.5756
3	61.5388	60.8399	46.8837	31.8319
4	120.3240	117.9210	79.6116	48.4153
5	198.2305	192.1182	114.9608	62.6941
6	293.0804	281.9967	151.3109	68.4167
7	408.1056	384.3208	187.5246	81.0305
8	539.8896	500.4664	220.7516	89.5881
9	688.3948	627.4645	238.7207	102.6107
10	853.3561	764.5730	250.6780	110.3643

TABLE 7.5
Computed Nondimensional Frequency Parameter ξ for the First 10 Flexural Bending Modes of a Rotating ($\bar{\Omega}=0.0877898$, Tip Tangential Velocity = 450 m/s) Cantilever Timoshenko Beam as a Function of Slenderness Ratio $\bar{d}$

	Slenderness Ratio $\bar{d}$			
	0.005	0.01	0.05	0.10
Mode No.	Nondimensional ξ	Nondimensional ξ	Nondimensional ξ	Nondimensional ξ
1	15.3513	8.3177	3.7538	3.3121
2	43.7037	28.9632	19.5034	14.6773
3	88.2832	68.6869	47.2828	31.9717
4	150.8615	127.7992	80.0909	48.5743
5	225.0000	200.3793	115.5232	62.8019
6	327.9073	290.1916	151.9531	68.5391
7	443.4582	393.9380	188.2302	81.1678
8	576.4866	510.3367	221.3706	89.7942
9	725.7231	637.6240	238.9201	102.7375
10	891.3798	775.0315	251.1273	110.6600

nondimensional frequency parameter with increasing slenderness ratio for the flexural and rotational mode frequencies are shown in Figures 7.8 and 7.9.

The graphical representation of ξ versus $\bar{d}$ illustrates the overall trend in both bending and rotational mode frequencies of the Timoshenko beam as we vary $\bar{d}$. These results are identical to the results reported by Zhou and Cheung [see Reference 8]. The dynamic buckling characteristics of the rotating radial beam with

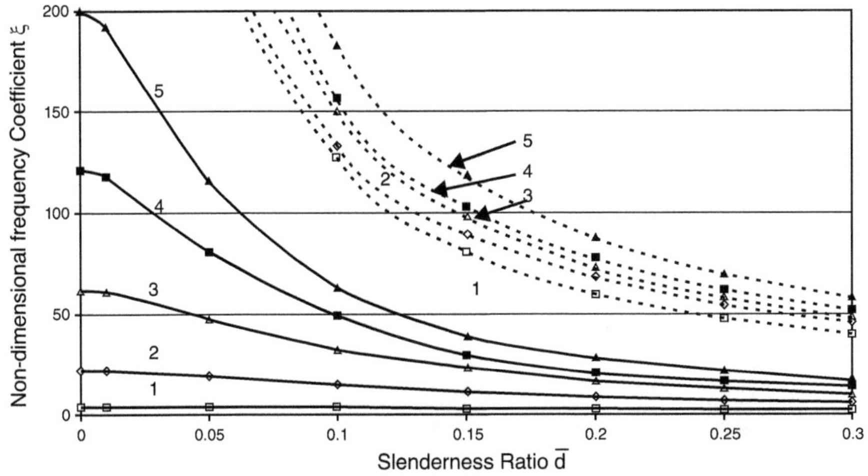

FIGURE 7.8 Drop in the first five flexural bending (———) and rotational (- - - - -) mode frequencies of a typical stationary ($\bar{\Omega}=0.0$) cantilever Timoshenko beam due to increasing slenderness ratio $\bar{d}$. (Reproduced with permission from Reference [1].)

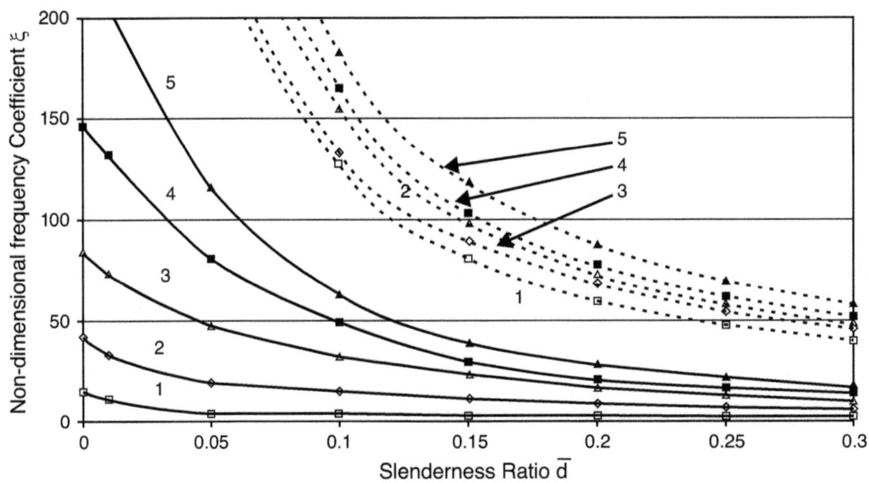

FIGURE 7.9 Drop in the first five flexural bending (———) and rotational (- - - - -) mode frequencies of a typical rotating ($\bar{\Omega}=0.1$) cantilever Timoshenko beam due to increasing slenderness ratio $\bar{d}$. (Reproduced with permission from Reference [1].)

its inner radius clamped at the top of the disk in terms of its frequency drop versus the compressive axial load are shown in Figures 7.10a and b. For the purpose of comparing the effect of spin angular velocity, Figure 7.10a shows the flexural bending mode frequencies for a stationary beam and Figure 7.10b shows the corresponding values for the same beam rotating at 300 radians/s.

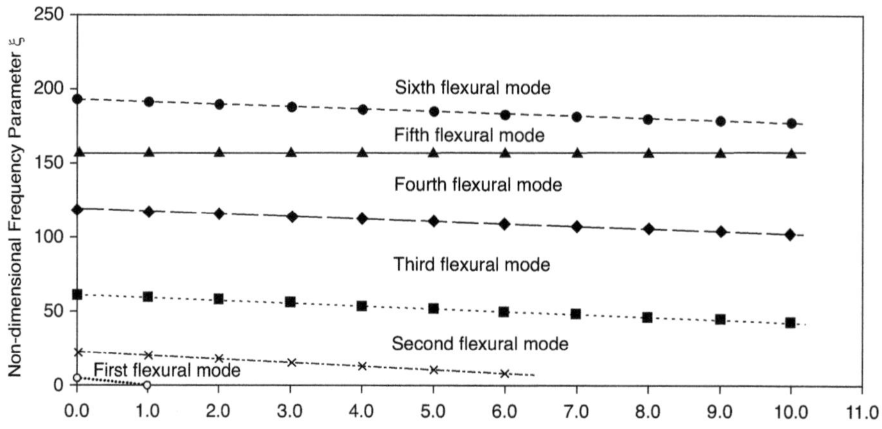

(a)

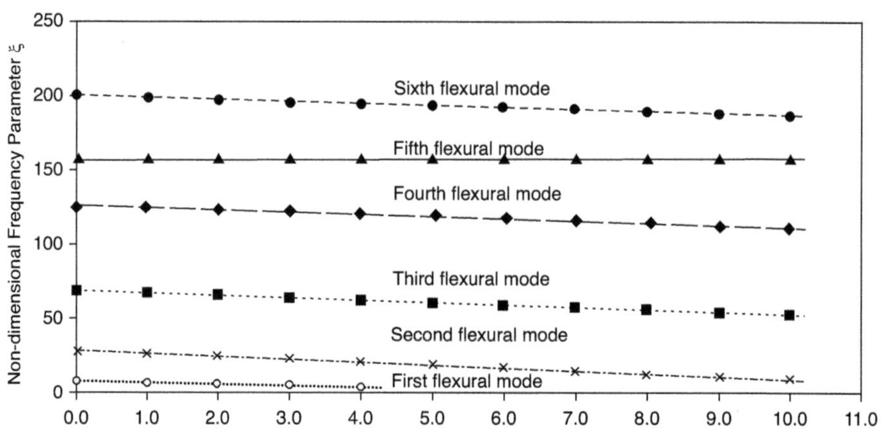

(b)

FIGURE 7.10 (a) Drop in the first six flexural bending mode frequencies of a typical stationary (angular velocity $\Omega = 0$ radians/s) cantilever Timoshenko beam (slenderness ratio $\bar{d} = 0.01$) under increasing compressive axial load F_a at the tip. (Reproduced with permission from Reference [1].) (b) Drop in the first six flexural bending mode frequencies of a typical rotating (angular velocity $\Omega = 300$ radians/s) Timoshenko beam (slenderness ratio $\bar{d} = 0.01$) under increasing compressive axial load F_a at the tip. (Reproduced with permission from Reference [1].)

By comparing these two plots, we observe that the rotational speed has a significant effect in raising the dynamic buckling load for the beam. For example, the current beam under consideration can withstand more than four times of the critical Euler Buckling load ($F_{max} > 4F_{cr}$) before collapsing in the first flexural mode under steady-state loading. Under the impact condition depending upon the duration of the axial load application, it can deform in some higher modes without actually collapsing. It can be seen in this plot that all the flexural mode frequencies drop due to an increase in the axial compressive load. However, the amount of drop is changing for different modes with the first and second flexural mode frequencies dropping significantly more than the higher order modes.

7.3.1.3 Parametric Tip-Rub Stability Analysis in the Frequency Domain

In the frequency domain, the dynamic stability of a rotating cantilever Timoshenko beam during the tip-rub event is under some kind of periodic axial loading in the longitudinal direction of the beam column. Theoretically, this can be analyzed by the harmonic balance method, initially suggested by Bolotin who formulated it as a Mahieu–Hill equation. The dynamic response of Timoshenko beam discussed in this section is based on the author's original research work published in 1989 in the *Journal of Sound and Vibration* [see Reference 9]. For the benefit of the readers, some of the results are being reproduced here with the permission of the publisher.

The current discussion is concerned with the dynamic stability problem of a Timoshenko beam subjected to a periodic sinusoidal axial force superimposed on a steady-state axial component of a constant magnitude. This analysis is an attempt to address this important practical problem and to identify the changing instability zones for different slenderness ratios $\bar{d}$ of the Timoshenko beam. To simplify the problem, for the current discussion purposes the Timoshenko beam under consideration is stationary such that the angular velocity $\Omega = 0$ radians/s. The mathematical basis of stability boundary determination using the harmonic balance method is a separate topic for a later discussion in this book. The governing equation of a stationary Timoshenko beam-column with a time-dependent axial force $F(t)$ can be written as follows:

$$EI\frac{\partial^4 \eta}{\partial s^4} - F(t)\frac{\partial^2 \eta}{\partial s^2} - \rho I\left(1+\frac{E}{\kappa G}\right)\frac{\partial^2 \ddot{\eta}}{\partial s^2} + \rho A\ddot{\eta} + \left(\frac{\rho^2 I}{\kappa G}\right)\frac{\partial^2 \ddot{\eta}}{\partial t^2} = 0 \quad (7.67)$$

Here the time-dependent axial force $F(t) = F_0 + F_t \cos(\omega_p t)$ has two components. Here, the first component F_0 is the steady-state constant load of fixed magnitude and F_t is the peak magnitude of the time-dependent component of the external axial load on the beam-column. It is an oscillatory excitation force vibrating at a frequency of ω_p radians/s such that $\omega_p = (2\pi f_p)$. For solving this problem, we introduce the following two non-dimensional parameters called $\alpha = (F_0/F_{cr})$ and $\beta = (F_t/F_{cr})$, such that in a cantilever beam they are defined as follows:

$$\text{Euler buckling critical load } F_{cr} = \left[\frac{\pi^2 EI}{4L^2}\right] \quad (7.68)$$

Beam first flexural bending mode frequency $\omega_N = \frac{\xi}{L^2}\sqrt{\frac{EI}{\rho A}}$ radians/s. ($\xi = 3.51602$)

The nondimensional excitation frequency $\bar{\omega}$ in this parametric resonance analysis problem is defined as follows:

$$\bar{\omega} = \frac{\omega_p}{\omega_N} \tag{7.69}$$

Other nondimensional parameters such as "slenderness ratio $\bar{d}$" are the same as defined earlier in the previous section.

$$\bar{d} = \sqrt{I/(AL^2)} \tag{7.70}$$

We also define the axial force parameter called "μ" as follows:

$$\mu = \frac{\beta}{2(1-\alpha)} = \frac{F_t}{2(F_{cr} - F_0)} \tag{7.71}$$

Timoshenko beam elasticity-shear characteristics in a nondimensional form are defined as follows:

$$\gamma = \frac{E}{\kappa G} \tag{7.72}$$

As per the method described in earlier sections, the partial differential equation outlined in Eq. (7.67) is converted into a matrix form. Similar to Galerkin's method described earlier for coupled partial differential equations Eqs. (7.15a)–(7.15c), the previous equation (7.67) is converted in a matrix form of Mahieu–Hill equation with terms like $[\boldsymbol{M}]$, $[\boldsymbol{C}]$, $[\boldsymbol{K}]$, and $[\boldsymbol{B}]$ matrices as shown below:

$$[\boldsymbol{M}]\{\ddot{\boldsymbol{Y}}(t)\}+[\boldsymbol{C}]\{\dot{\boldsymbol{Y}}(t)\}+([\boldsymbol{K}]-[\boldsymbol{B}]\cos(\omega_p t))\{\boldsymbol{Y}(t)\} = \{0\} \tag{7.73}$$

The deformation column vector $\{\boldsymbol{Y}(t)\}$ can be assumed in the form of an infinite Fourier series. However, for simplicity and demonstration purposes, here we will be considering only up to $m = 2$ terms of the series, as shown below:

$$\{\boldsymbol{Y}(t)\} = e^{\lambda t}\left[\frac{\{b_0\}}{2} + \sum_{m=1}^{2}\left(\{a_m\}\sin(2\pi m f_p t) + \{b_m\}\cos(2\pi m f_p t)\right)\right] \tag{7.74}$$

Substitution of the assumed series representation of $\{\boldsymbol{Y}(t)\}$ shown in Eq. (7.74) enables us to use the method of harmonic balance. The coefficients of Fourier series are put in the column vector, and the matrices can be separated as terms looking like mass matrix, damping matrix, and stiffness matrix terms. In this scheme, we collect the coefficients of $e^{\lambda t}$, $e^{\lambda t}\sin(m\omega_p t/2)$, and $e^{\lambda t}\cos(m\omega_p t/2)$. This system of equations can be rearranged with three different matrices, viz., $[\boldsymbol{P}]$, $[\boldsymbol{Q}]$ and $[\boldsymbol{R}]$ as the coefficients of ascending powers of λ in the following form:

$$\left([\boldsymbol{P}]+\lambda\ [\boldsymbol{Q}] + \lambda^2 [\boldsymbol{R}]\right)\{Y\} = \{0\} \tag{7.75}$$

Here, the column vector number of Fourier contains the unknown Fourier coefficients. In the present analysis, the numerical solution has been carried out by limiting the maximum coefficients "m" to 2. This implies that the solution of the assumed displacement function would have only two periodic components: one with the time period of $\left((2\pi)/\bar{\omega}\right)$ and second with the period of $\left((2\pi)/(2\bar{\omega})\right)$. The maximum number of instability zones, which can be determined by Eq. (7.75) in this manner, is equal to $2m$. In other words, for the present problem, one can predict up to four instability zones. Here, $\{\boldsymbol{Y}\}$ can be written as follows:

$$\{Y\} = \left\{ \left\{ \begin{matrix} b_0 & a_1 & a_2 & b_1 & b_2 \end{matrix} \right\}^T \left\{ \begin{matrix} b_0 & a_1 & a_2 & b_1 & b_2 \end{matrix} \right\}^T \right\} \tag{7.76}$$

In this relationship for $\{\boldsymbol{Y}\}$, the first set of Fourier coefficients are for Y_1 and the second set for Y_2, respectively. Then, the substitution of $\{\dot{Y}\} = \{X\} = \lambda\{Y\}$ in Eq. (7.75) makes the coefficient matrix completely asymmetric, which in the "state form" can be written as follows:

$$\begin{bmatrix} 0 & [I] \\ -[\boldsymbol{R}]^{-1}[\boldsymbol{P}] & -[\boldsymbol{R}]^{-1}[\boldsymbol{Q}] \end{bmatrix} \begin{Bmatrix} \{Y\} \\ \{X\} \end{Bmatrix} = \lambda \begin{Bmatrix} \{Y\} \\ \{X\} \end{Bmatrix} \tag{7.77}$$

The solution of Eq. (7.66) results in complex conjugate roots of eigenvalue λ, such that one can write $\lambda = \lambda_r \pm i\lambda_i$, where $i = \sqrt{-1}$; and eigenvalues λs have dimensions of per unit time (radians/s). The different signs of the real part of the eigenvalue (λ_r) determine the dynamic stability of the elastic system under consideration. These stability conditions are $\lambda_r > 0$ (unstable), $\lambda_r = 0$ (stable or oscillatory), and $\lambda_r < 0$ (asymptotically stable).

In dynamic stability problems, the usual interest is in identifying the unstable zones, i.e., to determine the boundaries of the principal instability regions in the frequency domain, where parametric resonance is most likely to occur. The boundary frequency equation (roots of ω_p) for the dynamic system governed by the Mathieu–Hill equation (see Eq. 7.73) is written as follows:

$$\text{Determinant} \begin{vmatrix} \left(\boldsymbol{K} + \frac{1}{2}\boldsymbol{B} - \frac{\omega_p{}^2}{4}\boldsymbol{M}\right) & -\frac{\omega_p}{2}\boldsymbol{C} \\ \frac{\omega_p}{2}\boldsymbol{C} & \left(\boldsymbol{K} - \frac{1}{2}\boldsymbol{B} - \frac{\omega_p{}^2}{4}\boldsymbol{M}\right) \end{vmatrix} = 0 \tag{7.78}$$

Finding the roots of ω_p for a large size determinant numerically speaking can become almost an impossible task. For example, even for current very simple non-rotating Timoshenko beam problem described by Eq. (7.67) with just $m = 2$ terms of the Fourier series, the size of determinant is 20×20. This computationally intensive task can be made a mathematically solvable problem by converting equation (7.78) into an eigenvalue problem. To that end, after some lengthy algebraic manipulations, Eq. (78) reduces to finding the eigenvalues of the following matrix:

$$\begin{bmatrix} 0 & 0 & 2[\boldsymbol{I}] & 0 \\ 0 & 0 & 0 & 2[\boldsymbol{I}] \\ [\boldsymbol{M}]^{-1}[2[\boldsymbol{K}]+[\boldsymbol{B}]] & 0 & 0 & -2[\boldsymbol{M}]^{-1}[\boldsymbol{C}] \\ 0 & [\boldsymbol{M}]^{-1}[2[\boldsymbol{K}]-[\boldsymbol{B}]] & 2[\boldsymbol{M}]^{-1}[\boldsymbol{C}] & 0 \end{bmatrix} \tag{7.79}$$

Since the matrices $[\boldsymbol{M}]$, $[\boldsymbol{C}]$, $[\boldsymbol{K}]$, and $[\boldsymbol{B}]$ are fully defined and known, either of the complex eigenvalue problems described in Eqs. (7.77) and (7.79), respectively, is easily solved by using IMSL-EISPACK routine [see Reference 10].

Determination of roots of ω_p in the above equation helps us in establishing the rub instability zones, shown in Figure 7.11a–c. These plots have been produced in a nondimensional form as axial force parameter μ vs nondimensional excitation frequency $\bar{\omega}$.

The computer plots of the unstable regions shown by cross-hatched areas in Figure 7.11a–c are for $\alpha = 0$ and $\gamma = 3$ and slenderness ratio of $\bar{d} = 0.01$, 0.1, 0.2, respectively. The stability boundary for $\lambda_r = 0$ shows three different zones of dynamic instability of a typical Timoshenko beam. It is evident that the two subharmonic zones are relatively narrow, however, one at $\bar{\omega} = 2$ for a thin beam of $\bar{d} = 0.01$ is very wide. It highlights the major danger zone of excitation frequency, this being at or around two times of the bending mode frequency of the beam, also called "principal boundary region", which becomes as low as $\bar{\omega} = 1.2832$ for $\bar{d} = 0.2$. The current results for $\bar{d} = 0.01$ are in close agreement with the values as obtained from the Mahieu equation derived from the Bernoulli–Euler beam theory. It shows that, as expected, the harmonic balance method analytical results for $\bar{d} \rightarrow 0.0$ does converge to other analytical results for thin beams.

7.3.1.4 Vibro-Impact Response Under Single Axial Pulse

The dynamic stability of a rotor blade subjected to pulse loading with any kind of general discontinuous distribution over the time axis can be checked in a time-domain solution by direct-time integration scheme discussed in this section. As an example, we start with the transient response of the rotating beam, which has been analyzed for a single axial pulse without any frictional force at the tip. In this simulation, we apply one single longitudinal pulse load of $F_{max} = 5F_{cr}$ lasting for $\bar{t}_c = 0.01$, i.e., $t_c = 0.209$ ms. The beam is allowed to vibrate in an undamped mode due to this single

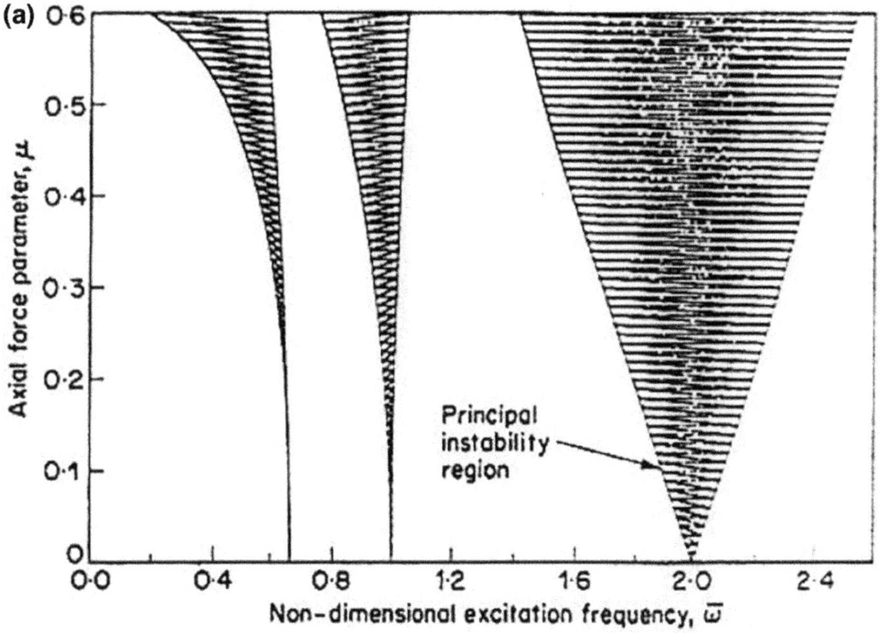

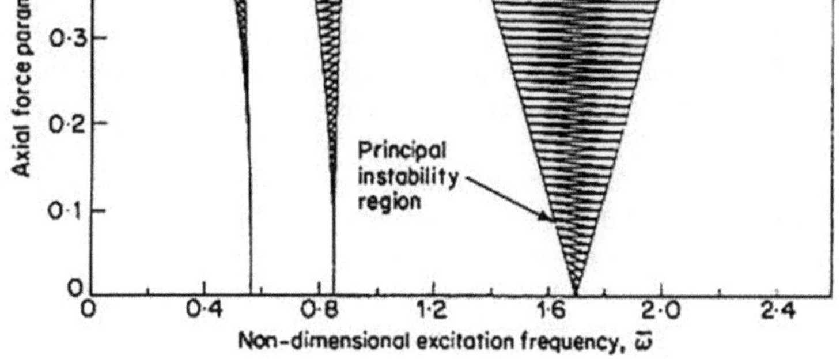

FIGURE 7.11 (a) Major dynamic instability region for $\alpha = 0$, slenderness ratio $\bar{d} = 0.01$ (similar to Bernoulli–Euler Beam). (Reproduced with permission from Reference [9].) (b) Dynamic instability region for $\alpha = 0$, slenderness ratio $\bar{d} = 0.1$. (Reproduced with permission from Reference [9].) (c) Dynamic instability region for $\alpha = 0$, slenderness ratio $\bar{d} = 0.2$. (Reproduced with permission from Reference [9].)

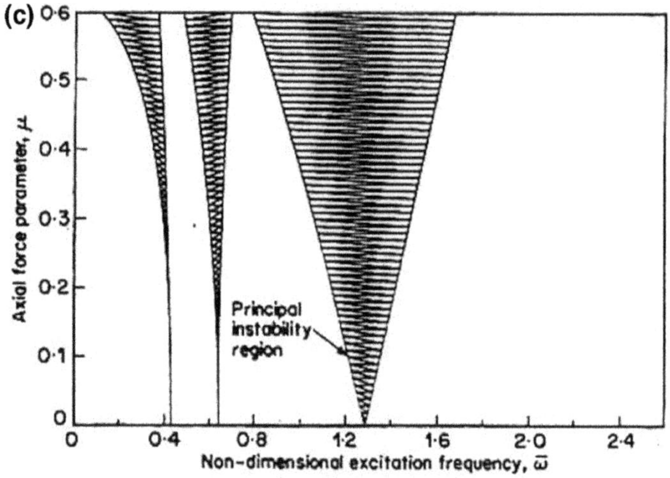

FIGURE 7.11 ***(Continued)*** (a) Major dynamic instability region for $\alpha = 0$, slenderness ratio $\bar{d} = 0.01$ (similar to Bernoulli–Euler Beam). (Reproduced with permission from Reference [9].) (b) Dynamic instability region for $\alpha = 0$, slenderness ratio $\bar{d} = 0.1$. (Reproduced with permission from Reference [9].) (c) Dynamic instability region for $\alpha = 0$, slenderness ratio $\bar{d} = 0.2$. (Reproduced with permission from Reference [9].)

pulse. An impulsive load excites all the modes in the beam simultaneously, but the higher order modes of the structure die down quickly leaving only the dominant mode active, which happens to be the first flexural mode of the beam. The Runge–Kutta integration scheme for the time-marching-forward solution has been carried out for 60 revolutions of the rotor, which is 1.26 seconds. The transient response in terms of the lateral tip deflection $\eta(L,t)$ has been plotted in Figure 7.12.

The stable nature of the practically undamped lateral response of ± 0.1 mm for 1.26 seconds shows the robustness and numerical stability of the integration scheme. It should also be noted that in the absence of tip friction ($\mu = 0$), the tip lateral deflection of the beam is the result of the coupling of axial motion with the lateral motion caused by the Coriolis effect. As expected, the undamped dynamic response of the beam is at 67.0 Hz, which matches very well with the first-flexural mode frequency of the rotating beam computed earlier as 67.8565 Hz. The corresponding tip displacement in the axial direction is shown in Figure 7.13a.

The beam tip oscillates about the steady-state deflection of 2 mm with the initial peak amplitude in the range of 1.2 mm $\leq \zeta(L,t) \leq$ 2.8 mm and then eventually settling down to 1.8 mm $\leq \zeta(L,t) \leq$ 2.2 mm. Figure 7.13b shows the zoomed-in view of the short-time response in the axial direction. As one can see from this plot that during the initial impact the beam responds at 1286 Hz. It is quickly observed that the longitudinal mode response frequency is much lower than the normal mode eigenvalues calculated from the stiffness and mass matrices. This subharmonic response is due to parametric resonance in the dynamical system. In the long-time response, the analytically computed time history of the dynamic deflection illustrates the presence of a small amount of numerical damping in the integration scheme, which

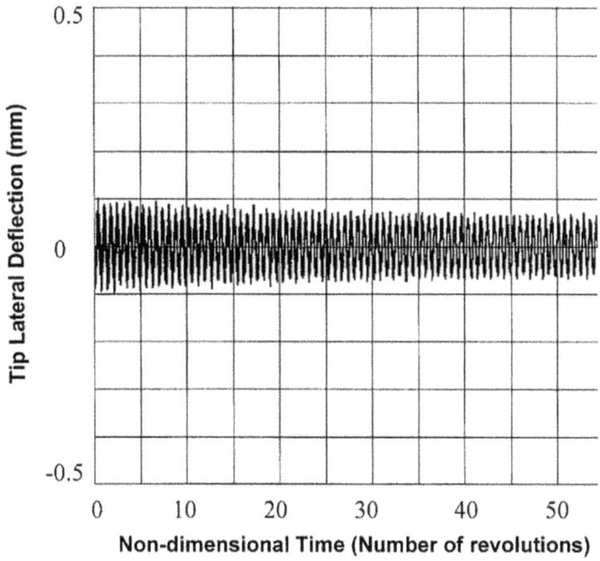

FIGURE 7.12 Analytically predicted transient lateral deflection of the tip after a single impulse of $\bar{F}_{max} = 5.0$ (rotational velocity $\Omega = 300$ radians/s). Full-time 50 revolutions transient response of the beam immediately after impact. (Reproduced with permission from Reference [1].)

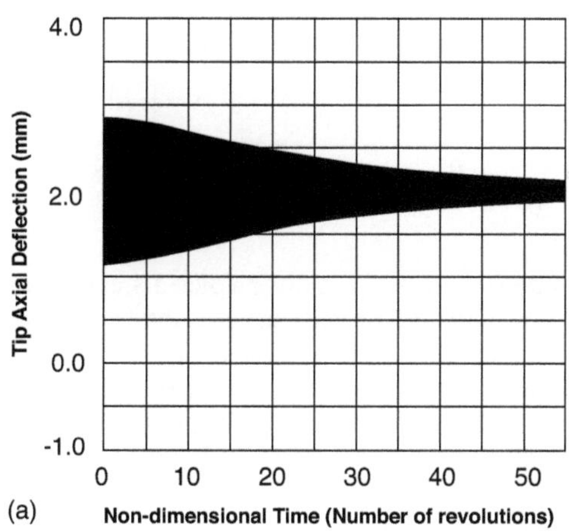

FIGURE 7.13 (a) Analytically predicted transient axial deflection of the tip after a single impulse of $\bar{F}_{max} = 5.0$ (steady-state tip deflection due to a rotational velocity of 300 radians/s = 2 mm): full-time 50 revolutions transient response of the beam immediately after impact. (Reproduced with permission from Reference [1].)

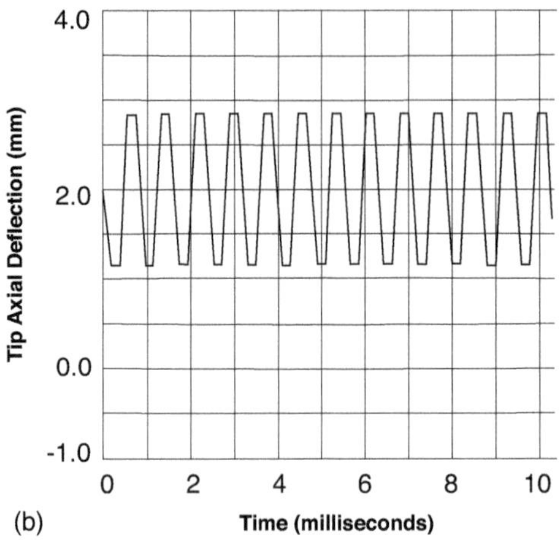

(b)

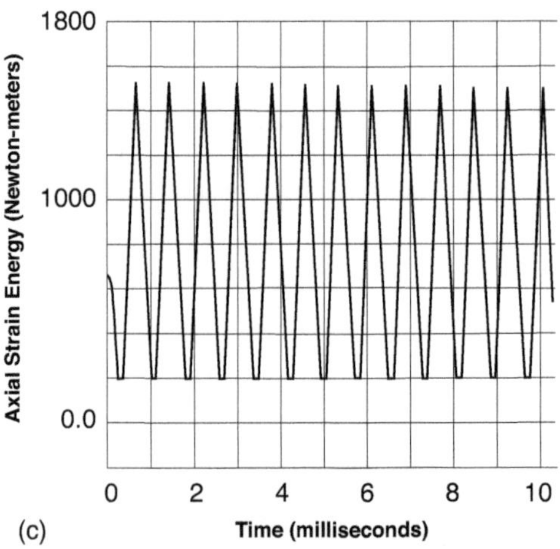

(c)

FIGURE 7.13 (*Continued*) (b) Analytically predicted transient axial deflection of the tip (zoomed-in view of longitudinal tip movement of the beam) after a single impulse of $F_{max} = 5.0$ (steady-state tip deflection due to a rotational velocity of 300 radians/s = 2 mm): zoomed-in 10 ms window immediately after impact. (Reproduced with permission from Reference [1].) (c) Analytically predicted transient response (zoomed-in view of axial strain energy in the beam) after a single impulse of $\bar{F}_{max} = 5.0$ (steady-state tip deflection due to a rotational velocity of 300 radians/s = 2 mm): zoomed-in 10 ms window immediately after impact. (Reproduced with permission from Reference [1].)

is inevitable and makes the transient response to appear somewhat damped. But, the most important fact is that there is no numerical instability associated with the time-step size t even for 50–60 revolutions of the rotor. To demonstrate the numerical stability and the robustness of the integration scheme, we have also looked into the strain-energy history plot of the dynamical system, which is shown in Figure 7.13c. It is obvious from this plot that there is no spurious build-up of strain-energy in the system as the integration marches forward along the time axis.

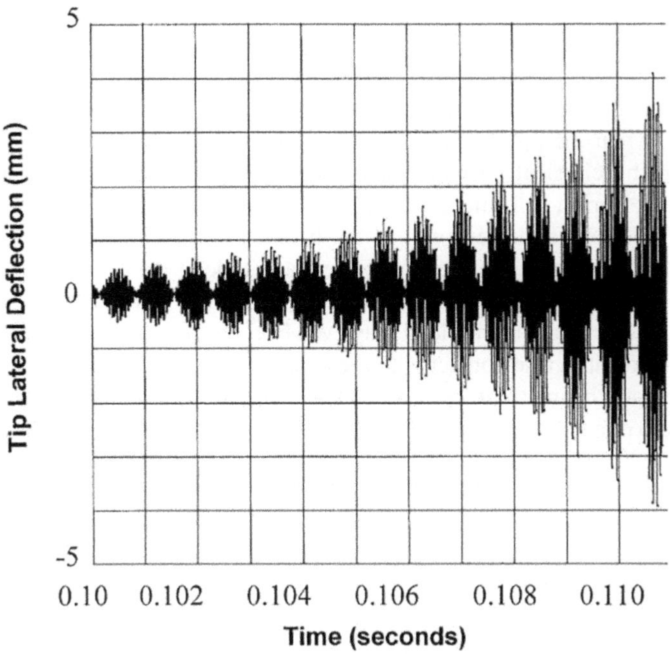

FIGURE 7.14 Autoparametric vibration instability due to energy transfer between lateral and axial modes (rotational velocity = 300 radians/s). (Reproduced with permission from Reference [1].)

The current time-marching-forward scheme captures all the complex dynamic characteristics of a nonlinear system very well (see Figure 7.14). If the dynamical system is undamped, depending upon the excitation frequency and the duration of the axial load $F(t)$, the transient response exhibits the characteristics of a typical parametric resonance.

Here, due to the energy transfer between the flexural and axial modes, the autoparametric vibration instability initiates at 0.1 second and grows to ± 4.5 mm in 10 ms. In this example, the increasing amplitude modulation frequency showing the spurt of growing large transverse motions of the tip of the Timoshenko beam is approximately 1400 Hz. This high-frequency modulation in the beam response is again the result of coupling between the longitudinal mode and the flexural mode deflection

As pointed out earlier, the interaction between the axial motion and lateral motion is caused by the Coriolis effect, which are coupled through the skew-symmetric terms in the coefficient matrix $[\mathrm{C}]_G$ to account for velocity-dependent circulatory forces. This term is directly proportional to the rotational angular velocity Ω. Thus, in other words, there is no coupled dynamic response in the transverse direction due to some prescribed axial motion in a stationary Timoshenko beam. Rather, it is a typical dynamic characteristic of a rotating beam only, where the interaction effect is directly proportional to the magnitude of rotational angular velocity Ω. In a stationary beam-column, theoretically in the absence of dynamic buckling, a pure axial impulse will not produce any lateral movement of the beam along the span.

In the majority of practical applications of a periodic tip-pulse problem of a cantilever beam, the blade is repeatedly deformed due to local interference with the outer case, when the tip clearance is closed. This problem is relatively more nonlinear than the classical periodic pulse-loading situation in the sense that it is a vibro-contact-impact deformation-controlled scenario rather than a known periodic-force-controlled dynamic response of the beam. In an interference-induced deformation-controlled tip-loading scenario, the magnitude of the external tip load is not known in priori; rather, the load is dependent on the amount of deformation. Being a transient contact load, the tip load may even become zero, if the dynamic response of the tip makes it to lose contact with the outer case. The current numerical scheme can handle both the interference-induced load and the longitudinal load momentarily going to zero due to contact separation.

7.3.1.5 Vibro-Impact Response Under Repeated Axial Impact

In the analytical discussion of the nonlinear transient response of a Timoshenko beam to a repeated pulse loading, we concentrate on the effects of the following three parameters:

i. Dynamic coefficient of friction at the contact surface between the tip and the outer case "μ".
ii. Magnitude of the peak pulse loading, i.e., "$F_{\max}$". The axial load at the blade tip is varied in steps such that $\bar{F}_{\max} = 1.0$, 2.0, 5.0, and 10.0. Figure 7.15 shows

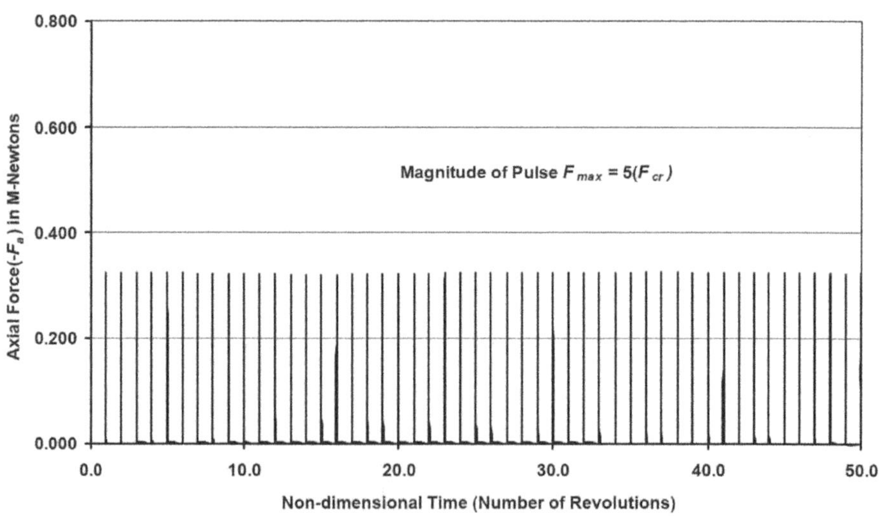

FIGURE 7.15 Time history of the axial load with repeated pulse loading (1 pulse/revolution, rotational velocity $\Omega = 300$ radians/s) with $\bar{F}_{max} = 5.0$.). (Reproduced with permission from Reference [1].)

the typical load history of the axial load $F(t)$ applied on the tip of the rotating beam, which clearly illustrates the characteristics of repeated pulse.

iii. Duration of rub (effect of increasing contact load time $\bar{t}_c$), keeping the peak axial load constant at $\bar{F}_{\max}=5$. In this example, we keep the dimensionless load parameter constant at $\bar{F}_{\max}=5$, which turns out to be $F_{\max}=0.3236\,\text{MN}$.

The resulting lateral deflection at the tip $\eta(L,t)$ is plotted against the nondimensional time, expressed as the number of revolutions for two values of the coefficient of friction, i.e., "μ"=0.0 (see Figure 7.16) and 0.2 (see Figure 7.17), respectively. A comparison of these two figures highlights the effect of nonlinearity in the dynamical system and leads to the following observations:

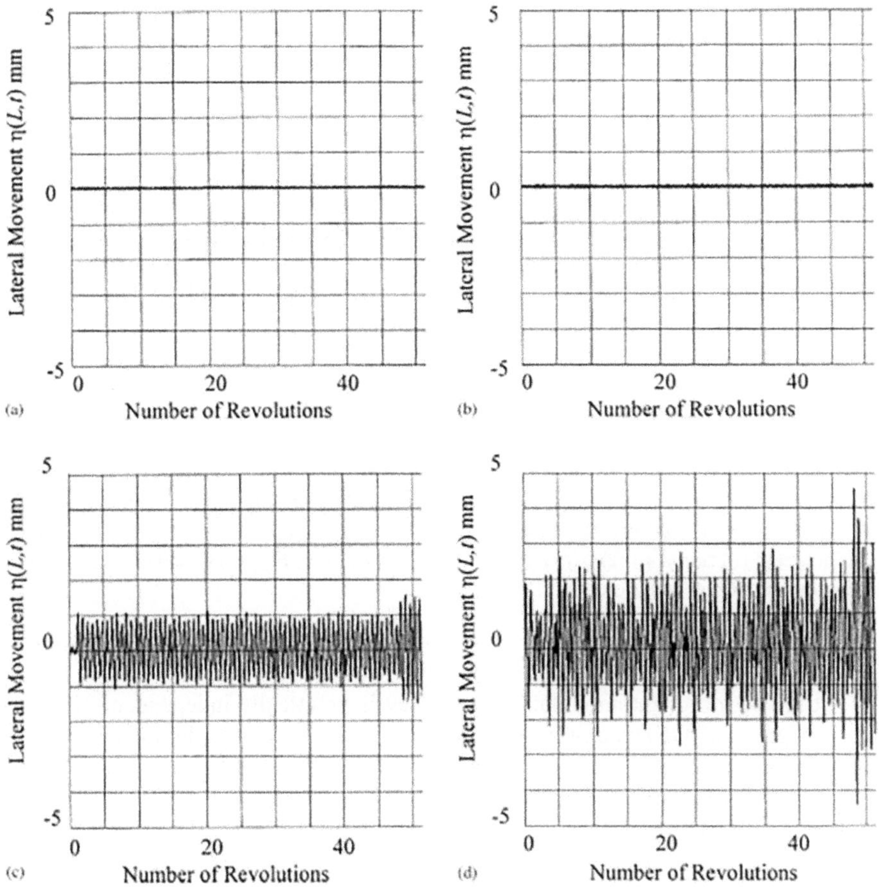

FIGURE 7.16 Transient response of the tip lateral deflection for repeated impulse (1 pulse per revolution, rotational velocity $\Omega=300$ radians/s) with increasing $\bar{F}_{max}=1.0, 2.0, 5.0, 10.0$ (coefficient of friction $\mu=0$): (a) $\bar{F}_{max}=1.0$, (b) $\bar{F}_{max}=2.0$, (c) $\bar{F}_{max}=5.0$, and (d) $\bar{F}_{max}=10.0$). (Reproduced with permission from Reference [1].)

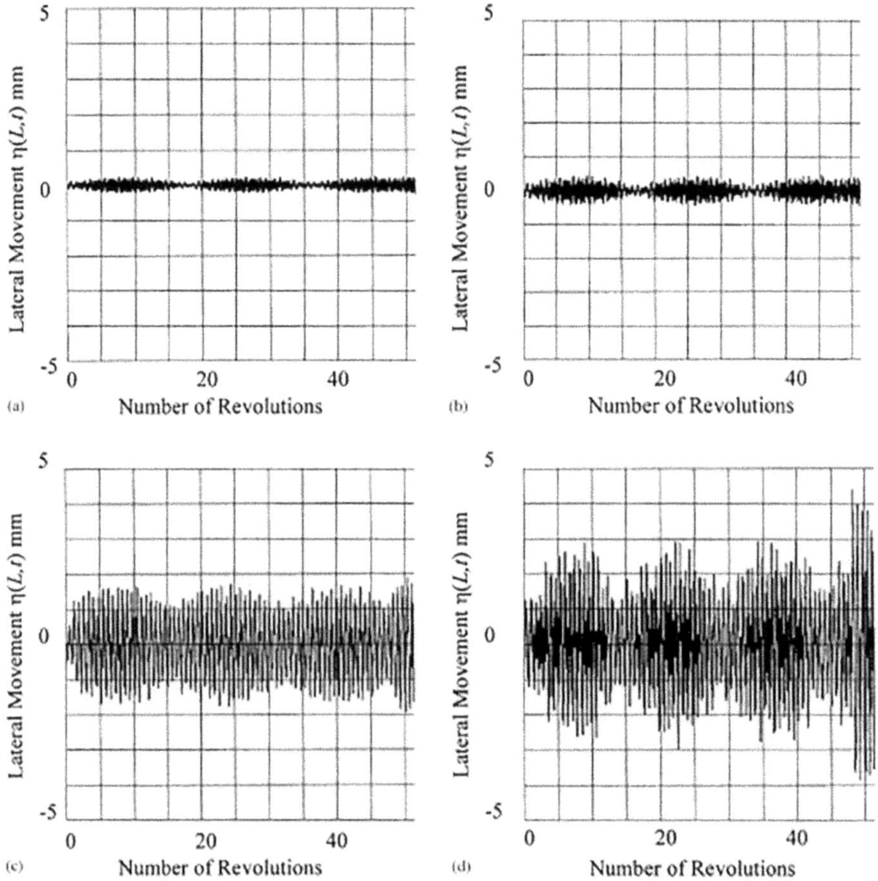

FIGURE 7.17 Transient response of the tip lateral deflection for repeated impulse (1 pulse per revolution, rotational velocity $\Omega=300$ radians/s) with increasing $\bar{F}_{max}=1.0, 2.0, 5.0, 10.0$ (coefficient of friction $\mu=0.2$):): (a) $\bar{F}_{max}=1.0$, (b) $\bar{F}_{max}=2.0$, (c) $\bar{F}_{max}=5.0$, and (d) $\bar{F}_{max}=10.0$.). (Reproduced with permission from Reference [1].)

a. Presence of friction at the blade-tip contact surface creates an amplitude-modulated response cycle for every 16 revolutions of the rotor. The time period of the modulation frequency is practically independent of the magnitude of the blade-tip axial load. In Figure 7.16, this modulation characteristic is completely absent, when the coefficient of friction $\mu=0$. However, after 48 revolutions, a bifurcation takes place and the dynamic behavior changes with the sudden jump in the amplitude of oscillations. This happens for all axial loads of magnitude $\bar{F}_{\max} \geq 5.0$. This chart shows that with a very low or no friction at all, for lower axial loads up to $\bar{F}_{\max} \geq 2.0$, there is practically no observable lateral motion of the blade tip, but for large axial loads $\bar{F}_{\max} \geq 10.0$, the response is extremely noisy and grows rapidly.
b. Figure 7.17 shows the response with coefficient of friction $\mu=0.2$. Here, with the increasing $\bar{F}_{\max}$, the modulation frequency remains constant for

every 16 revolutions of the rotor, but the peak transverse amplitude of the modulated response increases almost linearly with increasing $\overline{F}_{max}$ from ± 0.25 mm (for $\overline{F}_{max}=1$) to ± 2.90 mm (for $\overline{F}_{max}=10$). By harmonic balance, the typical transverse dynamic response with amplitude modulation for $\overline{F}_{max}=10$ is computed as follows:

$$\eta(L,t)\approx 2.90\,\sin(10t)\sin(320t)$$

i.e.,

$$\eta(L,t)\approx (2.90/2)\left(\cos(310t)-\cos(330t)\right) \tag{7.80}$$

Thus, the transient dynamic response in the lateral direction $\eta(L,t)$ with friction at the tip is a combination of two frequencies close to each other at 310 and 330 Hz. This particular amplitude-modulated response shows typical nonlinear characteristics caused by the presence of gyroscopic forces in the dynamical system. In an eigenvalue solution, it is characterized by two distinct roots of λ_i separated due to the presence of skew-symmetric terms in the coefficient matrix $[C]$ affecting the velocity-dependent forces in the system.

c. The timing of the bifurcation point appears to be a function of the load application time $\overline{t}_c$. For a given peak axial load $\overline{F}_{max}$, the contact load time $\overline{t}_c$ has the most significant effect in terms of complex nonlinear dynamic behavior. At some threshold value of $\overline{t}_c$, the response quickly becomes chaotic. The changing tangential tip lateral movements $\eta(L,t)$ are shown in Figure 7.18.

It can be seen that for the low values of $\overline{t}_c=0.01$ when the axial load is very close to a true impulse function, the dynamic response shows the characteristics of bifurcation. But, as the $\overline{t}_c$ increases, at first the magnitude of lateral tip displacement goes down from ± 1.0 to ± 0.5 mm. Then, it starts increasing with the motion becoming completely chaotic for $\overline{t}_c=0.2$. However, at $\overline{t}_c=0.5$, it becomes again periodic with an amplitude-modulated response frequency of 10 revolutions for each modulation cycle. Determination of threshold $\overline{t}_c$, keeping $\overline{F}_{max}$ constant, where the periodic motion becomes fully chaotic is an interesting research topic for further investigation.

The chaotic response of lateral deflection for varying contact-load time led us to look into the phase-plane trajectory plot in the axial direction (see Figure 7.19). The corresponding $\dot{\zeta}(L,t)$ versus $\zeta(L,t)$ plots are shown for $\overline{t}_c=0.01$, 0.1, 0.3, and 0.5, which illustrate the typical pattern. One can see the changing of two basins of attraction for $\overline{t}_c=0.01$ to six basins of attraction for $\overline{t}_c=0.5$. The other fact is also observed that while maximum axial deflection $\zeta(L,t)$ is peaking at 3 mm for all cases, the axial velocity is changing from 20 m/s on one extreme ($\overline{t}_c=0.01$) to 0.7 m/s ($\overline{t}_c=0.5$) on the other extreme. It is counterintuitive that for bigger impulse the peak axial velocity of the beam is actually lower than that for the lower impulse.

The section presents an analytical method to illustrate the nonlinear dynamic effect of repeated tip impact of a rotating Timoshenko beam, which simulates the intermittent rub of radial blades against the rigid outer case commonly occurring in a rotating machinery. Because the discontinuity of parameters affecting the coefficients

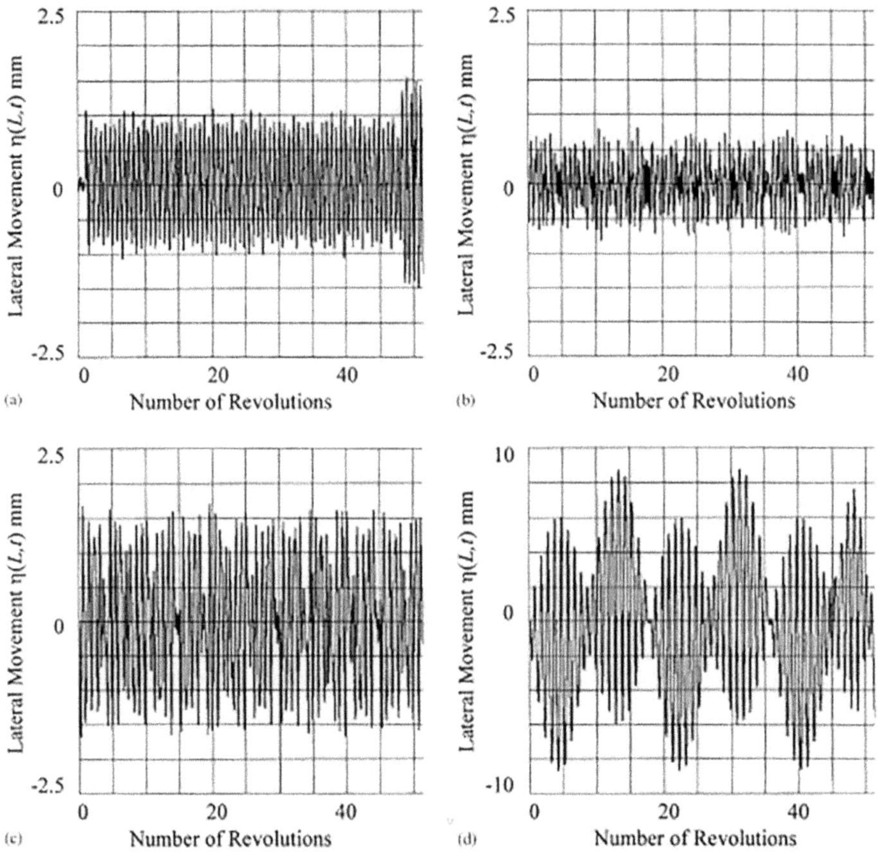

FIGURE 7.18 Transient response of the tip lateral deflection for repeated impulse (1 pulse per revolution, rotational velocity $\Omega = 300$ radians/s) with increasing contact load time $\bar{t}_c = 0.01$, 0.1, 0.3, 0.5 (coefficient of friction $\mu = 0$, $\bar{F}_{max} = 5.0$): (a) $\bar{t}_c = 0.01$ (circumf. 3.6°), (b) $\bar{t}_c = 0.10$ (circumf. 36°), (c) $\bar{t}_c = 0.30$ (circumf. 108°), and (d) $\bar{t}_c = 0.50$ (circumf. 180°). (Reproduced with permission from Reference [1].)

exists in the differential equations of vibro-impact systems, the transient response is more suitable for the exhibition of complicated dynamic behaviors than to do the traditional frequency domain eigenvalue analysis for the differential equations of the motion. The biggest drawback of any time-marching-forward explicit-integration scheme is the need to have a very small time-step size Δt to keep the solution process numerically stable, which makes the computation very expensive and extremely slow. In an impact and large-deformation problem, the difficulty of numerical instability is further complicated because the time-step size has to be adjusted as the event progresses. As the structure deforms more severely under impact loads, the numerically stable time-step size Δt keeps on getting smaller and smaller. In the present methodology, with the sixth-order Runge–Kutta–Fehlberg method, we have been successful in using a reasonable size time step, so that we can capture all the nonlinear characteristics such as bifurcation and chaotic behavior of a real-life multiple degrees-of-freedom dynamical system for 60 revolutions of the rotor. It is realized

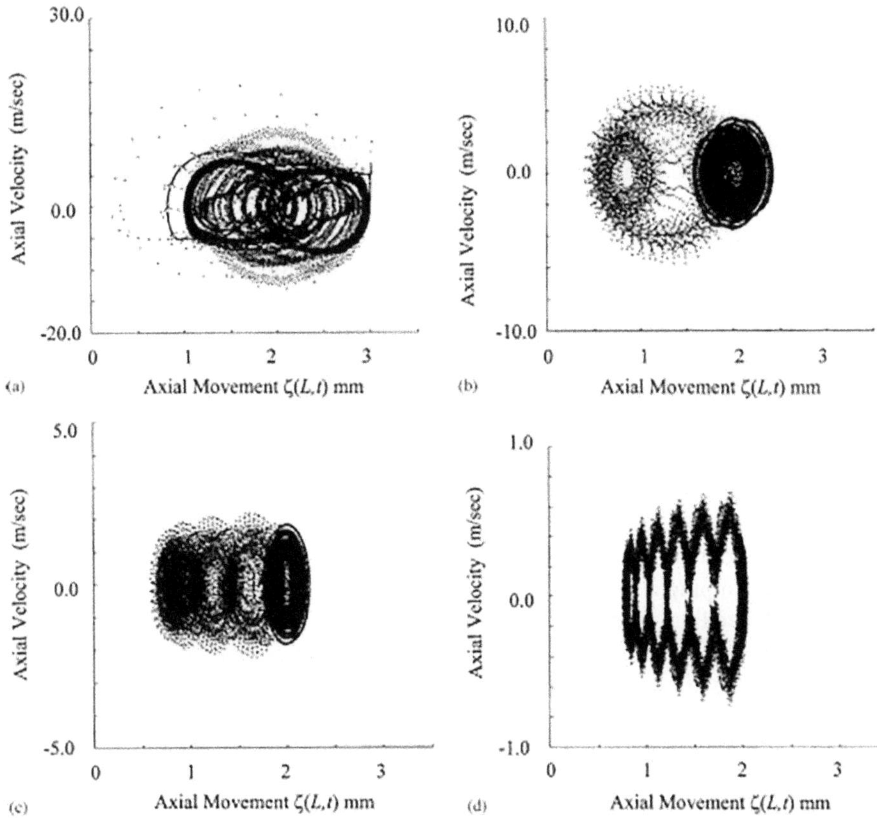

FIGURE 7.19 Phase plane trajectory for chaotic motion of the tip (at $s = L$) in the axial direction subjected to repeated impulse (1 pulse per revolution, rotational velocity $\Omega = 300$ radians/s) with increasing contact load-time $\bar{t}_c = 0.01$, 0.1, 0.3, 0.5 (coefficient of friction $\mu = 0$, $\bar{F}_{max} = 5.0$): (a) $\bar{t}_c = 0.01$ (circumf. 3.6°), (b) $\bar{t}_c = 0.10$ (circumf. 36°), (c) $\bar{t}_c = 0.30$ (circumf. 108°), and (d) $\bar{t}_c = 0.50$ (circumf. 180°). (Reproduced with permission from Reference [1].)

that the numerically integrated transient results of any theoretical investigation, even though it is stable, can be open for discussion until it can be verified with some other independent means. The only other way we can ensure the accuracy of any numerical scheme is to solve some simplified problem, for which case the closed-form solution is either known or it can be established by comparing the computed results with some test data. Right now, only a very limited amount of test data is available for comparison with the analytical model.

7.4 ANALYTICALLY PREDICTED TRANSIENT RESPONSE VS. TEST DATA

The materials covered in this section are based upon the author's original research reported in the May 2007 issue of the *Journal of Applied Mechanics*. For the benefit of the readers, some of the findings of that research are being reproduced here with the permission of ASME, as published in the journal.

To establish the accuracy of the current analytical model for its time-domain results, the numerically computed transient dynamic response of a typical high-pressure compressor blade is compared with the measured strain-gage data from a rig test [see Reference 11]. In this especially developed experimental rig, it was observed that during a controlled periodic rub scenario each rub-impact produced somewhat different transient dynamic characteristics than the one preceding the rub event until the rub-induced vibration of the blade reached a limit-cycle response under repeated rubs. For this part of the analysis using the sixth-order Runge–Kutta scheme, a direct-time integration of equations of motion defined in Eq. (7.27) is performed. The aspect ratio (L/c) of the test blade is 1.51745 with the span length L equal to 4 cm, and it is rotating at $N = 16{,}500$ rpm ($\Omega = 1727.775$ radians / s) with the tip tangential velocity of 406 m/s. The root radius of the blade airfoil is 19.5 cm, and the blade-tip radius $R = 23.5$ cm. This blade is made up of Inco718, and the other relevant dynamic characteristics used in this analysis are as follows:

Young's modulus $E = 200$ GPa
Poisson's ratio $\nu = 0.3$
Mass density $\rho = 7.833$ g/cc
Tip coefficient of friction $\mu = 0.02$
Dashpot damping at tip (nondimensional) $= 2.0\%$
Flexural bending mode fundamental frequencies (from the current analytical model)

	Stationary	Rotating
a. First flexural bending mode:	762.25 Hz	1112.46 Hz
b. Second flexural bending mode:	4735.59 Hz	5119.26 Hz

The radial lean angle is 0° for this blade, and the stagger angle is $\beta_r = \beta_R = -45°$. The bade tip rubs against a 72° circumferential rub zone with the contact-impact tip loading applied at the rate of one pulse per revolution with 0.1 mm radial interference. In the rig test, the 72° forced rub zone is created by inserting a partial sector of a circumferential shoe in the path of the moving blade tip. The transient analysis has been carried out for four repeated impacts. The measured dynamic data from a spanwise strain gage at the fillet of the airfoil root is compared with the numerically computed strain time history near the clamped end of the corresponding Timoshenko beam model using the current analytical technique. From the two sets of plotted data illustrated in Figure 7.5, it can be seen that the transient analytical results predict the dynamic characteristics and the resulting strain time history in the rubbing blade very well.

Numerically computed response shows highly nonlinear behavior of the airfoil root strains. However, in terms of frequency response, the analytical model responds at a slightly lower frequency than the test data. This can be attributed to the nonlinearity in the boundary conditions at the tip during the actual rub event. In the rig test, the tip is partially constrained during the rub, whereas in the analytical model it is considered free with rub-related forces as external loads on the system. In addition, both the analytical and the test data illustrate as to how the magnitude of the response builds up after the first rub, until it stabilizes after about the third rub. On this plot, the dynamic response of the blade as it passes through the 72° circumferential rub zone is shown by rectangular shaded areas with legends as the first, second, third, and fourth rubs. As the spinning blade tip comes out of the forced-rub zone, the extensional wave in the blade travels up-and-down its longitudinal axis with very high velocity, giving rise to large Coriolis forces, which are oscillatory in nature. Mathematically, this Coriolis force is a distributed load, which is represented by the terms such as $-2\rho A\Omega\cos\beta\dot{\zeta}$ shown in Eq. (7.42), and its value at the blade root is computed as the integrated sum over the span length of the blade given by

$$\text{Coriolis force at the airfoil root} = -2\rho A\Omega \int_0^L \cos\left(\beta_r + \beta' s\right)\dot{\zeta}\,ds \tag{7.81}$$

The transient characteristics of this Coriolis force are shown in Figure 7.20, and the magnitude of which at the airfoil root could be as high as 3200 G. For this test blade, whereas the first-flex bending mode frequency is 1600 Hz, the longitudinal wave frequency is about 33,400 Hz. These longitudinal stress waves, frequency of which is more than 20 times higher than the flexural bending waves, can generate intense heat at the mating surface of the blade root (dovetail) with the disk, which has been observed to result in local welding and fretting of the mating surfaces as well as severe bearing damage even after short-duration heavy rubs. It should be noted

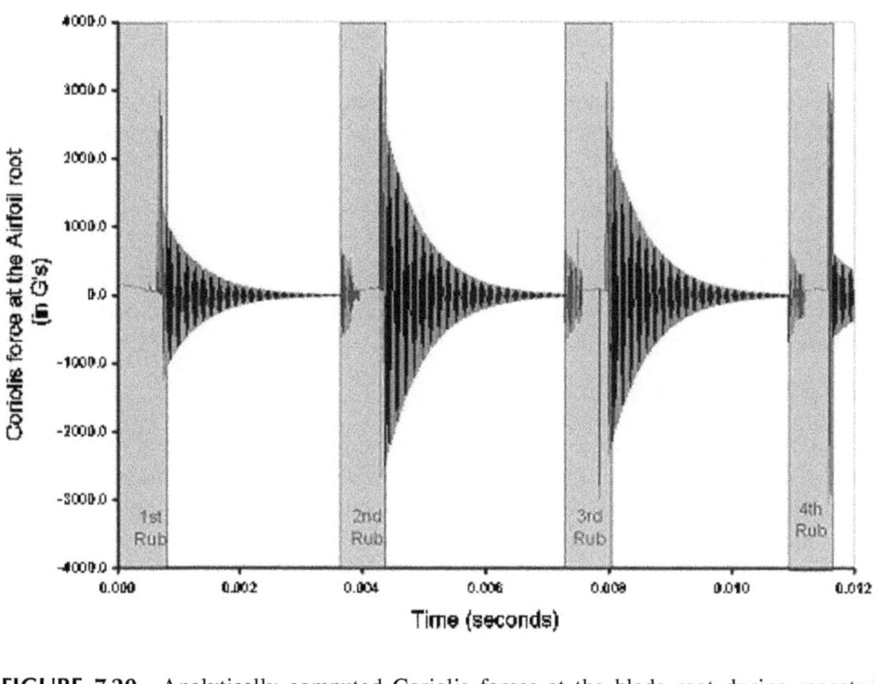

FIGURE 7.20 Analytically computed Coriolis forces at the blade root during repeated radial incursion of 0.1 mm at the blade tip with 72° circumferential rub zone for the first four rubs). (Reproduced with permission from Reference [2].)

that the decay in the magnitude of the Coriolis force, outside the imposed rub zone during the free vibration of the blade, is not due to damping in the system rather it is caused by the transfer of kinetic energy associated with the longitudinal motion into the lowest vibrational mode frequency, which invariably corresponds to its first flexural bending mode motion. The transient vibratory dynamic stresses in a rubbing airfoil are due to the interaction of longitudinal motion (hyperbolic wave) with the lateral motion (dispersive wave) of the beam, which in the case of rub-induced dynamic instability results in fatigue-type damage to the blade. Depending upon the eccentricity of the rub location at the blade-tip cross section, these rub-related damages can range from local tip curl due to plasticity to complete separation of the airfoil at the blade root.

7.4.1 Effect of Radial Incursion on Blade Tip Deformation

In this subsection, once the accuracy of the Timoshenko beam-column model has been established, we can use it confidently for some parametric studies about the blade tip deformation during rub. To that end, we introduce a radial incursion of a rub shoe in the path of the spinning blade. To keep the results applicable to any blade size, we nondimensionalize all the parameters related to length with respect to the reference "span length $L = 4$ cm" of the blade. For example, a nondimensional radial incursion $\overline{\delta}_r$ of 0.002 used in this sample study will amount for actual "radial incursion δ_r" determined as follows:

$$\delta_r = 0.002 \times 4 \text{ cm} = 0.008 \text{ cm} = 0.003149 \text{ inches} = 3.1 \text{ mils} \cong 3 \text{ mils}$$

Similarly, another nondimensional radial incursion value of 0.0067 used in this sample investigation will amount for actual "radial incursion δ_r" being computed as follows:

$$\delta_r = 0.0067 \times 4 \text{ cm} = 0.0268 \text{ cm} = 0.01055 \text{ inches} = 10.5 \text{ mils} \cong 10 \text{ mils}$$

We also introduce another parameter called nondimensional time $\overline{t}$ being defined as follows. Here, one unit of nondimensional time $\overline{t}$ = time in seconds to complete one revolution of the rotor, i.e.,

$$\text{Nondimensional time } \overline{t} = t\left(\frac{\Omega}{2\pi}\right)$$

In the above context, first we start with the analytically computed radial force, which the rub shoe needs to apply and push the tip inward in the longitudinal direction of the blade toward the engine axis of rotation. We make time-history plots of radial force shown in Figure 7.21 for two revolutions of the fan rotor under two varying amounts of nondimensional radial incursion $\overline{\delta}_r$ of the rub zone (0.002 and 0.0067). The analytical model predicts the peak magnitude of the radial force as follows:

a. For nondimensional $\overline{\delta}_r = 0.002$ $\rightarrow F_{max} = 1800$ Lbs $\cong$ 8007 Newtons
b. For nondimensional $\overline{\delta}_r = 0.0067$ $\rightarrow F_{max} = 7000Lbs\cong$31,137Newtons

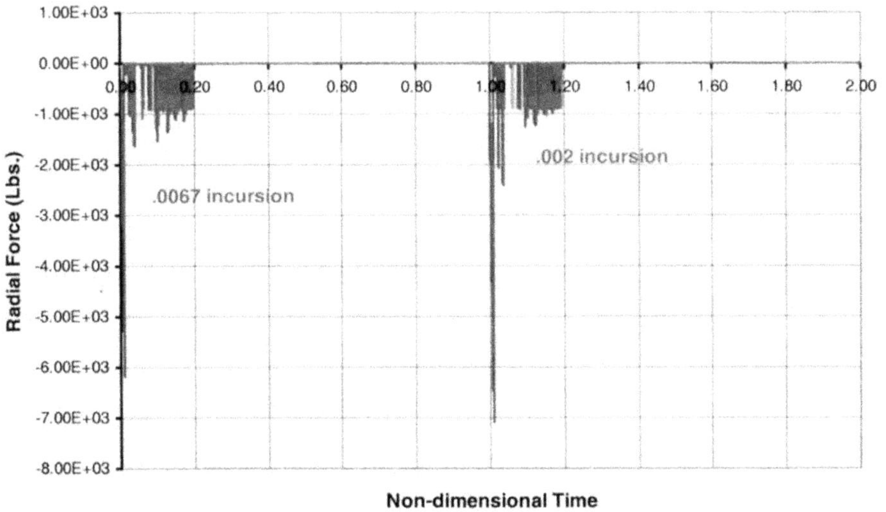

FIGURE 7.21 Comparison of analytically computed radial force at the blade tip due to rub at 16,500 rpm and subjected to radial incursions of $\bar{\delta}_r = 0.002$ and 0.0067 by a 72° circumferential rub zone.

It should be noted that the first peak of F_{max} is large, but it is also extremely so sharp that most of the measuring devices such as strain gage, load cells, or accelerometer may not be able to pick up the dynamic load of this magnitude.

Then, we make time history plots of nondimensional blade tip deformations in the longitudinal and lateral directions shown in Figures 7.22 and 7.23, respectively, for two revolutions of the fan rotor under two varying amounts of radial incursion of the rub zone.

A quick observation of the above plot highlights the following two characteristics of the rub event. First observation is that the blade-tip membrane compression relieves itself as the blade starts deflecting in the lateral direction due to dynamic flexural buckling. Another quick observation is that in a repeated rub sequence, each rub event is somewhat different from the previous one due to changes in the dynamic response of the blade.

Even in the same rub event, the magnitude of the lateral deflection is about 20 times larger than the peak longitudinal deflection. Figure 7.24 shows the same effect on lateral deflection magnitude for eight repeated rubs for three different radial incursions ($\bar{\delta}_r = 0.002$, 0.0067, and 0.0107).

It is observed from Figure 7.24 that peak lateral deflections of the rub blade take place in the sixth rub cycle for all three values of radial incursions $\bar{\delta}_r$ considered in this analytical study. The respective magnitudes of tip deformations are as follows:

a. For nondimensional $\bar{\delta}_r = 0.0020$ $\bar{\delta}_{\text{lateral}} = 0.030 = 1.20$ mm.
b. For nondimensional $\bar{\delta}_r = 0.0067$ $\bar{\delta}_{\text{lateral}} = 0.115 = 4.60$ mm.
c. For nondimensional $\bar{\delta}_r = 0.0107$ $\bar{\delta}_{\text{lateral}} = 0.141 = 5.64$ mm.

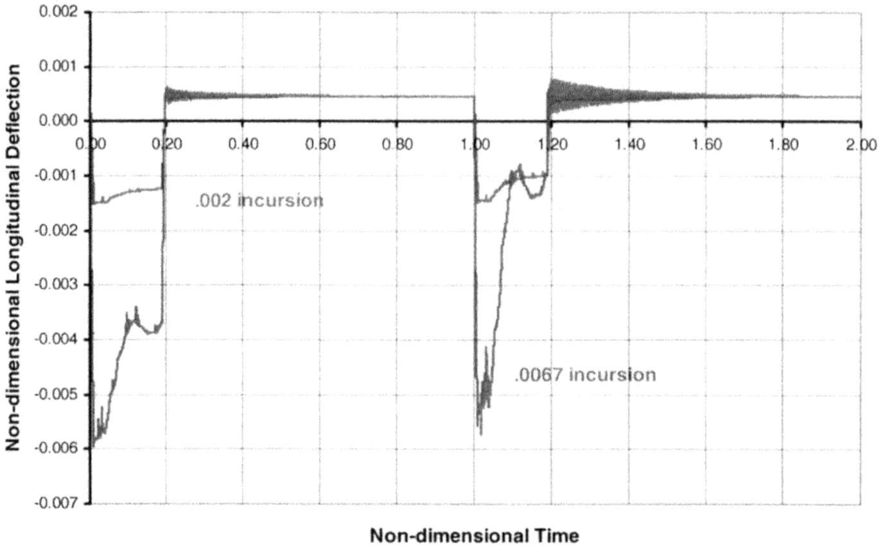

FIGURE 7.22 Comparison of analytically computed longitudinal deflection $(\delta)_{tip-longitudinal}$ time history of the blade-tip spinning at 16,500 rpm and subjected to radial incursions of δ_r = 0.002 and 0.0067 by a 72° circumferential rub zone.

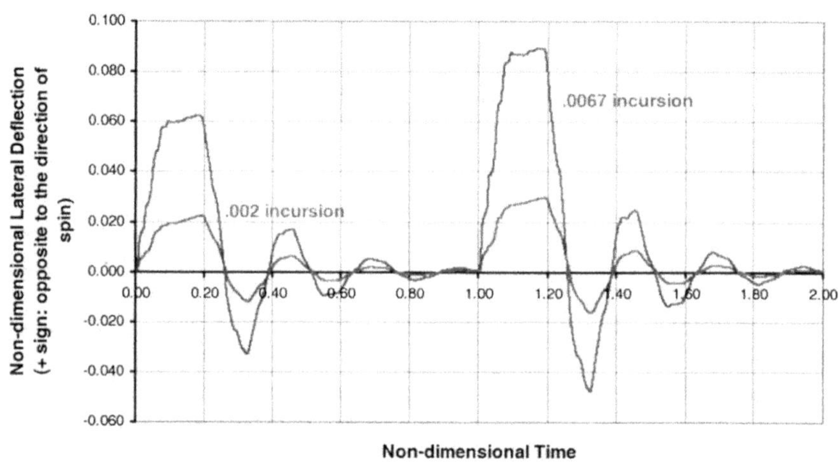

FIGURE 7.23 Comparison of analytically computed lateral deflection time history of the blade tip due to flexural bending spinning at 16,500 rpm and subjected to radial incursions of $\overline{\delta}_r$ = 0.002 and 0.0067 by a 72° circumferential rub zone.

It should be obvious that for a blade of span length of $L = 4$ cm $= 40$ mm, the lateral tip deflection of 5.64 mm is extremely huge and should be sufficient to plastically deform the tip of the blade in the form of a permanent tip curl. Now, we will study the transient dynamic stresses being developed during a repeated rub event.

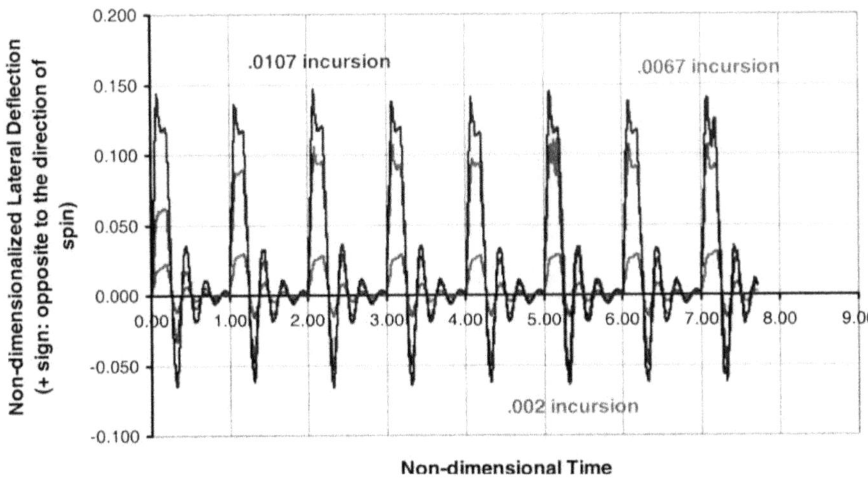

FIGURE 7.24 Comparison of analytically computed lateral deflection time-history of the blade tip due to flexural bending spinning at 16,500 rpm with eight repeated rubs (Radial incursion $\bar{\delta}_r$ = 0.002, 0.0067, and 0.0107).

With that objective in mind in Figure 7.25, we present a time-history plot of airfoil root stresses due to flexural bending during the rub event with blade spinning at 16,500 rpm for four repeated rubs (nondimensional radial incursion $\bar{\delta}_r = 0.002$). Based upon the numerical results of Figure 7.25, the peak value of bending stress near the airfoil root is $\sigma_{max} = 35$ ksi. Thus, one can linearly extrapolate for a quick estimate of the peak bending stresses at the airfoil root for some other values of radial incursion as shown below. Assuming that other impact and geometrical conditions remain identical, the peak flexural bending stresses during the rub event are as follows:

a. For nondimensional $\bar{\delta}_r = 0.0020$ $\quad \sigma_{max} = 35.0$ ksi.
b. For nondimensional $\bar{\delta}_r = 0.0067$ $\quad \sigma_{max} = 117.3$ ksi.
c. For nondimensional $\bar{\delta}_r = 0.0107$ $\quad \sigma_{max} = 187.3$ ksi.

It should be noted that the above peak stress values are only an estimate based upon the assumption of linear elastic deformation only. In reality, the contact-impact event especially with deep radial incursions is highly nonlinear in its dynamic response characteristics and with plastic deformation of tip curling including blade cracking failure may become inevitable.

7.5 EFFECT OF PRETWIST ON THE TIP RUB

In this section, we will apply the current analytical model to investigate the transient response of a twisted blade as opposed to a similar blade but without any pretwist. For this investigation, we will use the blade parameters the same as outlined in Section 7.3 with two different values of the total twist, that is $(\beta_R - \beta_r) = 0°$

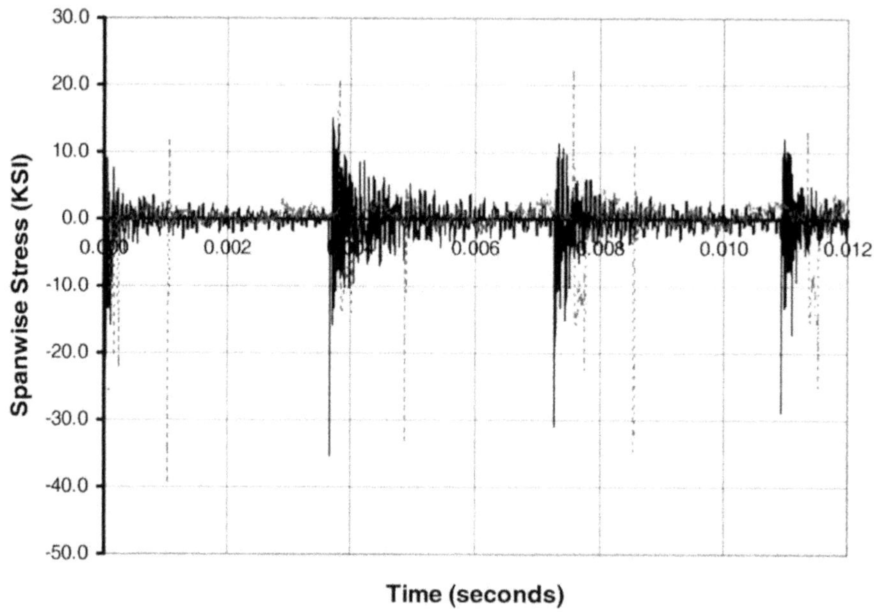

FIGURE 7.25 Analytically computed airfoil root stresses time history due to flexural bending during the rub event with blade spinning at 16,500 rpm for four repeated rubs (nondimensional radial incursion $\overline{\delta}_r = 0.002$).

and $(\beta_R - \beta_r) = -45°$. Here, we have compared the dynamic responses of these two blades, subjected to the same contact-impact loads during a typical rub event of one pulse per revolution, in terms of their nondimensional lateral tip deflection (η/L). The main difference in external load application between the results shown in the previous section and the current section is that the previous section results were generated for the rig-test conditions with displacement-controlled radial incursion of 0.1 mm, whereas in this section the results are for outer casing-imposed radial force of $F_{max} = 0.1$ times of the Euler critical buckling load applied at the free end of the beam. The numerical technique to implement the controlled radial incursion or longitudinal pulse of controlled magnitude in a transient simulation has been discussed in detail in the author's previous work [see Reference 1] on the related topic. The respective results of the lateral tip deflection (η/L) are plotted in Figure 7.26.

These time-history plots up to 25 ms clearly show that for a periodic contact-impact load of the same magnitude, the response of an untwisted blade is monotonically increasing, whereas for blade twisted at 45° the dynamic response shows a beating pattern. The beating pattern of a quasi-periodic nature indicates that this dynamic system is responding simultaneously at two different frequencies, which are very close to each other. It is observed that for both twisted and untwisted blades, the dynamic response until the fourth contact-impact pulse is almost identical. In addition, after each pulse loading, the twisted blade responds at a slightly higher frequency than an untwisted blade.

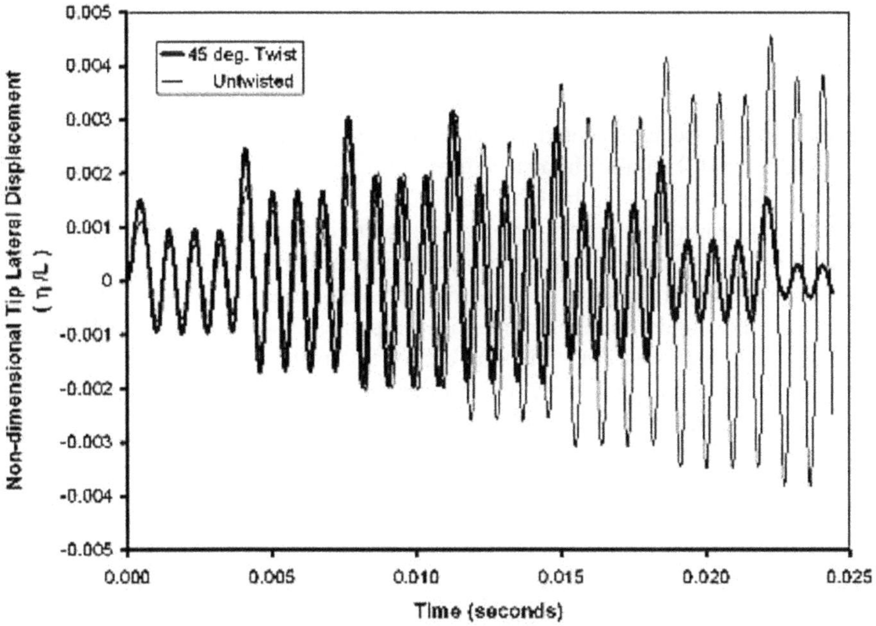

FIGURE 7.26 Comparison of analytically computed transient lateral tip displacements for an untwisted beam (—-------) versus a 45° twisted beam (...........) during repeated rubs (one pulse per revolution) at the blade tip with a periodic contact force of magnitude $F_{max} = 0.1 \times F_{cr}$). (Reproduced with permission from Reference [2].)

7.6 DYNAMICS OF THE PRETWISTED RECTANGULAR PLATE

The topics covered in this chapter clearly illustrate that there is a need to do more rigorous research to fully understand the complex physics of this vibro-impact problem. Until now for all our tip-rub discussion purposes, the blade has been modeled as a pretwisted Timoshenko beam. In this section, we will refine the blade model to consider a more realistic and relatively advanced blade modeling methodology. One of the major shortcomings of the beam formulation has been its inability to predict a very important blade vibrational mode associated with the chordwise bending (1EB) also called 2-stripe mode or coupled bending–bending vibrational mode. The coupling of spanwise bending with the chordwise bending is an important effect that a beam model will not be able to handle. However, blade designers have recognized it for a long time that the pretwist in the airfoil affects the chordwise bending characteristics of the blade in a profound way and therefore cannot be ignored. As a result, researchers have recently tried to apply advanced plate and shell theory in formulating the dynamic behavior of twisted airfoil as well, which also has the tendency to introduce untwisting of the airfoil cross section during spinning. The present discussion is an extension of the author's previous research [see Reference 2] on a similar topic, which dealt with the twisted-beam derivation versus twisted-plate formulation considered in the present work. In the current paper, a first attempt has been made to consider the contribution of the warping terms in von Karman's strain–displacement formulation and to couple them with the dynamical equations, which takes place due to the coupling of the twist and bending of a pretwisted plate of noncircular cross section. Warping deflection during the twist of a noncircular cross-sectional beam is a well-established effect in elastic deformation; however, its application in plate formulation has not been presented in any book until now. Furthermore, consideration of any additional flexibility in a structure will inherently bring down its computed natural frequency from the baseline value. This discussion quantifies the warping effect in reducing the natural frequency in a pretwisted plate with and without the centrifugal force field for a wide range of aspect-ratio configurations.

In the present formulation (see Figure 7.22), the shape of the blade is approximated by a rectangular plate of an identical in-plane aspect ratio (L/C) and equivalent cross-sectional area, "A". The one end of this plate is clamped to a rigid hub of radius "r"with a stagger angle (twist angle at the blade root also called "Setting angle" in literature) of θ_r and the free end is located at an outer radius of "R", making it behave like a cantilever beam in the radial direction of the rotating "tangential–axial–radial" coordinate system $(\hat{\boldsymbol{e}}_t, \hat{\boldsymbol{e}}_a, \hat{\boldsymbol{e}}_r)$ attached to the engine centerline. Here, $\hat{\boldsymbol{e}}_t$ points in the direction of rotation, $\hat{\boldsymbol{e}}_a$ is along the engine axis, and $\hat{\boldsymbol{e}}_r$ coincides with the span direction of the blade. For the mathematical derivation purposes, the plate shown in Figure 7.22 has a special orientation of setting angle θ_r being zero such that the local "thin shell" plate coordinate system $(\hat{\boldsymbol{e}}_{x0}, \hat{\boldsymbol{e}}_{y0}, \hat{\boldsymbol{e}}_{z0})$ at the blade root relates to the previously defined "tangential–axial–radial" $(\hat{\boldsymbol{e}}_t, \hat{\boldsymbol{e}}_a, \hat{\boldsymbol{e}}_r)$ system as $\hat{\boldsymbol{e}}_t = -\hat{\boldsymbol{e}}_{z0}$, $\hat{\boldsymbol{e}}_a = \hat{\boldsymbol{e}}_{y0}$, and $\hat{\boldsymbol{e}}_r = \hat{\boldsymbol{e}}_{x0}$. This orientation, and the setting angle θ_r being zero, is consistent with the majority of turbofan blade designs, toward which this discussion is mainly focused. However, for some other applications, such as helicopter blades, it may be more convenient to have the setting angle be 90°.

Per Reference [12], the dynamic characteristics of a twisted blade are also affected by the contributions of the rotary inertia terms as well as the warping inertia terms depending upon the amount of pretwist present in the typical noncircular airfoil cross section. It enables us to write the complete twisted blade dynamic equilibrium equation with a radial tip-rub load F_R at the cantilevered free end as follows:

$$D\nabla^4 w + D(\theta')^2\left[3\frac{\partial^2 w}{\partial x^2} + \frac{\partial^2 w}{\partial y^2}\right] - D(\theta')^4\left[x^2\frac{\partial^2 w}{\partial x^2} + \frac{(1-\nu)}{2}x^2\frac{\partial^2 w}{\partial y^2} + 2x\frac{\partial w}{\partial x}\right]$$

$$-\frac{F_R}{C}\left[\frac{\partial^2 w}{\partial x^2} - (\theta')^2 w\right] - h\frac{\partial(\sigma_{cf})_x}{\partial x}\frac{\partial w}{\partial x} - h(\sigma_{cf})_x\frac{\partial^2 w}{\partial x^2} - h(\sigma_{cf})_y\frac{\partial^2 w}{\partial y^2} -$$

$$\rho h\Omega^2\cos^2(\theta_r + \theta' x)w$$

$$= -\rho h\frac{\partial^2 w}{\partial t^2} + \rho\frac{h^3}{12}\frac{\partial^2}{\partial t^2}\left(\frac{\partial^2 w}{\partial x^2} + \frac{\partial^2 w}{\partial y^2}\right) - \rho h(\theta' x)^2\left[\frac{\partial^2 w}{\partial t^2} - \nu\left(\frac{Cy}{L}\right)^2\frac{\partial^4 w}{\partial y^2\partial t^2}\right] + q(x,y,t)$$

where

$$D = \frac{Eh^3}{12\left(1-\nu^2\right)} \tag{7.82}$$

In the above Eq. (7.82), the right-hand-side terms contain multiple inertia terms representing three different forms of inertia effects being considered in the current formulation, viz.,

a. Inertia due to transverse or lateral motion of the plate: $-\rho h \ddot{w}$

b. Rotary inertia due to cross-sectional rotation: $\left(\frac{\rho h^3}{12}\right)\left(\frac{\partial^2 \ddot{w}}{\partial x^2}+\frac{\partial^2 \ddot{w}}{\partial y^2}\right)$

c. Warping inertia due to pretwist in the plate: $-\ \rho h(\theta' x)^2\left[\ddot{w}-\nu\left(\frac{Cy}{L}\right)^2\frac{\partial^2 \ddot{w}}{\partial y^2}\right]$

As per the method described in earlier sections, the partial differential equation outlined in Eq. (7.82) is converted into a matrix form. Similar to Galerkin's method described earlier for coupled partial differential equations (Eqs. 7.15a–7.15c), the previous equation (7.67) is converted into a matrix form of vibrational equation with terms like $[\boldsymbol{M}]$ and $[\boldsymbol{K}]$ matrices as shown below. Similar to Eq. (7.57), we obtain the following eigenvalue formulation, which yields the following homogeneous equation:

$$[\boldsymbol{M}]\left\{\ddot{f}(t)\right\}+[\boldsymbol{K}]\left\{f(t)\right\}=\{0\} \tag{7.83}$$

For detailed results of the current plate model, the reader is referred to the author's original research paper of 2011 published in the *Journal of Sound and Vibration* [see Reference 12]. For the benefit of the reader and to demonstrate the accuracy of the plate model, some of the results of that study are being reproduced here with the permission of the publisher of the journal.

7.6.1 Finite-Element Results from the ANSYS® Versus Twisted Plate Model

To demonstrate the numerical accuracy of the current analytical model, first we will consider a set of twisted blades of rectangular cross section, which can be classified as "thin plates" with the ratio "C/h" being 20. For this purpose, we have developed two sets of finite-element models on a parallel basis with chord length $C = 25.4$ cm, thickness $h = 1.27$ cm, and the ratio L/C being 3 and 1, respectively. Following the modal analysis approach, the first six eigenvalues are computed using the commercially available code called ANSYS®. The reader is advised to make a note of a slight notation change, when we are discussing the results of a plate model being presented in the current section vs. the beam model results being presented until now. These notation changes are as follows:

a. Nondimensional frequency parameter ξ in Eq. (7.66) for Timoshenko-beam model is defined as

$$\xi = \omega_n L^2 \sqrt{\frac{\rho A}{EI}} \tag{7.84}$$

whereas, in the current pre-twisted rectangular plate formulation, we introduce a similar nondimensional frequency parameter called $\bar{\omega}$, which is defined as

$$\bar{\omega} = \omega_n \sqrt{\frac{\rho h L^4}{D}}$$

where

$$D = \frac{Eh^3}{12\left(1-\nu^2\right)} \tag{7.85}$$

b. Until now, in the Timoshenko-beam model we have used β_r and β_R as the twist angles of the beam cross section at the airfoil root radius r and at the fan blade outer radius R, respectively.

Whereas, in the current pretwisted rectangular plate formulation, we introduce a similar twist angle parameter called θ, such that θ_r and θ_R are the twist angles of the beam cross section at the airfoil root radius r and at the fan blade outer radius R, respectively (see Figure 7.27). Thus, we can write it as follows:

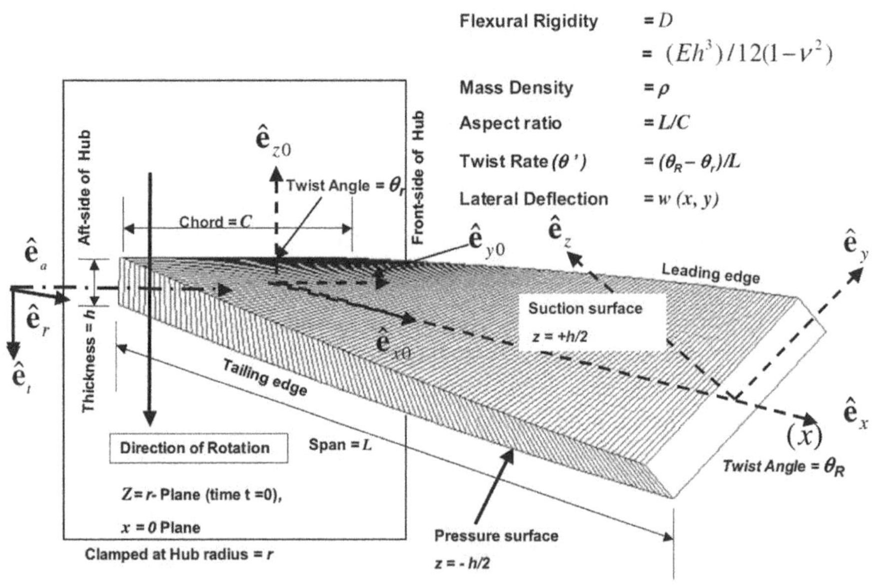

FIGURE 7.27 A typical pretwisted cantilever rectangular plate to simulate an airfoil blade and the respective thin local plate coordinate system (plate orientated with root stagger angle of $\theta_r = 0$). (Reproduced with permission from Reference [12].)

$$\text{Total twist angle in the rectangular plate} = (\theta_R - \theta_r) \text{ radians} \quad (7.86)$$

For checking the accuracy of the current analytical formulation, the twisted plate with an aspect ratio of $L/C=3$ and $C/h=20$ is a very important case, as Kielb et al. [see Reference 13] have provided the best set of test data for the first three modes (1F, 1T, and 3F) in the published literature. The ANSYS® models are made up of exact cube-shaped solids with eight-noded brick elements. Using an automatic mesh generator, a new mesh of an identical number of nodes and elements can be generated very quickly for a prespecified total twist value of $(\theta_R - \theta_r) = 0°, 15°, 30°, 45°$, and 60°, respectively. For validating the model, the nondimensional frequency parameters $\bar{\omega}$ coming out of ANSYS, as well as the test data from Kielb's paper [see Reference 13], have been plotted side-by-side and are shown in Figure 7.28.

The corresponding analytical results from the current solution for the first five modes (1F, 2F, 1T, 3F, and 2T) are shown with continuous solid lines in Figure 7.23. The test data has been shown with dotted lines and the respective ANSYS® results can be identified by a discrete set of filled circles. The corresponding ANSYS® generated first six-mode shapes are shown graphically in Figure 7.24.

As is evident from these graphical illustrations of the nondimensional frequency parameters $\bar{\omega}$, the results of the current analytical model are shown to capture the data trend for all the first five modes (1F, 2F, 1T, 3F, and 2T) of the plate vibration very well. The majority of previously published twisted beam or plate model results start showing significant deviation and scatter in trends for anything beyond the first two vibrational modes [see Reference 13]. In fact, the numerical values of the nondimensional frequency parameter $\bar{\omega}$ for the first two fundamental modes 1F and 1T, computed by the current analytical model, are in much closer agreement with Kielb's

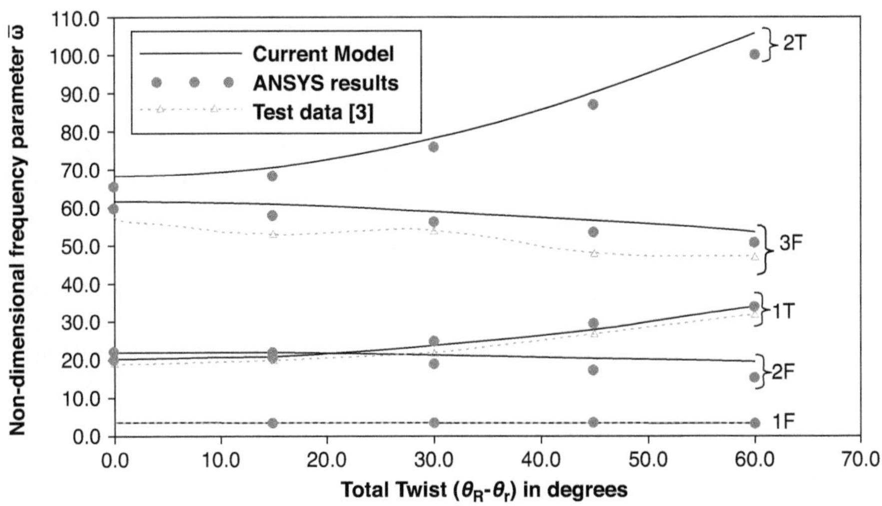

FIGURE 7.28 First five nondimensional frequency parameters $\bar{\omega}$ for cantilever plates with $L/C = 3$, $C/h = 20$ and varying total twist angle $(\theta_R - \theta_r)$. (Reproduced with permission from Reference [12].)

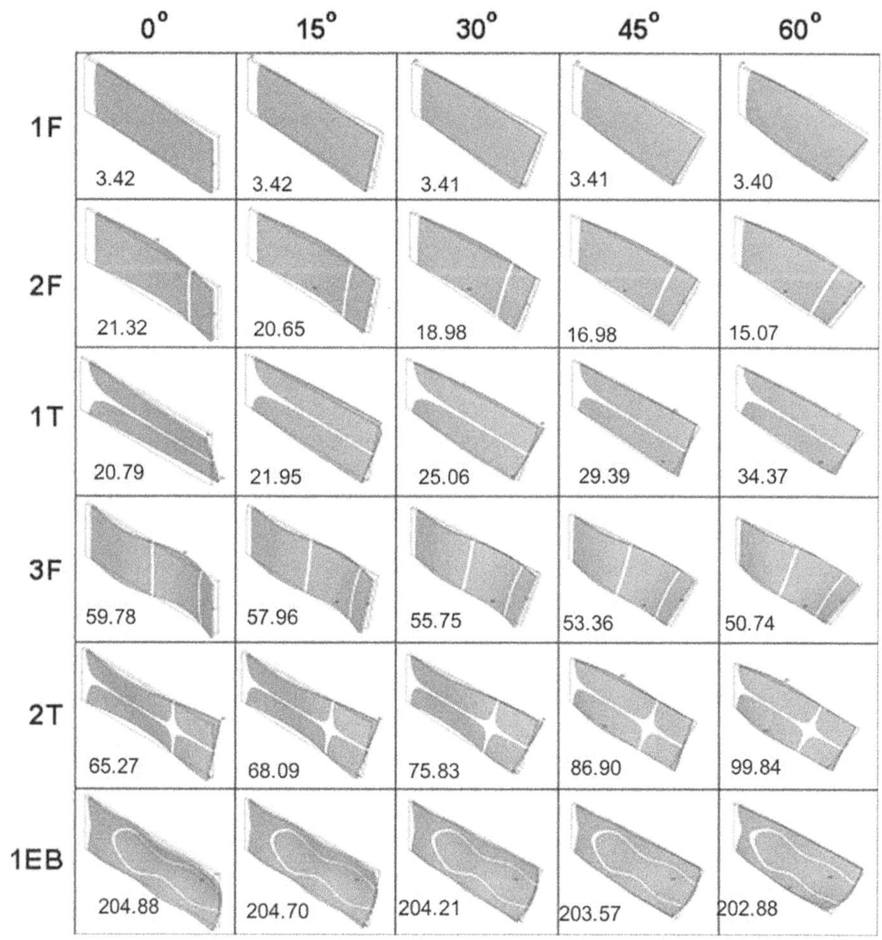

FIGURE 7.29 ANSYS® generated first six mode shapes and nondimensional frequency parameters $\bar{\omega}$ for cantilever plates with $\boldsymbol{L/C} = 3$, $\boldsymbol{C/h} = 20$ and varying total twist angle $(\theta_R - \theta_r) = 0°$, 15°, 30°, 45°, and 60°). (Reproduced with permission from Reference [12].)

actual test data than those produced by the detailed ANSYS® finite-element model. This is the first time in the published literature that a twisted blade analytical model has produced such a consistent set of results without any contradictions (Figure 7.29).

To quantify the accuracy of the $\bar{\omega}$ computed by the current analytical technique, numerical results obtained from the model are compared with Kielb's published test data in Table 7.6.

The summarized results in a tabulated form include Kielb and his coworkers' experimental values [see Reference 13] limited to three vibrational modes, viz., 1F, 1T, and 3F, which are the only available test data in their 1985 paper. The computed difference of $\bar{\omega}$ shown in Table 7.6 highlights an overall good agreement between Kielb's experimentally measured data versus the analytically computed results from the physics-based dynamical model for the entire range of twist angles. For the first

TABLE 7.6
Nondimensional Frequency Parameters $\bar{\omega}$ for the First Three Modes in the Case of a Thin Rectangular Plate ($L/C = 3$, $C/h = 20$, and $\nu = 0.3$) with Varying Twist Angles $(\theta_R - \theta_r) = 0°$, 15°, 30°, 45°, and 60°

Mode	Comparison with Kielb's Ref. 13	Total Twist $(\theta_R - \theta_r)$ 0°	15°	30°	45°	60°
1F	Current	3.51606	3.57503	3.64323	3.55711	3.26543
	Ref. 13	3.31000	3.35000	3.40000	3.40000	3.26000
	Difference	6.2%	6.7%	7.2%	4.6%	0.2%
1T	Current	20.13806	20.93168	23.29004	27.19427	32.69526
	Ref. 13	19.53000	20.50000	23.00000	27.50000	32.00000
	Difference	3.1%	2.1%	1.3%	1.1%	2.2%
3F	Current	61.69397	61.02262	59.19023	56.60553	53.67544
	Ref. 13	57.10000	53.20000	54.20000	48.50000	47.00000
	Difference	8.0%	14.7%	9.2%	16.7%	14.2%

two fundamental modes 1F and 1T commonly used for benchmarking all twisted beam and plate analyses, the largest difference is within 6.7%, and for the 3F bending mode the highest difference is 16.7%, which is considered reasonably good with respect to the results of many other previous models.

With this discussion, we have shown that the Timoshenko-beam model analysis procedure of coupling aspects of the longitudinal component of deformation considered in Section 7.2 can also be extended to the pretwisted plate model. The implementation of this methodology involves a relatively complex mathematical formulation. However, that model is expected to provide better insight and more accurate answers to the highly complex physics of blade tip rub inside a deformable containment structure.

7.7 ADVANCEMENTS IN BLADE TIP-RUBBING ANALYTICAL MODELING

The topics covered in this chapter clearly illustrate that there is a need to do more rigorous research in order to fully understand the complex physics of this vibro-impact problem. Since the publication of original groundbreaking research work of the author, discussed in this chapter, there have been some recent advancements in this field. Here is a comprehensive summary of some recent contributions of other follow-up research in the area of tip-rub numerical simulations. Specifically, the following observations require further investigations:

a. *Rubbing blade interaction with the rotor shaft at the root and the casing at the tip:*
   During the tip-rub event, a flexible blade deforms at the tip under the effect of the rub forces, and eventually following the load path, the dynamic loads

get transferred to the rotor shaft and to the engine structure. This is a very complex, highly nonlinear, transient dynamic event, and the entire rotor system can become unstable due to multiple contact locations in the load path. This contact effect on the rotor stability was first studied by Lesaffre et al. in 2007 [see Reference 14]. In 2016, Ma and his colleagues [see Reference 15] published a very comprehensive set of transient dynamics results considering all the complexities of contact interactions in their mathematical model. Later in 2019, Ivanov et al. [see Reference 16] considered the effect of the same set of parameters in their reduced finite element model to analyze the dynamic response of a load reduction device during the Fan Blade out event. In 2020, Guo and his colleagues at Northeastern University, Shenyang facility [see Reference 17] extended the previous modeling effort by considering more parameters affecting the rub.

b. *The blade-tip interactions with the abradable material.*
During the last 20 years, this interaction effect has been studied as a separate discipline both experimentally and theoretically. Most of the experimental studies have been carried out at the Ohio State University Gas Turbine Laboratory facility. One notable reference in this field is the January 2011 *ASME Journal of Turbomachinery* paper published by Padova, Dunn, and his team [see Reference 18]. They published several follow-up papers investigating different aspects of tip rub. French researchers led by Goldspiegel and his team [see Reference 19] gave a theoretical basis for blade vibration build-up during the repetitive rub with abradable coating material. In 2024, Zhang and his team [see Reference 20] studied the rub-induced vibration in their investigation to get a better insight into the role of flexibility during the rub interaction event between the coated casing and the blade tip. To investigate the characterization of the blade-coating contact force, Lin et al. [see Reference 21] established a dissipative contact force model considering plastic deformation based on the principles of energy conservation and momentum conservation. Thiery and Chandran [see Reference 22] developed a simplified bladed rotor model using a Jeffcott rotor with blades represented as damped elastic pendulums. This is a relatively simple model, which may be helpful for doing some quick parametric studies related to blade tip rub.

c. The intermittent rub may not be a periodic event, rather it is a random event inside the containment structure. Another field which has recently got some attention is to determine the effect of crack initiation due to tip rub. To that end, the recent paper by Shen and his team [see Reference 23] studied the dynamic characteristics of blade-casing rub-impact coupling fault considering breathing crack. Yu and his team [see Reference 24] carried out the dynamic modeling and failure analysis for a compressor blade in aero-engine undergoing blade-casing rubbing events.

d. Another curious observation is that the peak axial velocity is smaller for a larger impulse and larger for small impulse, which raises obvious questions about the reasons for this interesting phenomenon. Further work is needed to make sure that it is not a numerical artifact, and if what is the physics behind this.

The current analytical methodology is limited to a constant rotational speed, but it can be extended to a nonconstant angular velocity. Furthermore, to make the results more realistic and useful in design applications, the effect of damping on the transient dynamic response should also be investigated. Analytically predicted magnitude of undamped response may make numerical results to appear much worse and severe than what the actual test might show and vice versa. It can guide the implementation of mitigation solutions to prevent worst-case failure scenarios and help perform adjustments to either operational or design parameters.

REFERENCES

1. S. K. Sinha (2005) "Non-linear Dynamic Response of a Rotating Radial Timoshenko Beam with Periodic Pulse Loading at the Free-End", *International Journal of Non-linear Mechanics*, Vol. 40, No. 1, pp. 113–149. https://doi.org/10.1016/j.ijnonlinmec.2004.05.019
2. S. K. Sinha (2007) "Combined Torsional-Bending-Axial Dynamics of a Twisted Rotating Cantilever Beam with Contact-Impact loads at the free End", *Journal of Applied Mechanics*, Vol. 74, pp. 505–522. https://doi.org/10.1115/1.2423035
3. S. Timoshenko, D. H. Young and W. Weaver (1974) *Vibration Problems in Engineering*, Fourth Edition, Wiley, New York, pp. 432–434.
4. L. Meirovitch (1989) *Analytical Methods in Vibrations*, The Macmillan Company, Collier-Macmillan Limited, London, pp. 225–235.
5. H. E. Lindberg and A. L. Florence (1987) *Dynamic Pulse Buckling: Theory and Experiment*, Martinus Nijhoff Publishers, Dordrecht, pp. 112–140.
6. T. N. Farris and J. F. Doyle (1989) "Wave Propagation in a Split Timoshenko Beam", *Journal of Sound and Vibration*, Vol. 130, No. 1, pp. 137–147. https://doi.org/10.1016/0022-460X(89)90524-5
7. V. V. Bolotin (1964) *The Dynamic Stability of Elastic Systems*, Holden-Day, San Francisco, pp. 272–278.
8. D. Zhou and Y. K. Cheung (2001) "Vibrations of Tapered Timoshenko Beams in Terms of Static Timoshenko Beam Functions", *ASME Journal of Applied Mechanics*, Vol. 68, No. 4, pp. 596–602. https://doi.org/10.1115/1.1357164
9. S. K. Sinha (1989) "Dynamic Stability of a Timoshenko Beam Subjected to an Oscillating Axial Force", *Journal of Sound and Vibration*, Vol. 131, No. 3, pp. 509–514. https://doi.org/10.1016/0022-460X(89)91010-9
10. B. T. Smith, J. M. Boyle, J. J. Dongarra, B. S. Garbow, Y. Ikebe, V. C. Klema and C. B. Moler (1976) *Matrix Eigensystems Routines—EISPACK Guide*, Second Edition, Springer, New York, pp. 2–21.
11. C. Padova, J. Barton, M. G. Dunn, S. Manwaring, G. Young, M. Adams and M. Adams. (2005) "Development of an Experimental Capability to Produce Controlled Blade Tip/Shroud Rubs at Engine Speed", *ASME Journal of Turbomachinery*, Vol. 127, No. 4, pp. 726–735. https://doi.org/10.1115/1.1934429
12. S. K. Sinha and K. E Turner (2011) "Natural Frequencies of Pre-twisted Blade in a Centrifugal Force Field", *Journal of Sound and Vibration*, Vol. 330, No. 11, pp. 2655–2681.https://doi.org/10.1016/j.jsv.2010.12.017
13. R. E. Kielb, A. W. Leissa and J. C. MacBain (1985) "Vibrations of Twisted Cantilever Plates—A Comparison of Theoretical Results", *International Journal for Numerical Methods in Engineering*, Vol. 21, pp. 1365–1380. https://doi.org/10.1002/nme.1620210802

14. N. Lesaffre, J.-J. Sinou and F. Thouverez (2007) "Contact Analysis of a Flexible Bladed Rotor", *European Journal of Mechanics, A/Solids*, Vol. 26, No. (3), pp. 541–557. https://doi.org/10.1016/j.euromechsol.2006.11.002
15. H. Ma, F. Yin, Z. Wu, X. Tai and B. Wen (2016) "Nonlinear Vibration Response Analysis of a Rotor-Blade System with Blade-Tip Rubbing", *Nonlinear Dynamics*, Vol. 84, No. 3, pp. 1225–1258. https://doi.org/10.1007/s11071-015-2564-5
16. I. Ivanov, V. Myasnikov and B. Blinnik (2019) "Study of Dynamic Loads Dependence on Aircraft Engine Mount Variant after Fan Blade-out Event", *Vibroengineering Procedia*, Vol. 26, pp. 1–6. https://doi.org/10.21595/vp.2019.20800
17. X. Guo, J. Zeng, H. Ma, C. Zhao, X. Yu and B. Wen (2020) "A Dynamic Model for Simulating Rubbing Between Blade and Flexible Casing", *Journal of Sound and Vibration*, Vol. 466, 115036. https://doi.org/10.1016/j.jsv.2019.115036
18. C. Padova, M. G. Dunn, J. Barton, K. Turner, A. Turner and D. DiTommaso (2011) "Casing Treatment and Blade-Tip Configuration Effects on Controlled Gas Turbine Blade Tip/Shroud Rubs at Engine Conditions", *ASME Journal of Turbomachinery*, Vol. 133, p. 011016. https://doi.org/10.1115/1.4000539
19. F. Goldspiegel, G. Portemont and J. Berthe (2022) "On the Divergence Mechanisms Analysis of a Vibrating Blade Subjected to Repetitive Rubs with an Abradable Coating", *ASME Journal of Turbomachinery*, Vol, 144, No. 5, p. 051008. https://doi.org/10.1115/1.4053018
20. Y. Zhang, S. Yang, X. Tai, H. Ma, H. Guan, Q. Mu, L. Qu and X. Ding (2024) "Study on Rubbing-Induced Vibration Characteristics Considering the Flexibility of Coated Casings and Blades", *Machines*, Vol. 12, No. 7, p. 481. https://doi.org/10.3390/machines12070481
21. J. Lin, B. Wu, J. Zhang, Z. Wei, X. Zhang and H. Dai (2024) "Blade-Coating Rub-Impact Force Analysis Using a Dissipative Contact Force Model with Plastic Deformation", *Journal of Mechanical Science and Technology*, Vol. 38, No. 12, pp. 6731–6745. https://doi.org/10.1007/s12206-024-1127-4
22. F. Thiery and P. Chandran (2024) "On the Operational Similarities of Bladed Rotor Vibrations with Casing Contacts", *International Journal of System Assurance Engineering and Management*, Vol. 15, No. 8, pp. 1–15. https://doi.org/10.1007/s13198-024-02455-z
23. X. Shen, G. Chen, F. Liu, R. Zhang and Y. Chi (2025) "Study on Dynamic Characteristics of Blade-Casing Rub-Impact Coupling Fault Considering Breathing Crack", *Nonlinear Dynamics*, Vol. 113, pp. 32091–32115. https://doi.org/10.1007/s11071-025-11784-y
24. P. Yu, X. Tao, J. Zhang, Z. Xiang, Y. Jin and J. Chen (2025) "Dynamic Modelling and Failure Analysis for a Compressor Blade in Aero-Engine Undergoing Blade-Casing Rubbing Events", *Engineering Failure Analysis*, Vol. 177, p. 109700. https://doi.org/10.1016/j.engfailanal.2025.109700

8 Advances in Impact Simulations and Shock-Resistant Structural Designs

The contents of this chapter are focused on practical applications of all the mathematical theory and physics behind the impact dynamics problems covered in Chapters 1 to 7. Until now, we have discussed numerical simulations of various impact dynamics problems related to aerospace applications. To that end, our focus has been essentially in the area of propulsion system transient dynamics situations during normal operating conditions of the aircraft, viz., bird strike, ice shedding, blade containment and fan blade out (FBO), blade tip rub, engine rotordynamic response, and mount loads. These modeling procedures and recommended best practices for LS-DYNA® analytical simulations have been developed over many years to meet and comply with the certification requirements of regulatory agencies like FAA (Federal Aviation Administration). We will also be showing with other examples as to how the LS-DYNA® model can be improved and build confidence in its predictive capability to produce credible transient dynamics results based on the physics of the problem alone. Having said that, in this chapter, we will be discussing another class of aerospace civil aviation issues related to flight safety and aircraft survivability under extreme conditions such as blast simulations, missile attack, and aircraft crashworthiness. The materials presented in the following section are based on the author's unique contribution and pioneering academic work published in the public domain [see Reference 1] in the 2013 issue of the *AIAA Journal of Aircraft*. Here for the benefit of the readers, some of the results, findings, and illustrations of that investigation are being reproduced with the full permission of the journal publishing organization AIAA.

Later on in this chapter, we will also be discussing specific challenges about LS-DYNA® modeling issues related to analysis of pressure pulse transmission through the airspace due to blast and detonation, failure of bolted joints, Johnson–Cook plasticity model, Mooney–Rivlin Rubber-like hyperelastic materials, honeycomb and foam characteristics, modeling of oil sloshing and other flammable liquids, etc. Due to the unique capability of LS-DYNA® as the only commercially available solver code in simulating various kinds of impact scenarios, the topics being discussed in this chapter will be tailored toward the capability, numerical accuracy, contact algorithms, and formulations of different material models inside LS-DYNA®.

DOI: 10.1201/9781003749752-8

8.1 RECENT APPLICATIONS OF HIGH-VELOCITY IMPACT MODELING

The contents of this section, already available in the public domain since its September 2013 publication in the academic *Journal of Aircraft* [see Reference 1], are being reproduced in their entirety almost verbatim with the full permission of the journal publishing organization AIAA. Lately, aircraft survivability and flight safety have been a concern in light of recent events of missile attacks on commercial aircraft. Large wide-body military and commercial aircraft are particularly susceptible to man-portable air defense systems (MANPADS) missiles during takeoff and landing due to large infrared emissions, slow speed, predictable flight paths, unencrypted air traffic communications, and publicly available commercial schedules. Propulsion systems on these aircraft are the largest source of infrared energy and, as such, have a high probability of being the impact point, if the missile successfully intercepts the aircraft.

In this chapter, we report a hypothetical scenario, whereby a shoulder-fired missile hits a running engine in-flight. This section presents an in-depth study of the analytically predicted damage to the engine, as well as the transient loads that are developed at the engine mount system.

The analytical simulation involved modeling the engine and the missile and the correct representation of the entire physics of a high-velocity impact with target penetration. There have been many experimental investigations as well as theoretical investigations in the area of target penetration with a sharp-nosed ballistic impactor. In the present investigation, we also considered the additional damage caused by a detonating warhead. The explosion-related, traveling pressure wave was modeled separately by blast modeling. The finite-element mesh of the engine included the details of the outer structure, which most likely affected the missile trajectory and bore the brunt of the impact. For example, parts of the turbine casing that have a honeycomb structure, which may absorb some of the impact energy, were considered in the finite-element mesh. The elastic–plastic material modeling of various engine components included strain-rate-dependent material properties, as well as high-temperature characteristics, especially in the turbine area. Results of such a study can help us in planning a representative live-fire test under controlled conditions. For example, an analysis of peak dynamic loads on the mount system can help in determining the reliability of the engine attachment system with the aircraft wing.

8.1.1 Details of the Analytical Model

Analytical modeling was performed with LS-DYNA® and included a full representation of low-pressure (LP) and high-pressure (HP) rotors. The missile model was developed independently of the engine model. The goals of the current investigation are listed as follows:

1. Develop an engine–MANPADS modeling procedure that is capable of producing high-fidelity predictions of damage on a large turbofan engine.
2. Apply the modeling procedure to develop credible damage predictions for the hypothetical engagement scenarios.

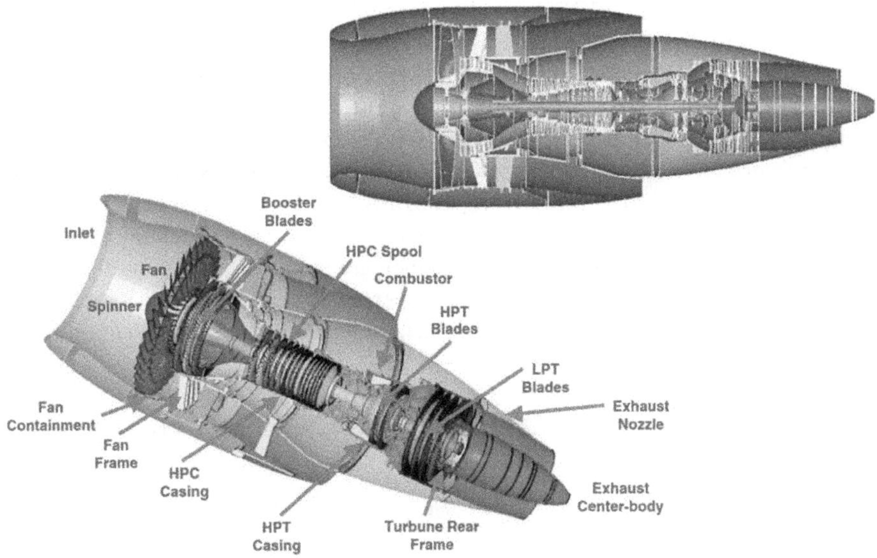

FIGURE 8.1 Two views of the full-engine LS-DYNA® model. (Reproduced with permission from Reference [1].)

8.1.1.1 Details of the Engine Model

Figure 8.1 shows two different views of the engine model. The model includes the stator and rotor components typical of jet engine construction. The full-engine LS-DYNA® model consists of about 1.5 million nodes and elements, in which the majority of elements are eight-noded solid brick elements. About 100,000 shell elements are used to represent the stator and rotor airfoil blades, stator vanes, and nozzles. In the engine model, the LP rotor rotates at roughly one-third the speed of the HP rotor. Because the intention of the engine finite-element model is to study missile impact and detonation-related damage, the aft-end components of the engine were represented using an extremely refined mesh. The containment of damaged stator components can be analyzed by modeling the full 360° sector of the relevant casing structure.

In this approach from the rotor side, one models all rotor blades that are separated in the circumferential direction by an angle, which is determined by using the following formula:

$$\text{Circumferential separation in degrees} = (360°)/(\text{number of blades on the rotor}) \quad (8.1)$$

The fan blades, as well as the fan containment structure, are modeled in sufficient detail using eight-noded brick elements called *ELEMENT_SOLID. With a constant strain element, one must use a minimum of three elements through the thickness everywhere. The blades are rotated about the engine axis using *BOUNDARY_PRESCRIBED_MOTION_NODE for the root nodes and

*INITIAL_VELOCITY_NODE for all other nodes. It is recommended that ahead of the LS-DYNA® transient analysis, initial stresses for the rotating blades must be done using an implicit code, such as ANSYS®, or the implicit version of LS-DYNA® itself. All inner surfaces of the containment surface, as well as any other filler material surfaces, should be specified in separate sets of *SET_SEGMENT. The contact interface between the blades and inner surfaces of the fan case must be defined using *CONTACT_ERODING_NODES_TO SURFACE. Furthermore, it is imperative that due to very high strain rates involved in these simulations, one uses strain-rate-dependent properties whenever they are available. One good resource in the public domain for the high-strain-rate properties of typical aerospace materials is [see Reference 2] written in 1980. The material model for both blades and fan casing should be either

*MAT_STRAIN_RATE_DEPENDENT_PLASTICITY

or

*MAT_PIECEWISE_LINEAR_PLASTICITY.

All material models in LS-DYNA® must be used with the FAILURE option "ON". One does not have to use any specific hourglass control parameter or penalty factor, other than the default values that are built into the code already. If LS-DYNA® results indicate the presence of hourglass instability, it is best handled by making the mesh finer, rather than using different types of hourglassing parameters. The remeshing should be done in such a way as follows:

1. Point load situations (throughout the entire duration of the transient analysis) are eliminated by using a finer mesh option, such that concentrated loads are always shared by multiple surrounding nodes.
2. Slave side segments are finer than the master side segments.

8.1.1.2 Details of the Missile Model

The MANPADS missile finite-element model was constructed in detail for use with the LS-DYNA® code and it comprised seeker, warhead, guidance and control, and rocket motor sections (see Figure 8.2).

The MANPADS missile model also contains detailed data on section properties, exterior dimensions, joint construction, joint strength, component construction, material properties, mass properties, and rocket motor case strength. The current version of the MANPADS missile model consists of shell and solid elements. The shell elements used are based on the Hughes–Liu formulation and are mostly quadrilaterals, although some triangles are also used. The solid elements used are all hexahedrons. There are currently 59,576 nodes, 39,725 shell elements, and 9809 deformable solid elements. The fragments are modeled with 4032 rigid elements. Materials in the structural model are varied and include glass, phenol, aluminum, and steel. In each case, nonlinear material models are employed. The majority of these nonlinear

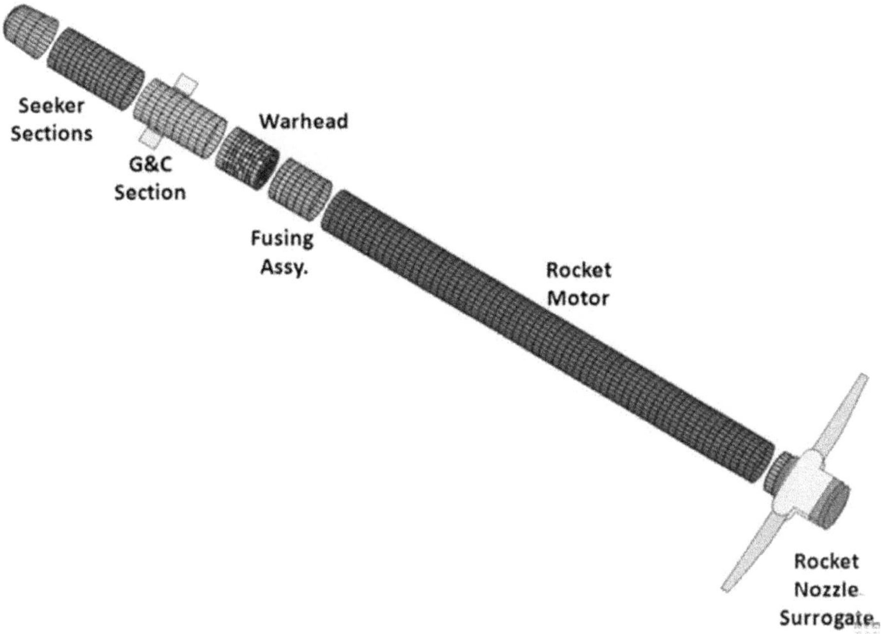

FIGURE 8.2 The MANPADS model. (Reproduced with permission from Reference [1].)

material models are of the piecewise linear plasticity type. In this type of material model, density, elastic modulus, yield stress, tangent modulus, and failure strain are specified, as well as a nonlinear stress–strain curve.

8.1.1.3 Modeling Methodology of Blast Pressure Characteristics

The warhead on the missile contains high-explosive materials that produce a significant blast pressure. Shattering the warhead casing generates fragments as well. To obtain the correct fragment pattern and velocity distribution, a coupled fluid-structure interaction technique was chosen. In this technique, the explosive and surrounding air were modeled as fluids in an Eulerian domain. Warhead structures, including fragments and end caps, were modeled in the Lagrangian domain. Fragment accelerations for the first several microseconds were extracted from this analysis and combined with the fragment masses in order to obtain fragment force vs time curves.

This expensive coupled analysis was performed once, and the force curves were generated. These curves were then used in a much more economical, all Lagrangian model of the MANPADS model. Blast characteristics of the warhead were introduced into the model using a parametric representation that was developed for the Conventional Weapons Effects (CONWEP) code [see Reference 8.3]. The warhead explosive charge was mass-scaled to match the peak pressure and time of arrival of the pressure pulse as a function of range. Figure 8.3 shows a comparison of the fragment velocities from the model with data obtained experimentally in static arena tests.

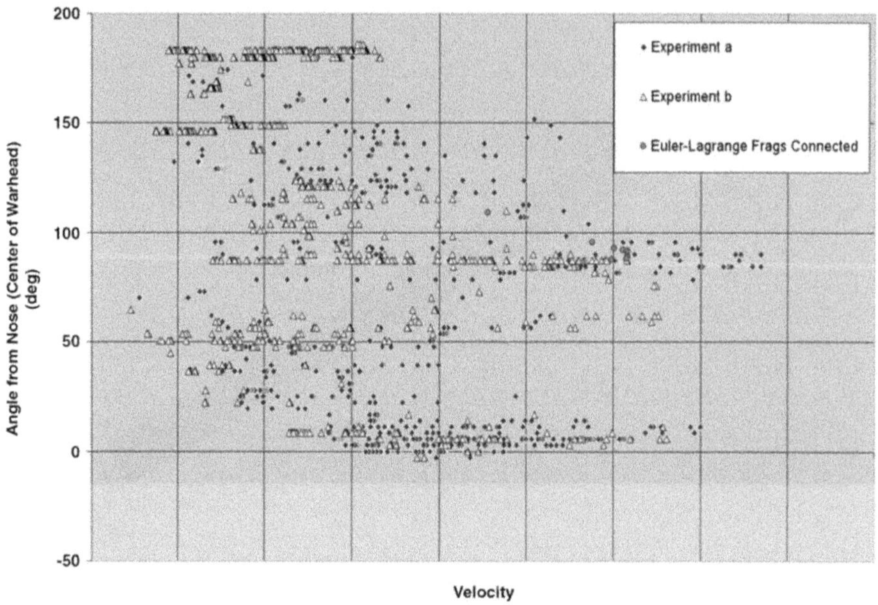

FIGURE 8.3 Fragment velocities compared with test data. (Reproduced with permission from Reference [1].)

Note that the experimental data contain fragments having very small masses. No attempt was made to model these small fragments in the finite-element model. Emphasis was placed on the modeling fragments associated with a side spray above a certain mass and velocity threshold. This is why there is limited model data (see the data labeled "Euler–Lagrange frags connected") shown in Figure 8.3. The thresholds were chosen to create a credible model that had reasonable mesh density, which led to reasonable runtimes.

Figure 8.4 shows the model-generated peak pressure vs range that is compared with experimentally obtained data. The peak pressures shown in the figure are "side-on" pressures. The comparison with the side-on pressures is shown, because that is what was measured in the experiments, which were performed in static arena tests. The actual application of pressure in the finite-element model incorporates the pressure contributed from the dynamics of the blast wave. Fragment drag was accounted for by assuming a flat trajectory. Equation 8.2 gives the solution of the flat trajectory problem, where V_f is the fragment final velocity at distance x, given an initial velocity of V_i, with mass, m, in the air with density ρ. The drag coefficient is given by C_D and the presented area by A:

$$V_f = V_i e^{\left(-\frac{\rho C_D A}{2m}x\right)} \tag{8.2}$$

Experimentally, many of the parameters are unknown, so Eq. (8.2) is rewritten with those parameters lumped into a single parameter, k,

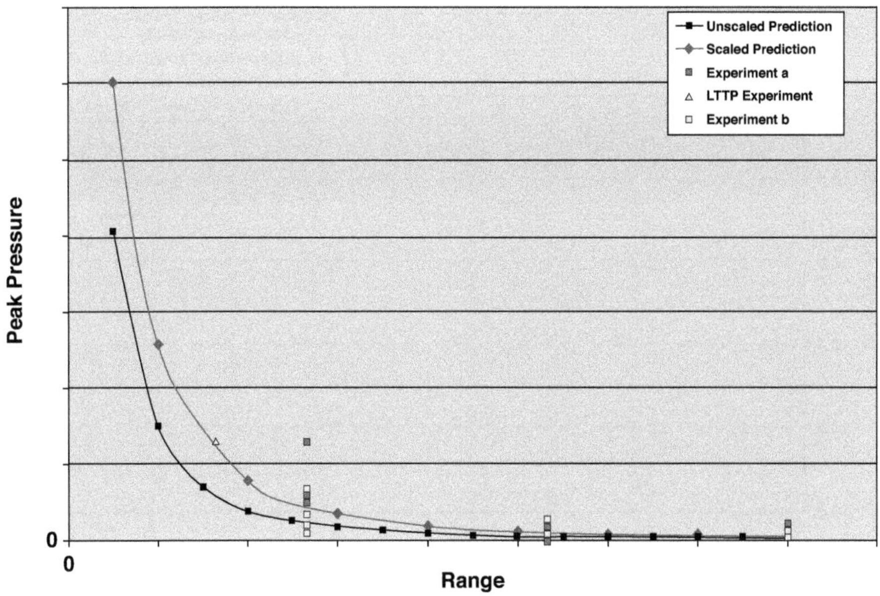

FIGURE 8.4 Modeled peak pressure compared with the test data showing the explosive charge mass scaling comparison. (Reproduced with permission from Reference [1].)

$$V_f = V_i e^{(-kx)} \quad (8.3)$$

By experimentally measuring fragment velocities in static arena tests at several distances, x, reasonable values for k were determined. Note that the current arena test procedures do not track individual fragments. Therefore, the individual fragment masses and areas are not associated with individual fragment velocities. Statistical methods are used to estimate the association. This shortfall in arena testing is currently the topic of intense research. The local velocity V_f, thus, calculated can be used to determine the magnitude of the pressure pulse through the air medium as follows:

$$\text{Magnitude of local pressure at a distance } x \text{ in a moving shock wave} = \frac{1}{2}\rho\left(V_f\right)^2 \quad (8.4)$$

Implementing the effects of drag on fragments in the model was done using a mass damping option in LS-DYNA®. Although this option is generally used to damp low-frequency structural modes, it can also damp rigid-body modes, which is how it was used to simulate the rigid-body motion of fragments in the model. Damping parameters are scaled according to standard atmospheric density in order to simulate varying altitudes at the target. Figure 8.5 shows the quality of using this method to reproduce fragment drag at two different altitudes.

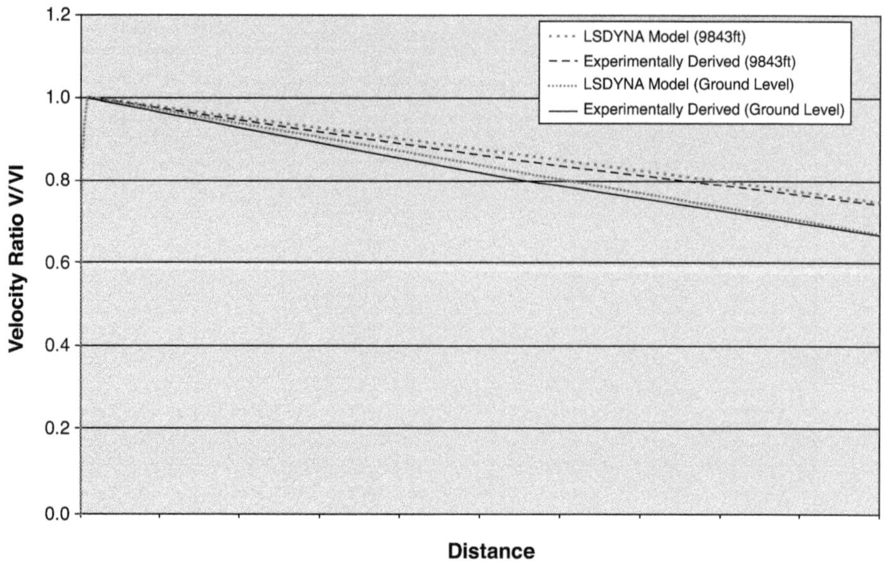

FIGURE 8.5 Fragment velocity vs range showing drag effects of velocity degradation. (Reproduced with permission from Reference [1].)

Fragment drag is generally not included in close-in detonation simulations such as this. However, because the authors have an interest in the applications of detonation at greater distances, the drag contribution was derived and included in this work.

8.1.1.4 Effect of Missile Orientation and Shotlines on the Engine

In this analysis, two different aft-end shotlines are considered (see Figure 8.6). In the first shotline, the incoming missile enters the engine at the exhaust cone, and in the second shotline it goes toward the core of the engine. Figure 8.6a schematically shows the engine and the orientation of the incoming missile for shotline 1. Similarly, Figure 8.6b schematically shows the engine and the orientation of the incoming missile for shotline 2. Note that the missile and engine are shown in the same scale.

In the analytical simulation, the engine model described earlier in Section 8.1.1.1 is coupled with the MANPADS missile model described in Section 8.1.1.2 of this

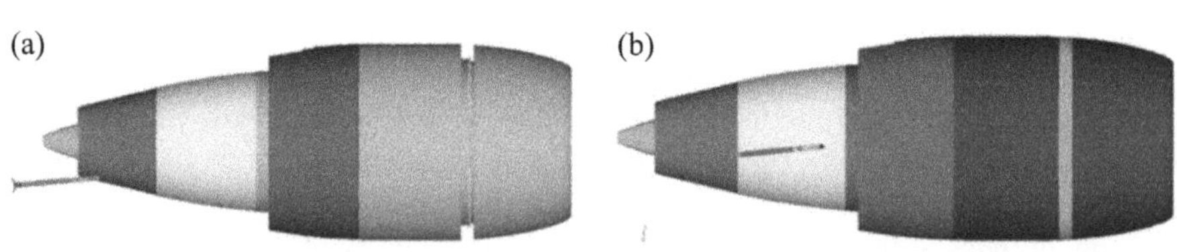

FIGURE 8.6 Missile model coupled with engine model for the two shotlines' orientations: (a) Shotline-1 and (b) Shlotline-2. (Reproduced with permission from Reference [1].)

presentation. The transient dynamic analysis of this combined model has been run successfully using LS-DYNA® as the solver, which is carried out for 15 ms from the time of initial contact of the detonating missile with the engine.

When coupling the missile model with that of the engine, it is not necessary to consider airflow, gas loads, and thermal gradients. Because liberated blade fragments after impact have a tendency to go forward for forward-end shotlines, it is necessary to include the engine inlet. Similarly, the aft-end of the stator structure needs to be included in the analytical simulation. Depending upon the design, one may model bolts and bolt holes in the flanges explicitly or one can use the *CONTACT_SURFACE_TO_SURFACE interface definition between flanges.

One needs to model the fully bladed rotor, including the low-pressure turbine (LPT) rotor and disk for simulating the rotor response in LS-DYNA® and running it for a sufficient length of time (in the 15–100 ms range), such that one can observe the rotor response due to unbalanced forces. These models are large in terms of the number of nodes and elements involved, but they are capable of simulating the real engine dynamics as close as possible to the real thing. The main challenge concerns modeling the bearing surface interface between the rotating shaft and the outer race bearing surface that is connected to the stator structure. In the jet engines, there are two types of bearings: (i) ball bearings (which can take axial loads) and (ii) roller bearings (which cannot take axial loads). In LS-DYNA®, these bearings can be simulated using a *CONTACT_NODE_TO_SURFACE type of interface. The stator side of the ball bearing is represented by a U-shaped groove for the rotor side elements sitting inside two axial walls, preventing their axial movement. The stator side of the roller bearing is represented by a cylindrical surface, such that the rotor side elements are free to move axially. In this modeling approach, one represents the core of the engine with two concentric rotors involving the fan LPT rotor as the inner rotor and the high-pressure compressor–high-pressure turbine (HPT) rotor as the outer rotor. The model is capable of capturing deformation in the shaft due to the relative rubbing of two concentric rotating shafts. In this modeling approach, other than the fan blades, all other blades are modeled using shell elements with the ELEMENT_SHELL option. It is very important from the dynamic results accuracy point of view that individual blades on both rotors are modeled in such a way that their deformation due to rubbing or clashing with any stator components is properly captured in the LS-DYNA® model. Because of the typical dimensions of the parts, the disks are usually represented by eight-noded brick elements. It should be pointed out that one must follow a geometrically correct procedure for combining shell nodes with six degrees of freedom on one side with a brick node having three degrees of freedom on the other side. For aft-end shotlines (the main focus of this report), one must include the following critical components in the LS-DYNA® analytical model: (i) HPT casing, (ii) LPT casing, (iii) exhaust nozzle, (iv) exhaust cone center body, (v) turbine stator vanes/nozzles, (vi) cowling and nacelle, and (vii) forward and aft-mounted systems.

Other items, such as the oil tank and auxiliary gearbox, may be represented by separate rigid masses. If one is interested in assessing the potential for fire, fuel, and hydraulic lines will have to be included in the analysis. Because there are multitudes of engine parts that may come into contact with the missile, it is better to use *CONTACT_AUTOMATIC_SINGLE_SURFACE, which is also efficient in terms of computational time. This type of contact algorithm is widely used in impact simulations. In the engine model, the LPT, turbine rear frame, and exhaust/cowl have multiple surfaces that can come into contact with the missile. When the missile penetrates through the exhaust cowling, liberated cowl fragments could potentially contact the remaining undamaged portions of the exhaust cowl. Also, LPT blades could come into contact with adjacent LPT stator vanes in the same stage or into contact with the LPT casing. All default options are used in this *CONTACT_AUTOMATIC_SINGLE_SURFACE input, as this impact event has metal-to-metal contact.

In addition, the time-deletion limit option was used in *MAT_PIECEWISE_LINEAR_PLASTICITY, which activates the deletion of highly distorted elements; otherwise, the model's time step would have to be extremely small. It should be noted that in this modeling approach, it is usually sufficient to model all of the blades with a relatively coarse mesh representation.

The following steps are involved in coupling the full-engine model with the missile model:

1. Redefining the missile nodes with respect to the engine nodes in the global frame of reference. This process also requires redefining the initial nodal velocities of each node in the LS-DYNA® model.
2. Definition of segment sets using *SET_SEGMENT.
3. Using the *CONTACT_ERODING_NODES_TO_SURFACE option.

The initial conditions of the engine and missile are incorporated into the simulation by imposing stresses and velocities onto the finite elements of the rotating parts of the engine, positioning the missile model at the surface of the impact point, and specifying all nodes associated with the missile and their initial velocities.

8.1.2 Results of the Analysis

As with any impact scenario, the majority of the missile-related, analytically predicted damage to the engine is confined in the near vicinity of the initial contact. This is true for both shotlines 1 and 2 and because the time duration of both analyses was limited to 15 ms from the time of the first contact.

Because LS-DYNA® is an explicit time integration code, there is a minimum time step size that is imposed by the Courant condition of the model, which is dependent on the smallest deformable element. Because many of the elements were quite small, the minimum time step was also small (less than 1.0×10^{-6}s). To maintain solution stability over many time steps, as well as to complete the computations in a reasonable time while capturing the majority of damage resulting from the event, the simulation duration was limited to 15 ms.

Figure 8.7 shows the damage predictions to the test engine. In these plots, the red color fringe patterns highlight the areas of large plastic deformation in various engine components.

The damage shown in Figure 8.7 is a result of the accumulation of damage resulting from the forces that are imposed on the model elements from all sources.

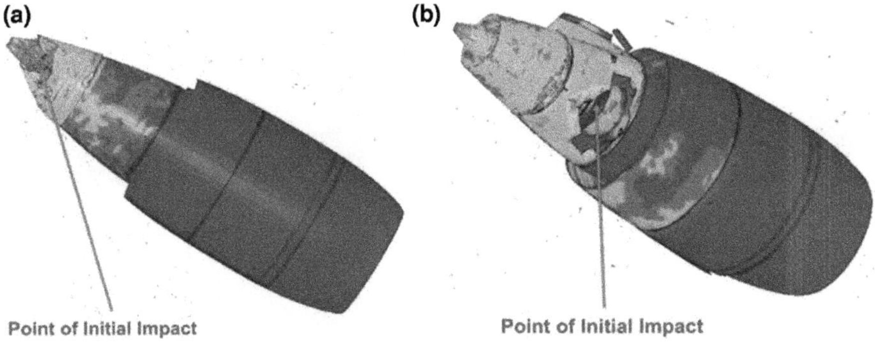

FIGURE 8.7 Damage predictions to the engine for two different shotlines. (a) Damage prediction for shotline-1. (b) Damage prediction for shotline-2. (Reproduced with permission from Reference [1].)

These sources include the blast pressures and impacts of fragments from the missile warhead and body, as well as the impacts from the engine parts contacting one another. The damage is simulated by using nonlinear (and in some cases rate-dependent) material properties for all of the components and imposing failure criteria that have been previously determined for each of the material types used.

The simulation is started, and a global solution is found at each time step. Also at each time step, each element is checked to determine whether its failure criterion has been reached. The material model *MAT_24 (piecewise linear plasticity) is used with the failure subroutine invoked. If it has failed, it is eroded out of the model before the next time step is taken. This process continues until the final time is reached. Thus, one can see that the total simulation entails the solution of many problems, each of which can be very different from its neighbors in time.

Generally speaking, one can see that the loading due to blast pressures and very-high-velocity warhead fragments occurs early in the simulation. The loading due to the kinetic energy of the rest of the missile and to deformations and contacts of the static and rotating parts of the engine results in damage later in the simulation. In the numerical simulation, the damage caused by the pressure-blast wave can be characterized by a wide area plastic buckling of skin in the vicinity of the impact, whereas piercing and tearing of mangled metal at the hit location is the sign of a direct ballistic impact by the main body of the missile (see Figure 8.8b). In the same way, the secondary fragmentation caused by the shrapnel is indicated by isolated holes appearing farther away from the shotline trajectory.

In this section, sample results of the LS-DYNA® analysis are presented for certain engine components, which exhibit most of the visible damage in the aftermath of the missile shot. Analytically predicted damage results to these engine components are compared with the test observations. Figures 8.8a,b and 8.9a,b show the predicted damage vs the test damage and illustrate the LS-DYNA® capability to credibly predict engine damage ahead of the test. Damage predictions correlate reasonably well with test results.

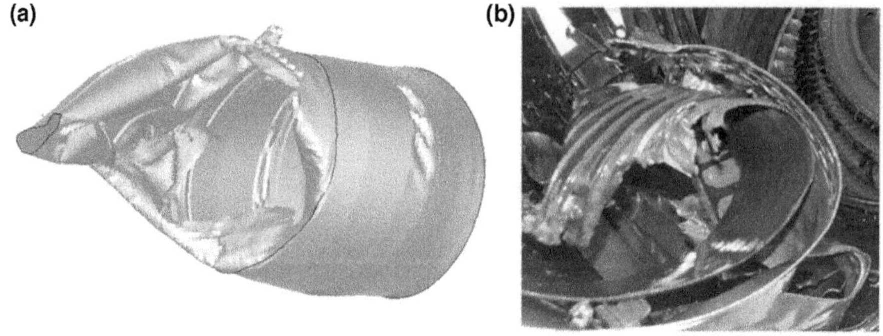

FIGURE 8.8 Damage to the exhaust cone center body. (a) Analytical pretest prediction. (b) Test result during shotline-1. (Reproduced with permission from Reference [1].)

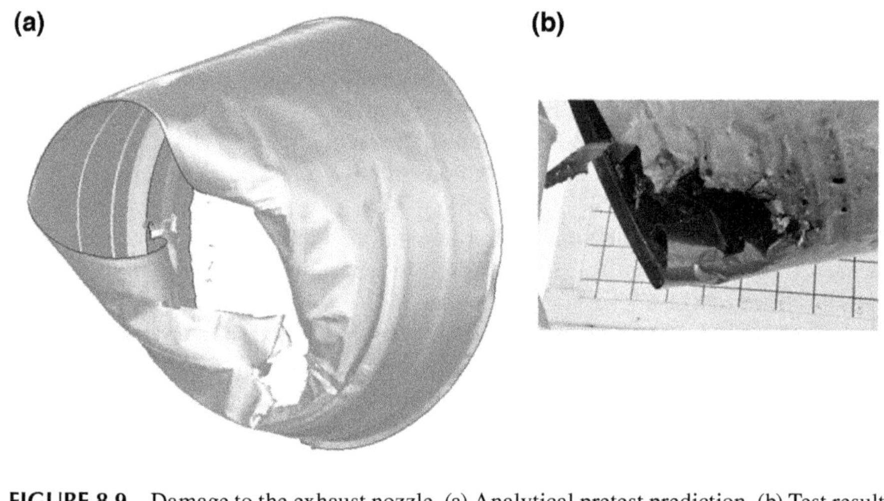

FIGURE 8.9 Damage to the exhaust nozzle. (a) Analytical pretest prediction. (b) Test result during shotline-1. (Reproduced with permission from Reference [1].)

Here, Figure 8.8 shows the damage to the exhaust cone center body, and Figure 8.9 provides a zoomed-in view, highlighting the damage to the exhaust nozzle. It can be seen that the pretest analytical prediction of the damage to the engine was very consistent with the post-test observations. The excellent finding of this project clearly demonstrates the predictive capability of a physics-based LS-DYNA® model in analyzing a very complex high-velocity nonlinear transient dynamic event involving multidisciplinary applications. In the next few sections, we will be showing some other applications and capabilities of LS-DYNA® in capturing a transient dynamic event about its numerical accuracy and precision. Section 8.2 will discuss one such application of LS-DYNA® simulation, where the rub blade would exhibit extremely large deformation during the transient rubbing of the blade tip and the correct and detailed LS-DYNA® captures it very well.

8.2 TRANSIENT LARGE BENDING DEFORMATION DURING BLADE TIP RUB

Figure 8.10 shows the comparison of actual transient dynamic deformation of a blade observed through a high-speed movie in a rig test (Figure 8.10a) versus LS-DYNA® simulation (Figure 8.10b). One can clearly see the relatively very large lateral deformation δ_{rub} of the blade tip due to its rubbing action against the fan containment ring. In that process, the free end of the cantilevered blade near the tip is deformed due to contact-rub loading acting in the tangential direction of the fan case and opposite to the direction of rotation of the fan rotor.

In the actual rig test, based upon the high-speed movie taken during the test, the blade exhibited large deformation of the tip in the direction opposite to the direction of rotation of the fan blade. The magnitude of such a large deformation by as much as almost 30% of the blade span was unexpected by any static calculation, but was in excellent agreement with the LS-DYNA® computed value for the entire duration of the transient dynamic analysis. For comparison purposes about the accuracy of numerical schemes in our earlier theoretical discussion related to the partial differential equation approach for similar blade tip-rub lateral deformation, the reader should refer to Figure 5.32. In Figure 5.32, the deformed blade shown schematically due to the interference-induced rub at the tip due to clearance closure with the outer case was based upon the numerical solution of the Timoshenko beam problem discussed in Section 7.2. For similar calibration purposes, referring to Figure 7.24, the Timoshenko beam model analytical results predicted 14% lateral deflection due to 1.07% of radial incursion of the casing rub surface.

The above example demonstrated an excellent numerical accuracy of LS-DYNA® results for any typical transient dynamics contact-impact blade tip-rub situations.

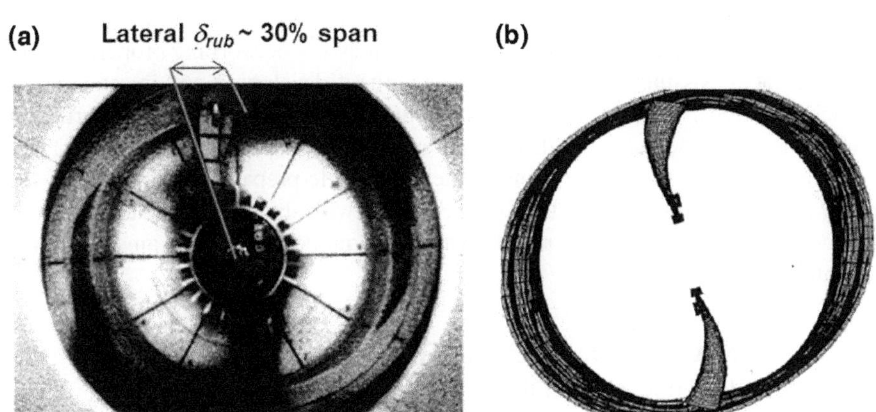

FIGURE 8.10 Comparison of transient dynamic blade lateral deformation during tip rub. (a) In the test ≈ 30% of blade span length deformation opposite to the direction of rotation. (b) LS-DYNA® analytical results (showing 1X deformation without any magnification). (a) Test picture of blade rubbing. (b) LS-DYNA simulation.

Such LS-DYNA® models, when properly meshed with good aspect-ratio brick elements and an appropriate choice of *CONTACT surface definitions of mating surfaces such as abradable coating and honeycomb, are fully capable of producing accurate numerical results prior to the actual test. A proper selection of the correct *CONTACT option is very critical including the MASTER and SLAVE sets depending upon the impact conditions. In general, the SLAVE side mesh must be finer than the MASTER side mesh, which is generally coarser. Depending upon the design configuration and surrounding structural components, the types of appropriate contact surfaces could be as follows. The following is a short list of the most commonly used interface *CONTACT definitions, from a large number of additional options offered in LS-DYNA®.

1. *CONTACT_AUTOMATIC_SURFACE_TO_SURFACE
2. *CONTACT_AUTOMATIC_NODES_TO_SURFACE
3. *CONTACT_AUTOMATIC_SINGLE_SURFACE
4. *CONTACT_ERODING_SURFACE_TO_SURFACE
5. *CONTACT_ERODING_NODES_TO_SURFACE
6. *CONTACT_ERODING_SINGLE_SURFACE
7. *CONTACT_TIED_ SURFACE_TO_SURFACE
8. *CONTACT_TIED_ NODES_TO_SURFACE
9. *CONTACT_TIEBREAK_NODES_TO_SURFACE
10. *CONTACT_TIEBREAK_SURFACE_TO_SURFACE

Another additional point must be noted about sometimes the need of "SOFT" option to be turned on in certain interface conditions. Most of the time, the stiffness values of both MASTER and SLAVE sides of the interface are almost of the same order of magnitudes and there are no numerical issues in the solution, with the default value of the "SOFT" option being "OFF". However, when the relative stiffness values of the two sides of any contact interface are significantly different from each other, usually by an order of magnitude then the interface may not be working properly resulting in surfaces passing through each other without developing any interface load. In such situations, the user should turn the SOFT option "ON" for that particular interface through the input *.k file correction.

Using the above finer details and guidelines, while preparing the input file, the reader can use a physics-based LS-DYNA® model with standard material properties to produce credible analytical results with full confidence without any additional validation of test data.

8.3 DYNAMIC RESPONSE OF BOLTED JOINTS UNDER SHOCK LOADING

Bolted connections are key components of various joints at various locations of any typical jet engine. This is another very important application of LS-DYNA® to determine the dynamic behavior of the entire supporting structure of any high-bypass jet engine under extreme dynamic loads experienced during the FBO event.

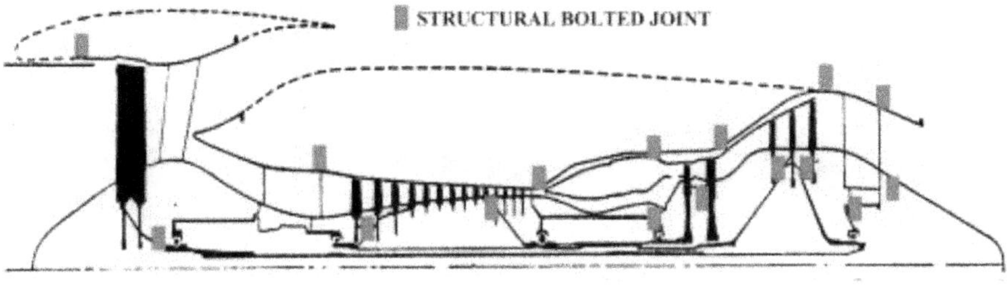

FIGURE 8.11 Structural bolted joints in a typical turbofan engine showing the side view. (Reproduced with permission from Reference [4].)

Czachor's ASME paper of 2005 [see Reference 8.4] is an excellent contribution in this field. The illustrations presented in this discussion have been directly taken from his papers with full permission of the journal publishing organization ASME. Depending upon different design philosophies as well as assembly considerations, a typical turbofan jet engine may have a combination of circumferential and split-line horizontal bolted connections in various locations. For example, Figure 8.11 shows schematically various engine axis locations, where circumferential flanges are connected through bolted joints.

These bolted joints (Figure 8.12) used to connect two circumferential flanges have basically three components: (i) a bolt made up of bolt head and a shank with threaded ends, (ii) a tightening nut at the threaded end, and (iii) washer, which may be a thin ring-shaped sheet metal plate just to distribute the load from the bolt head or the nut over a wider surface area of the flange or sometimes a plastically crushable long cylindrical washer to take dynamic loads during the FBO event. At the installation time, these bolts are preloaded by applying torque at the bolt head and by tightening the nut, which almost automatically makes the bolt shank to go into tension with the flange mating surfaces being pushed into compression against each other.

The typical plastically deformed shape of a bolted joint connection in bending under transverse loading is schematically shown in Figure 8.12a. It can be seen in Figure 8.12b as to the major mode of load transfer path from the bolt to the flange hole is through a surface contact between the outer surface of the bent bolt shank and the inside surface of the bolt hole of the flanges. It may be counterintuitive for most of designers trained in static load concepts under the assumption that the bolts are

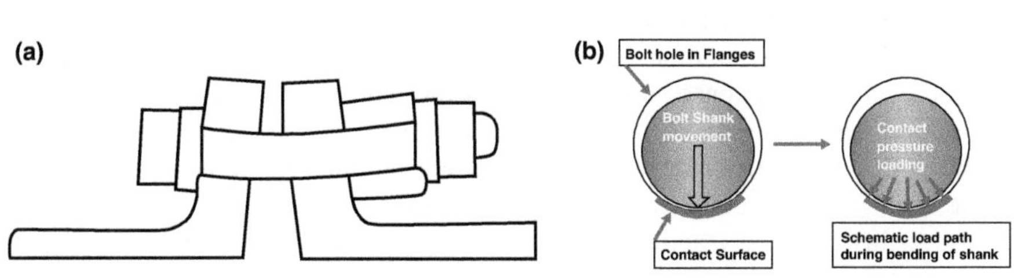

FIGURE 8.12 Typical deformation of bolt shank inside the bolt holes of the flanges. (a) Side view (Reproduced with permission from Reference [4]); (b) zoomed-in cross-sectional view.

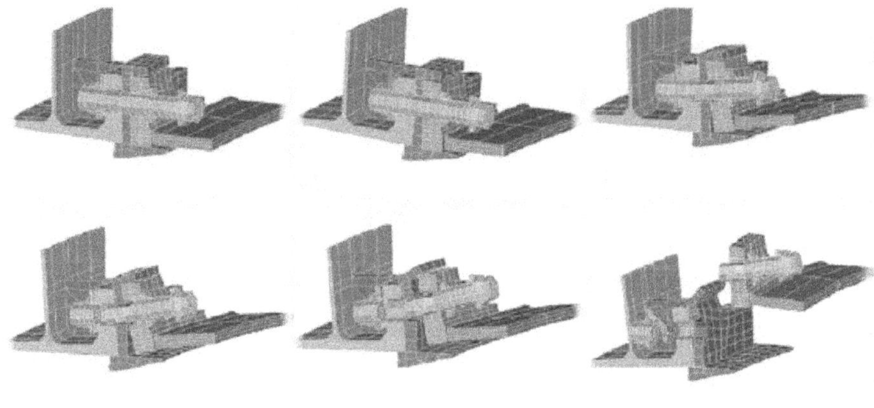

FIGURE 8.13 LS-DYNA® analysis of a bolted joint in the vicinity of a blade impact on the adjacent casing. (Reproduced with permission from Reference [4].)

mostly loaded in tension only, and they think that it may fail when the tensile stress in the shank would exceed the failure strength limit of the bolt-shank material capability. Under the impact and shock type of dynamic loading of bolted joint connections, which is the focus of the current discussion for LS-DYNA® simulation, the accurate and precise modeling of the geometry and contact details of the FE mesh becomes extremely important. It should be recognized that depending upon the accuracy desired and modeling complexity, there are various ways to represent a bolted connection in any typical LS-DYNA® simulation. It can vary from a simple spot weld representation between two mating flanges to an explicit modeling of the bolt-hole in the flanges and bolt shanks along with its pretension loadings (see Figure 8.13).

To pick up a right strategy to model a typical bolted joint, one must have an in-depth understanding of its behavior in transferring the load. A bolted joint subjected to bending load works as what is commonly known as "Toe-heel action". When two flanges are joined together by a bolted joint in the circumferential direction then on the tension side the joint would open up. Once the initial compression between the mating flange surfaces is used up, the bolt shank will go into tension. Thus, on the inner radius, the flanges will open up, but at the outer radius it will compress against each other. This characteristic of a bolted connection makes the joint more flexible due to local bending of the flanges at the root fillet. Furthermore, the bolt may be a loose bolt with the shank diameter is smaller than the flange hole diameter (see Figure 8.12b). In this situation, the bolt shank bends inside the hole and then transfers the load through contact forces between the flange-hole inner surface and the bolt shank. Another Figure 8.14 from Czachor's paper is also being reproduced for the benefit of the current readers, which highlights all the key dimensions of interest for designing a bolted connection with withstand the dynamic load without failure.

Using this "Toe-heel action", a fully detailed LS-DYNA® lever action model has been created [see Figure 8.13], and it is a relatively simple task to generate the load-deflection characteristics of any such bolted connection. Figure 8.15 shows one such load deflection plot for a lever flange at high load levels of the containment shell.

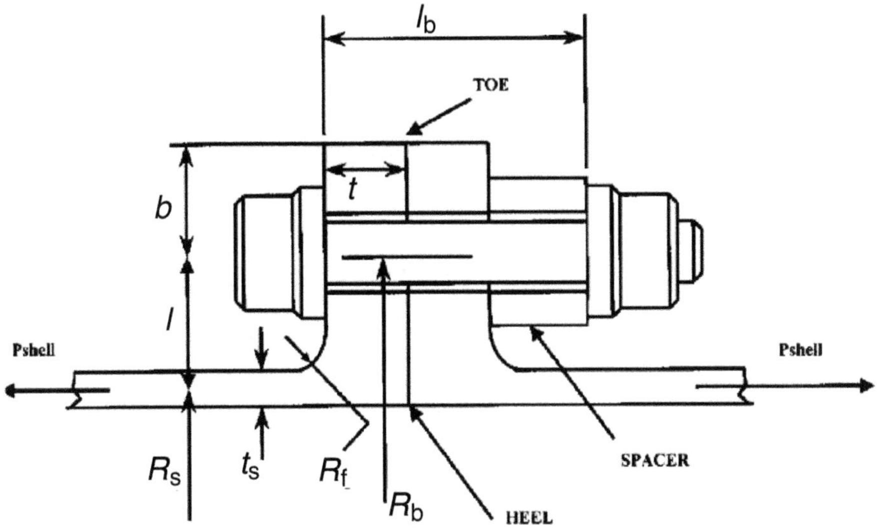

FIGURE 8.14 Fully detailed geometrical dimensions of a bolted joint flange connection. (Reproduced with permission from Reference [4].)

The results of the dynamic load-deflection characteristics of a typical bolted joint (see Figure 8.13), analyzed in a typical generic LS-DYNA® simulation, have been summarized in the form of schematic plot shown in Figure 8.15. In the next subsection, we discuss in detail about applying the pretension in the bolt shank before any transient dynamic analysis in LS-DYNA® can be performed.

8.3.1 Applying Initial Preloads in the Bolted Joint

In the LS-DYNA® modeling scheme, there are two common ways to apply the Bolt Preloads prior to the transient analysis run is started. It should be noted that bolted joint preloading is a self-equilibrating condition, in which all the internal contact forces in the vicinity of the bolt head, shank, and flanges are active in such a way that when the bolt shank is in tension the surrounding contact surfaces at the flanges and the bolt heads remain in compression.

1. ***CONTACT_INTERFACE**: This is an iterative process and requires multiple runs to tune the load in such a way to converge on a desired bolt prestress value. Any typical LS-DYNA® model with a very large number of bolted connections can use significant amounts of computational resources during the iteration and is not a recommended approach.
2. ***INITIAL_STRESS_SECTION:** *INITIAL_STRESS_SECTION will preload a cross section of solid elements to a prescribed stress value. In this method, the bolt stress is applied directly and no guess work is needed. Being a noniterative process, it is a better and preferred way to apply the bolt

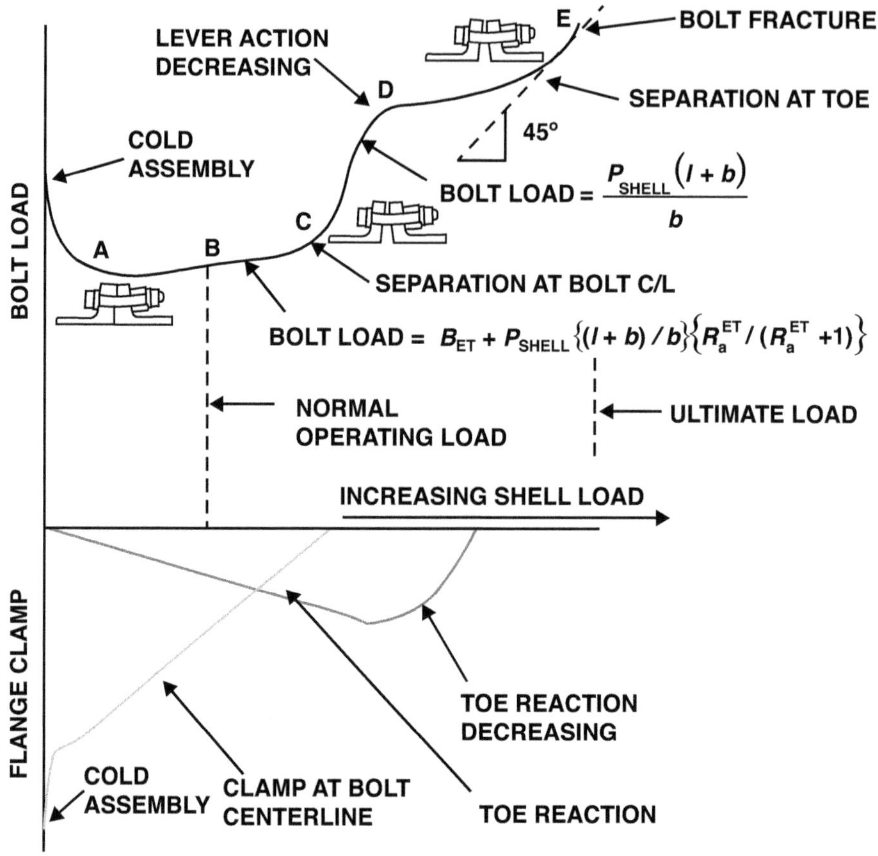

FIGURE 8.15 The joint diagram for a lever flange at high load levels. (Reproduced with permission from Reference [4].)

preloads, specifically in any detailed meshing scheme using eight-noded brick elements to model the bolted connections explicitly, as discussed in the previous subsection.

The implementation of this method in the LS-DYNA® model involves the following steps, as discussed here. Let us consider a bolted connection shown in Figure 8.16 with all its details of LS-DYNA® modeling. All the potential contact surfaces should be using *CONTACT_AUTOMATIC_SURFACE_TO_SURFACE.

- Preload stress (normal to the cross section) is defined via *DEFINE_CURVE (stress versus time). Stress must start from zero and then can be ramped linearly.
- This curve is typically flagged with **SIDR=1**, invoking **Dynamic Relaxation** option for applying the preload.
- Physical location of cross section is defined via *DATABASE_CROSS_SECTION. As shown in Figure 8.16, the cross section of the bolt shank should be chosen as much as possible in the middle.

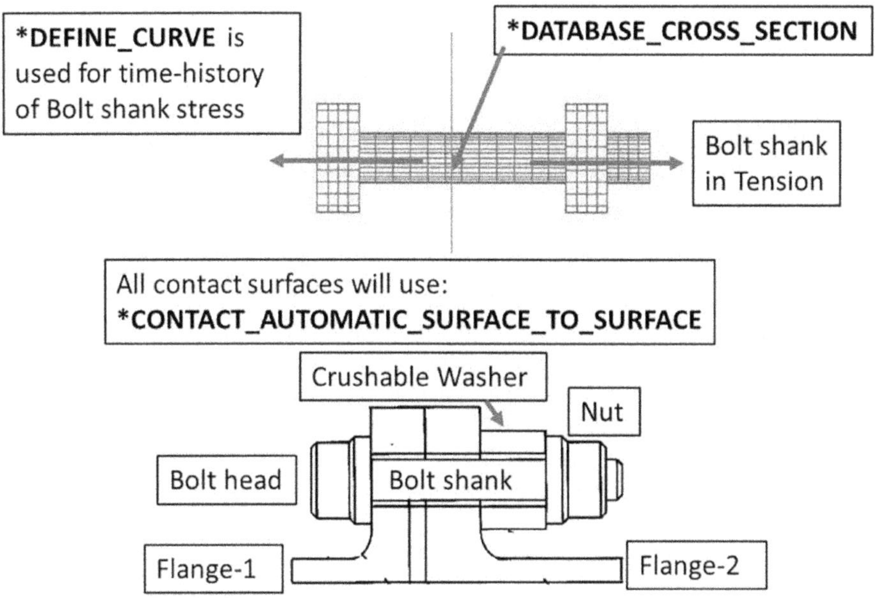

FIGURE 8.16 Steps needed for bolt preload application in LS-DYNA.

- A part set, together with the cross section, identifies the elements subject to the prescribed preload stress.
- Contact damping (VDC) and/or *DAMPING_PART_STIFFNESS may be required to attain convergence and reduce the magnitude of oscillatory stresses during the dynamic relaxation analysis. This step is needed to stop or to mitigate the effect of elastic oscillations during the transient analysis run of LS-DYNA® solver.

Using the recommended bolt modeling procedure in LS-DYNA®, described in Section 8.3, is extremely important in correctly predicting the outcome of the FBO event or a test rig, where some circumferential bolt failures essentially in the vicinity of the main impact zone of the containment ring are almost inevitable. With the bolt failure option turned "ON" via the material model, the containment analysis will show that the containment shell would ovalize and other blades on the fan rotor will start rubbing against the containment ring. That was the observation made during the rig-test discussed in Section 8.2.

8.4 HONEYCOMB AND FOAM STRUCTURAL DESIGN FOR CRASHWORTHINESS

Honeycombs are a type of very lightweight composite due to their sandwiched-type architecture. Because of its lightweight sandwich construction made up of outermost layers of very thin metallic or fiberglass sheet glued with Hexcel shells in the core, it is a very commonly used material in many aerospace applications. In addition to being a lightweight material, its mechanical properties are highly anisotropic. To such highly desirable quality of strength-to-weight ratio requirements, it is relatively strong in out-of-plane flexural bending deformation, but its crushing strength in lateral compression is very low. Once its threshold crushing strength is reached, it will crush practically under constant load until it becomes practically solid. For a more in-depth information about its configuration and properties, the readers are referred to a marketing-type technical document published by the Hexcel composites [see Reference 5]. This document is a comprehensive guide to standard Hexcel honeycomb materials, configurations, and mechanical properties.

In this section, with our goal of modeling details of LS-DYNA® simulations, we will focus on the technical details of its mechanical properties and how to use them in LS-DYNA® analysis and modeling schemes. Honeycombs can be represented by eight-noded brick elements with equivalent material density and their orthotropic properties. In that context, the LS-DYNA® choice of material models offers the following two options. The reader should refer to References [6,.7] for detailed information about the features and capabilities of these two honeycomb material models used in the recent versions of LS-DYNA®.

1. *MAT_026 or *MAT_HONEYCOMB
   This material model is for honeycomb and foam materials with real anisotropic behavior. This material model in LS-DYNA® formulation is basically similar to *MAT_002 *MAT_ORTHOTROPIC_ELASTIC, and MAT_002_ANIS, or *MAT_ANISOTROPIC_ELASTIC. Due to this reason, the user is better off staying with *MAT_002 option in terms of numerical accuracy and robustness of the results. *MAT_HONEYCOMB option enables the user to modify the effective values of directional Young's modulus according to the amount of compaction as per formula shown below. The following formulas based upon the volume fraction method are used to determine the effective anisotropic values of E_{aa}, E_{bb}, E_{cc}, G_{ab}, G_{bc}, G_{ca}, etc., which are updated as the material is compacted during the transient dynamic event.

$$E_{aa} = E_{aau} + \beta(E - E_{aau}) \tag{8.5}$$

$$E_{bb} = E_{bbu} + \beta(E - E_{bbu}) \tag{8.6}$$

$$E_{cc} = E_{ccu} + \beta(E - E_{ccu}) \tag{8.7}$$

$$G_{ab} = G_{abu} + \beta(G - G_{abu}) \tag{8.8}$$

$$G_{bc} = G_{bcu} + \beta(G - G_{bcu}) \tag{8.9}$$

$$G_{ca} = G_{cau} + \beta(G - G_{cau}) \tag{8.10}$$

where the dimensionless compaction parameter $\beta = \max\left[\min\left(\frac{1-V}{1-V_f},1\right),0\right]$ (8.11)

In the above equations, Young's modulus E and the shear modulus G are the isotropic elastic material properties in the fully compacted configuration, which are as follows.

$$G = \frac{E}{2(1+\nu)} \tag{8.12}$$

The relative volume, V, is defined as the ratio of the current volume to the initial volume, with typically, $V = 1$ at initial time $t = 0$.

The basic weakness or limitation of this material model is that it does not have any material FAILURE option to invoke. Another drawback of this material model is that it can introduce spurious numerical noise in the transient solution.

2. *MAT_126 or *MAT_MODIFIED_HONEYCOMB
   The major improvements in this material model are for aluminum honeycomb crushable foam materials with anisotropic behavior with the available options of three independent yield surfaces to be specified. However, as more options are available so is the need of additional test data, before anyone can use such a complex material model with any certainty. Additionally, more advanced material models invariably add to the computational cost as well as numerical instability issues of these added features. The reader is cautioned about the challenges involved with such complex material models, which again make the results more unreliable and less useful in any real-life applications or design decisions. The other weakness or limitation of this material model remains as similar to the previous *MAT_026 is that it does not have any material FAILURE option to invoke.

8.4.1 "Progressive Failure" Option for Crushable Honeycomb

There are many aerospace transient dynamics applications, where the FAILURE option must be invoked for the "progressive failure" of the elements to be enabled for the correct numerical simulation of the dynamics involved in a typical high-velocity impact event. In such events like a blade containment during the FBO event, aluminum or Nomex honeycomb being used in the fan containment structural system (see Figures 4.5 and 4.6) gets crushed and fails due to the moving release blade tip. The LS-DYNA® analytical model must be capable of allowing these honeycomb or foam elements to progressively fail and being eroded based upon the limiting "crushing strength" criteria in the material model in order for the transient analysis to continue for a long time. Since all the anisotropic material models available in LS-DYNA® to represent honeycomb such as *MAT_002, *MAT_026, and *MAT_126 lack this important feature, one must develop an alternate way to invoke this parameter of "Limiting Crushing Strength" in the existing LS-DYNA® model.

To that objective in consideration, we will use the "FAILURE ON" capability of *MAT_024. This is a very versatile powerful tool as a material model inside LS-DYNA® to make a brick element to fail and erode at any new time step $(t+\Delta t)$, as soon as that element meets or exceeds the limiting "FAILURE CRITERIA", usually as a strain limiting feature. Using the digitized stress–strain curve, one can invoke practically any type of failure criteria in the LS-DYNA® run. The only limitation of *MAT_024 failure option is that it is only applicable for isotropic material models. However, as a practical solution, in the majority of impact situations including anisotropic barriers in an impact event, this material model *MAT_024 performs very well. Using this capability, as an example we will demonstrate in a step-by-step fashion as how to use "MAT_024" to simulate the dynamic behavior of aluminum honeycomb in an FBO analysis using Hexcel [see Reference 5] public-domain data.

***MAT_024 for Honeycomb:** Among the commonly available material models in LS-DYNA with failure option, it can easily be represented by

*MAT_024 (*MAT_PIECEWISE_LINEAR_PLASTICITY).

It is an orthotropic elastic material, which can withstand a large compressive load in the direction of hexagonal shells. It is made up of 5052 aluminum alloy and each hexagonal shell is connected to the others. Thus, similar to other crushable foam-like materials, it can be compressed up to 95% strain at a constant compressive load called "crushing strength", but cannot take any tension. Its mechanical behavior can easily be captured by an elastic–plastic material model with the following condition and also for a positive slope of stress–strain curve. Similar to "volume fraction ratio methodology" introduced earlier in the composite Section 6.4.1, we will redefine the term called "solidity factor" in the context of aluminum honeycomb as follows:

$$\text{“Solidity factor” of aluminum honeycomb} = \frac{\rho_{\text{AL-Honeycomb}}}{\rho_{\text{Solid-AL}}} \tag{8.13}$$

The "solidity factor" is an important piece of information for developing equivalent elastic–plastic material characteristics data needed to represent the honeycomb properties in the form of *MAT_024 input parameters to simulate the "Progressive failure and Honeycomb elements erosion" during the transient analysis run. As a sample example, we will generate these equivalent material characteristics data in a step-by-step fashion for a typical 5052 Alloy Hexagonal Aluminum Honeycomb Hexcel 1/4–5052 –.0015 [Refer to page 16 of Hexcel manual, Reference 5], which are as follows:

Nominal **Density of 5052 AL-honeycomb** = 3.4 Lbs/cft = 0.00197 Lbs/inch^3 = 0.05446 gm/cc

Solid AL Density = 167.3 Lbs/cft = 0.0968 Lbs/inch^3 = 2.68000 gm/cc

Per formula (8.12), "Solidity Factor" of 5052 AL-Honeycomb = 0.05446/2.68 = 0.02032

Elastic modulus per page-16 data = 90 ksi = 90,000 psi

Crush strength per page-16 data = (Crushing strength)$_{\text{compression}} = \sigma_{\text{crush}} = 150.0$ psi.

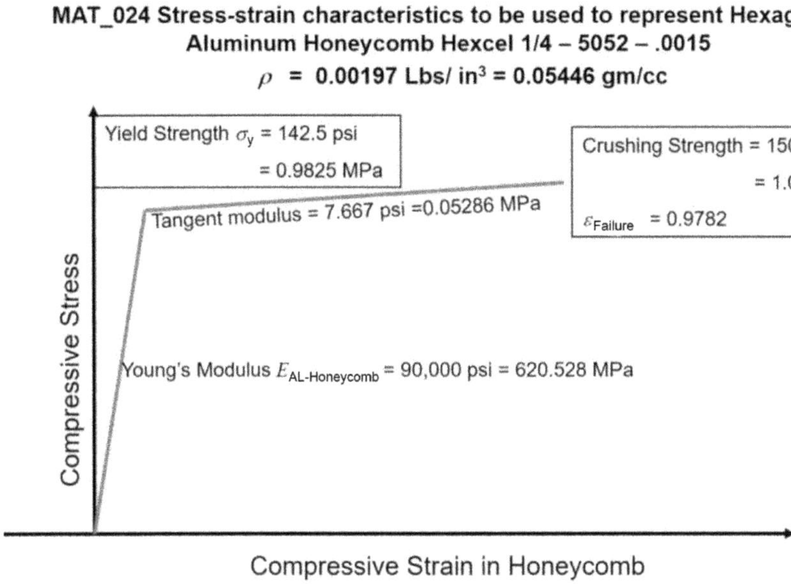

FIGURE 8.17 Equivalent bilinear stress–strain plotted data to be used as input with MAT_024 material model to represent crushing failure of a typical hexagonal aluminum honeycomb Hexcel 1/4–5052 –.0015 in LS-DYNA® analysis.

Estimated yield strength (for numerical stability and positive slope of the bilinear tangent modulus) = σ_Y = 95% of (crushing strength)$_{compression}$ = 0.95 × 150 psi= 142.5 psi.

Yield strain = Estimated yield strength/elastic modulus = 142.5/90,000 = 0.00 1583 = 0.1583%.

Total strain capability until AL-Honeycomb is crushed = 1- (Solidity Factor of Honeycomb) = 1–0.02023 = 0.97977.

Effective plastic strain at FAILURE $\varepsilon_{Failure}$ to be used in MAT_024 input = 0.97977–0.001583 = 0.9782 = 97.82%.

Tangent modulus of AL-Honeycomb $= (150.0 - 142.5) / 0.9782 = 7.667$ psi.

The above-calculated equivalent values for *MAT_024 input are graphically shown in Figure 8.17 through a schematic stress–strain curve. The reader can generate their own similar stress–strain bilinear curve starting with the Hexcel data given in Reference [5].

8.5 DYNAMIC RESPONSE OF RUBBER-LIKE MOONEY–RIVLIN MATERIALS

Rubbers are hyperelastic materials with viscoelastic characteristics, which have multiple usages as mechanical components to perform as dampers, shock absorbers, gaskets, bolted joint washers, etc. in many aerospace applications. In many parts of the engine, we use vibration isolators made up of rubber-like materials. These are viscoelastic materials, and as per classification of viscoelastic material characteristics, they are known as Mooney–Rivlin materials. Depending upon their critical locations in the load path, they become key components to be included in the LS-DYNA® model to carry out the transient dynamic analysis. LS-DYNA® provides the option of two different material models to properly describe the physics and dynamic behavior of such materials during the transient high-velocity impact event. For the benefit of the reader, the following details are as per the description of the LS-DYNA users' manual [see Reference 6].

1. ***MAT_027**: *MAT_MOONEY-RIVLIN_RUBBER (2 coefficients description)
   In this material model, the user needs to supply two coefficients called *A* and *B*, which are based on curve fitting of load-deflection test data collected in a coupon test up to 40% strain. The coefficients *A* and *B* are used in the following strain energy density function equation:

$$W = A(I-3)+B(II-3)+C\left(III^{-2}-1\right)+D\,(III-1)^2 \tag{8.14}$$

where

$$C = 0.5A + B$$

$$D = \frac{A(5\nu-2)+B\,(11\nu-5)}{2(1-2\nu)}$$

ν = Poisson's ratio

$$5(A+B) = \text{shear modulus of linear elasticity}$$

$$I,\ II, III = \text{invariants of right Cauchy – Green Tensor } C$$

Similar to the original definition of strain, the "Principal stretch ratio" in the uniaxial direction, λ_1, is defined as follows:

$$\lambda_1 = \frac{L_0 + \Delta L}{L_0}$$

Using the load-deflection data curve fitting, the input data for *MAT_027, needed as the coefficients *A* and *B* parameters, are generated. For details of this process, the reader is referred to Reference [6].

2. ***MAT_127**: *MAT_ARRUDA_BOYCE_RUBBER (6 coefficients stress–strain data.)
   Rubber is generally considered fully incompressible since the bulk modulus greatly exceeds the shear modulus in magnitude. In comparison to two coefficients input namely *A* and *B* for *MAT_027, in *MAT_127 the user

can specify up to six coefficients, which are based on a similar curve fitting technique of load-deflection test data collected in a coupon test for much bigger strain values. Again, it is the same challenge that more coefficients mean more number test data is needed for the curve fitting. The user must confirm that all six coefficients are positive, as the negative value of any of the coefficients may start numerical instability. To model the rubber as an unconstrained material, a hydrostatic work term, $W_H(J)$, is included in the strain energy functional, which is a function of the relative volume, which is as follows. The following strain energy density function equation is for reader's quick information only. For detailed information about the derivation and other backgrounds of the following strain energy density function equation (8.15), the readers must refer to Reference [6],

$$W(J_1, J_2, J) = nk\theta\left[\frac{1}{2}(J_1 - 3) + \frac{1}{20N}\left(J_1^2 - 9\right) + \frac{11}{1050N^2}\left(J_1^3 - 27\right)\right] +$$

$$nk\theta\left[\frac{19}{7000N^3}\left(J_1^4 - 81\right) + \frac{519}{673{,}750N^4}\left(J_1^5 - 243\right)\right] + W_H(J) \quad (8.15)$$

where the hydrostatic work term is in terms of the bulk modulus, K, and the third invariant, J, as

$$W_H(J) = \frac{K}{2}(J - 1)^2 \quad (8.16)$$

Without going into a lot of unnecessary complexities of *MAT_127 material model options to represent rubber, for simple rate effects, the relaxation function represented by six terms $(\alpha_0, \alpha_1, \alpha_2, \alpha_3, \alpha_4, \alpha_5)$ of the following Prony series is sufficient to capture the full physics of rubber deformation.

$$g(t) = \alpha_0 + \sum_{m=1}^{N} \alpha_m e^{-\beta t} \quad (8.17)$$

Here, in the above six-term equation, $N = 5$. In this form, this *MAT_127 model is effectively a Maxwell fluid, which consists of a damper and spring in series. We characterize this in the input by shear moduli, G_i, and decay constants, β_i. The viscoelastic behavior is optional and an arbitrary number of terms may be used. This simplified model adequate and sufficiently accurate for most of the practical aerospace applications.

8.6 ELASTIC–PLASTIC MATERIAL MODELS FOR METALS WITH HIGH DUCTILITY

We will start our discussion of this section with typical steps needed to compute the plastic strain for a ductile material in a transient dynamic analysis using the explicit time marching forward integration scheme. To that end for each node with nodal motions fully decoupled, the governing equations of motion in the explicit scheme can schematically be written as follows (see Eq. 2.6):

$$m\ \ddot{u}_{t+\Delta t} = F_{\text{external}} + F_{\text{contact}} - F_{\text{stiffness}}\left(\sigma_{xx},\, \sigma_{yy},\, \sigma_{zz},\, \tau_{xy},\, \tau_{yz},\, \tau_{zx}\right)_t \tag{8.18}$$

The above equations are vector equations, and as such represent three equations for each node in a three-dimensional spatial coordinate system. The corresponding nodal acceleration, velocity, and displacement in the global coordinate system at time $= t + \Delta t$ are

$$\ddot{u}_{t+\Delta t} = \frac{1}{m}\left[F_{\text{external}} + F_{\text{contact}} - F_{\text{stiffness}}\left(\sigma_{xx},\, \sigma_{yy},\, \sigma_{zz},\, \tau_{xy},\, \tau_{yz},\, \tau_{zx}\right)_t\right] \tag{8.19}$$

After the nodal accelerations $\ddot{u}_{t+\Delta t}$ have been computed in the simple integration scheme, we can write the equations for the corresponding nodal velocity $\dot{u}_{t+\Delta t}$, nodal displacements $u_{t+\Delta t}$, and the new nodal coordinates $x_{t+\Delta t}$ at the new time step of time $= t + \Delta t$ as follows:

$$\dot{u}_{t+\Delta t} = \dot{u}_t + \left(\ddot{u}_{t+\Delta t}\right)\Delta t \text{ (velocity)} \tag{8.20}$$

$$u_{t+\Delta t} = u_t + \left(\dot{u}_{t+\Delta t}\right)\Delta t \text{ (displacements)} \tag{8.21}$$

$$x_{t+\Delta t} = x_t + \left(\dot{u}_{t+\Delta t}\right)\Delta t \text{ (new nodal coordinates)} \tag{8.22}$$

The corresponding strain rate tensors in any element $\dot{\varepsilon}_{ij}$ at time $= t + \Delta t$ are

$$\dot{\varepsilon}_{ij} = \frac{1}{2}\left(\frac{\partial \dot{u}_i}{\partial x_j} + \frac{\partial \dot{u}_j}{\partial x_i}\right) \tag{8.23}$$

In the above equation, the strain rate tensors $\dot{\varepsilon}_{ij}$ at time $= t + \Delta t$ are determined by taking the partial derivative of the nodal velocities at the new time step of $(t + \Delta t)$ with respect to the nodal coordinates at time "t".

Let us assume that these strain rates are represented as d_1, d_2, d_3, d_4, d_5, d_6 for $\dot{\varepsilon}_{xx}$, $\dot{\varepsilon}_{yy}$, $\dot{\varepsilon}_{zz}$, $\dot{\varepsilon}_{xy}$, $\dot{\varepsilon}_{yz}$, $\dot{\varepsilon}_{zx}$, respectively, for each element. Then, the effective strain rate $\dot{\varepsilon}_{\text{effective}}$ is computed as follows:

$$d_{\text{average}} = -\frac{1}{3}\left(d_1 + d_2 + d_3\right) \tag{8.24}$$

Then, the three components of deviatoric strain rates $d1d$, $d2d$, $d3d$ in any typical element are

$$d1d = \left(d_1\right)_{\text{deviatoric}} = d_1 + d_{\text{average}} \tag{8.25}$$

$$d2d = \left(d_2\right)_{\text{deviatoric}} = d_2 + d_{\text{average}} \tag{8.26}$$

$$d3d = (d_3)_{\text{deviatoric}} = d_3 + d_{\text{average}} \tag{8.27}$$

Then, the effective strain rate $\dot{\varepsilon}_{\text{effective}}$ in any typical element is

$$\dot{\varepsilon}_{\text{effective}} = \sqrt{\frac{2}{3}\left[d1d^2 + d2d^2 + d3d^2 + \frac{1}{2}\left(d_4{}^2 + d_5{}^2 + d_6{}^2\right)\right]} \tag{8.28}$$

The effective strain rate $\dot{\varepsilon}_{\text{effective}}$ is converted into effective strain in any typical element, and the corresponding stress field is determined for that element depending upon its material properties. It should be noted that it is not necessary to compute the effective strain explicitly, rather it can be embedded in the stress-field calculation from the strain rate parameters, viz., d_1, d_2, d_3, d_4,d_5, d_6 multiplied by the time-step size $= \Delta t$. This is done in the following five steps:

a. A trial stress field is established to determine how much of this stress is elastic component and how much is plastic component.
b. For the plastic stress component, hydrostatic pressure component called $p(i)$ and the deviatoric stress components called $t_1(i)$, $t_2(i)$, $t_3(i)$ for σ_{xx}, σ_{yy}, σ_{zz} are computed.
c. Then, the von Mises effective stress for this 3-D stress field is computed.
d. After determining the equivalent uniaxial von Mises stress, the incremental plastic strain called $\text{depi}(i)$ accumulated during the time Δt is computed for this element.

$$\text{depi}(i) = \left[\dot{\varepsilon}_{\text{effective}}(i)\right]\Delta t \tag{8.29}$$

It should be noted that $\text{depi}(i)$ is a scalar number, always positive, and as such has no direction. At any given time during the analysis, the plastic strain in any element is the cumulative sum of incremental plastic strain $\text{depi}(i)$ accumulated during that period and will never go down. Plastic deformation is nonconservative and is similar to entropy in the thermodynamic sense. In this way, the total accumulated cumulative plastic strain is called $\text{ep}(i)$.

$$\left[\text{ep}(i)\right]_{t+\Delta t} = \left[\text{ep}(i)\right]_t + \text{depi}(i) \tag{8.30}$$

e. Once the incremental plastic strain $\text{depi}(i)$ has been determined, a radial return method is used to establish the new stress field at time $(t+\Delta t)$, which yields $\left(\sigma_{xx}, \sigma_{yy}, \sigma_{zz}, \tau_{xy}, \tau_{yz}, \tau_{zx}\right)_{t+\Delta t}$ for this element.

After this step is completed, the analysis continues for the next set of values to be computed at the new time step of $(t+\Delta t)$. If at any time during the integration cycle, the accumulated plastic strain in any element exceeds the failure limit $\varepsilon_{\text{Failure}}$ specified in the material input data entry through the corresponding mat_data.k file, that element will be eroded at that point and will be taken out of the analysis from that point onwards.

8.6.1 High Strain Rate Characteristics of Ductile Materials Under Impact

The readily available commercial code LS-DYNA® offers three main choices of elastic–plastic material model for most of the ductile materials commonly used in aerospace applications. The following is a list of three main classes of metals widely used in the aerospace industry with their typical mechanical properties based on the average of various public-domain sources.

a. Aluminum-based alloys, e.g., Aluminum 6061.
   Typical properties: Young's modulus = 68.9 GPa (10.0E+6 psi)
   Density = 2.7 gram/cc (0.0975 Lb/inche^3)
   Yield strength = 276 MPa (40.0E+3 psi)
   UTS = 290 MPa (42.0E+3 psi)
   Velocity of sound = 5056.32 m/s (16,589 ft/s)
b. Titanium-based alloys, e.g., TI 6-4.
   Typical properties: Young's modulus = 113.8 GPa (16.5E+6 psi)
   Density= 4.43 gram/cc (0.160 Lb/inch^3)
   Yield strength = 931 MPa (135E+3 psi)
   UTS = 1241 MPa (180E+3 psi)
   Velocity of sound = 5070.35 m/s (16,635 ft/s)
c. Steel-based alloys, e.g., INCO718.
   Typical properties: Young's modulus = 205.0 GPa (29.7E+6 psi)
   Density = 8.19 gram/cc (0.296 Lb/inch^3)
   Yield strength = 1035 MPa (150E+3 psi)
   UTS = 1375 MPa (199E+3 psi)
   Velocity of sound = 5001.46 m/s (16,409 ft/s)

From the above summarized material data, in commonly used aerospace applications for typical ductile metals with elastic–plastic behavior, one fact should be noted that regardless of the type of metal alloys, the velocity of sound is practically constant and varies only within a small 5000–5100 m/s range. This characteristic is extremely important and very useful in all explicit time-marching forward integration schemes, where the time-step size Δt is entirely dependent on essentially two parameters only (a) velocity of sound in the medium for stress wave propagation and (b) the smallest element size in the finite-element meshing scheme. Since the velocity of sound is practically constant, the actual Δt to be used in the analysis is controlled by only one parameter, the smallest characteristic length of the smallest element size typically expressed as "Δx". Thus, the stable time-step size value of Δt to be used during the transient analysis phase can be estimated, using the maximum velocity of sound to be 5100 m/s prior to the actual LS-DYNA® run, as follows:

$$\Delta t \leq \frac{\Delta x}{\text{Max. Velocity of Sound}} = \frac{\Delta x \text{ in mm}}{5100 \times 1000 \dfrac{\text{mm}}{\text{s}}} \tag{8.31}$$

To calibrate the user about the true numerical value of actual Δt in any typical mechanical component transient dynamic analysis, let us assume the dimension of the smallest element size in the FEA mesh $(\Delta x = 1\,\text{mm})$, then $\Delta t \leq 1.96E-7$ seconds. This calibration factor can be used for a quick estimate of Δt, depending on the actual value of the smallest element size Δx being used in the FEA mesh of the reader.

For representing ductile metals with elastic–plastic behavior, there are four material models, which are commonly used to capture the full stress–strain characteristics in most of the aerospace impact simulations including strain hardening and rise in the yield strength at high strain rates.

1. ***MAT_003: *MAT_PLASTIC_KINEMATIC**
   It is a very efficient, cost-effective, and extremely robust material model, and the most widely used elastic–plastic material model with the capability of eroding elements, based upon the limiting plastic strain criteria. By default, it uses bilinear stress–strain characteristics, the first one with Young's elastic modulus, and the second one for the plastic strain computation with a low slope value called "tangent modulus". To account for the strain rate effects, it uses Cowper Symonds strain rate model. Schematically, its stress–strain characteristics are shown in Figure 8.18.

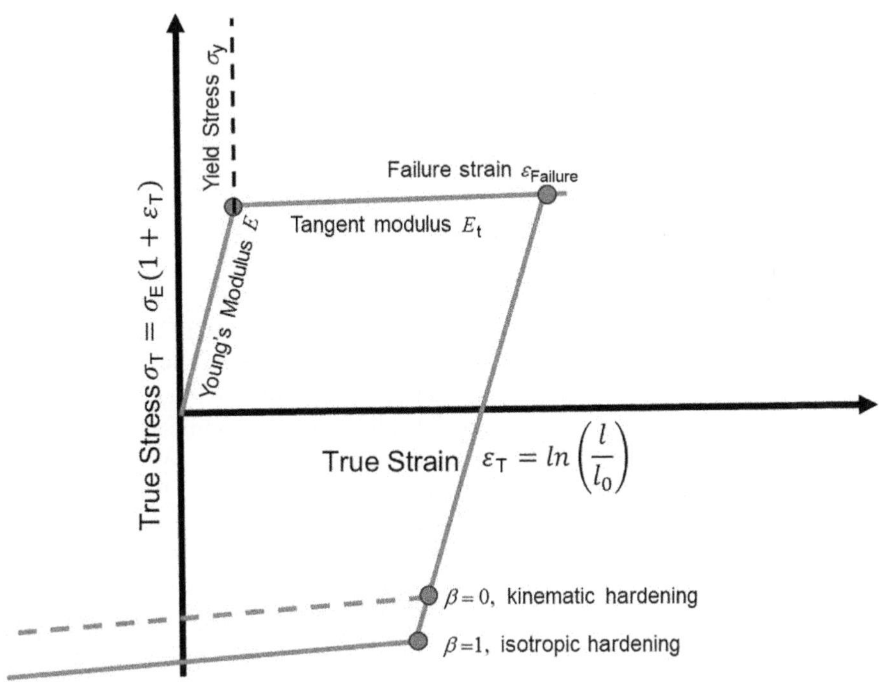

FIGURE 8.18 A bilinear stress–strain diagram schematically showing full loading and unloading characteristics of typical ductile metals with elastic–plastic behavior.

On studying Figure 8.18, the reader should notice that the stress–strain diagram uses the terms "True Stress σ_{True}" and "True Strain $\varepsilon_{\text{True}}$". It should be pointed out that during the large plastic deformation, the "True stress σ_T" and the "True strain ε_T" values are somewhat different from the conventional "Engineering Stress σ_E" and "Engineering Strain ε_E" values, which are related to each other as follows:

$$\text{True strain} = \varepsilon_{\text{True}} = \ln\left(\frac{l}{l_0}\right) = \ln\left(1+\varepsilon_{\text{Engineering}}\right) \tag{8.32}$$

$$\text{True stress} = \sigma_{\text{True}} = \sigma_{\text{Engineering}}\left(1+\varepsilon_{\text{True}}\right) \tag{8.33}$$

The main drawback of this model is that it considers the plastic strain a scalar quantity and does not distinguish between the plastic strain accumulated during the tension cycle of loading or the compression side of loading. Also, the elastic–plastic behavior is exactly similar both in the tension side of loading and in the compression side of the loading.

2. ***MAT_098: *MAT_SIMPLIFIED_JOHNSON_COOK**
The Material type 98 uses a Johnson-Cook strain rate sensitive plasticity formulation, which ignores the effect of temperature change and all thermal effects. It is used for problems where the strain rates vary over a large range. The model uses a parameter called PSFAIL, which is the effective plastic strain at failure. If the parameter PSFAIL is zero, the element failure is not considered. The Johnson and Cook material model expresses the flow stress as follows [see References 6,8,9]:

$$\sigma_y = \left(A + B\bar{\varepsilon}^{pn}\right)\left(1 + c\ln\dot{\varepsilon}^*\right) \tag{8.34}$$

where
A, B, c, n, and m = input constants
$\bar{\varepsilon}^{p}$ = effective plastic strain

3. ***MAT_024: *MAT_PIECEWISE_LINEAR_PLASTICITY**
The Material Type 24, which is an elasto–plastic material model with an arbitrary stress versus strain curve and the capability to arbitrarily define the strain rate dependency as well. This is the most versatile elastic–plastic material model and is highly recommended for various types of analyses and for numerical robustness in terms of its capabilities. It can handle even a negative slope of the stress–strain curve in the plastic deformation zone, without running into numerical issues. Also, element FAILURE and EROSION based on a plastic strain or a minimum time step size can be defined. This is the best material model with many enhanced capabilities and should be the first choice, when a lab-test data is available to characterize the material behavior even in extreme conditions. A complex stress–strain behavior can easily be supplied in a digitized columnwise data with stress values versus strain data either directly or through a load curve.

4 ***MAT_019: *MAT_STRAIN_RATE_DEPENDENT_PLASTICITY**
This is Material Type 19. A strain rate-dependent material can be defined. Its capability and formulation is similar to MAT_024. It is up to the user to decide as to which material model should be the first choice. For such applications, *MAT_024 is highly recommended.

8.7 MULTILAYERED CONTAINMENT DESIGN

Earlier in Chapter 4, we had discussed in a generic sense about the physics of fan blade containment in the context of FBO event as a softwall versus hardwall containment design. In this section, we will discuss some fine points related to LS-DYNA® modeling of a multilayered containment system in terms of meshing and other details necessary from accuracy of numerical result considerations from such analyses. Referring to Figure 4.2a and b, any containment system cross section has multiple layers of various materials such as an abradable layer, a septum, a metallic AL-Honeycomb or Nomex-Honeycomb, the main hardwall containment structure, or various concepts of Aramid materials like Kevlar-49 blankets for the softwall containment system. Regardless of various conceptual designs, all of them have one common characteristic; they all have multilayered configurations acting as a system made up of multiple barriers to stop the movement of a release blade and bring it to rest and fully nested inside the containment system without breaching the outermost layer and exiting its confined space.

8.7.1 Recommended FE Meshing Procedure

During the FE meshing process of a multilayered containment system, the user is faced with the following two conflicting requirements:

1. In any continuous structure, a flow-through meshing scheme is always a better and preferred approach than introducing a tied interface anywhere in the structure as much as possible.
2. Two sides with widely varying stiffness values should never share the same node. As such, the two sides can only be connected through a TIED surface-to-surface interface. However, there is an additional requirement that any interface to work correctly, the SLAVE side mesh must be FINER or at least the same size as the MASTER side mesh. Additionally, on a TIED interface, no NODES should have any interference whatsoever at time $t=0$. These conflicting requirements are mathematically impossible to satisfy due to the curvature of the hardwall metal containment ring in the circumferential direction (see Figure 8.19).

Referring to Figure 8.19, the outermost layer of elements in the containment structure could be solid metal with very high stiffness values. However, an adjacent layer could be made up of aluminum or Nomex honeycomb, which are relatively speaking extremely soft. The only way the two sides can be connected is through a

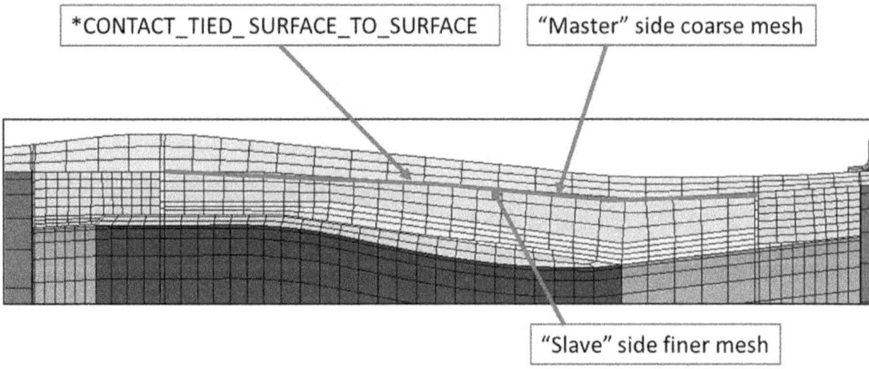

FIGURE 8.19 A typical theoretically CORRECT mesh from interface considerations of TIED surface-to-surface requirements, prone to numerical issues of surfaces passing through each other.

*CONTACT_TIED_ SURFACE_TO_SURFACE. However, due to a finer mesh on the Honeycomb side being SLAVE side, some of these nodes will protrude through the MASTER surfaces, which will definitely create numerical errors in the model during the transient analysis run.

To avoid such conflicting situations prone to numerical errors, highlighted through the meshing scheme shown in Figure 8.19, we will now present in a step-by-step fashion an alternate meshing scheme, which is schematically shown in Figure 8.20.

a. First, generate a flow through mesh of the entire containment system, starting from the innermost surface of abradable, septum, honeycomb, outermost barrier of containment ring, etc. as one continuous system as shown schematically in Figure 8.20.
b. Then separate out each layer based upon the respective materials and assign them separate PID numbers. Make sure that each separate layer represents the correct thickness such as abradable, septum, honeycomb, and outer

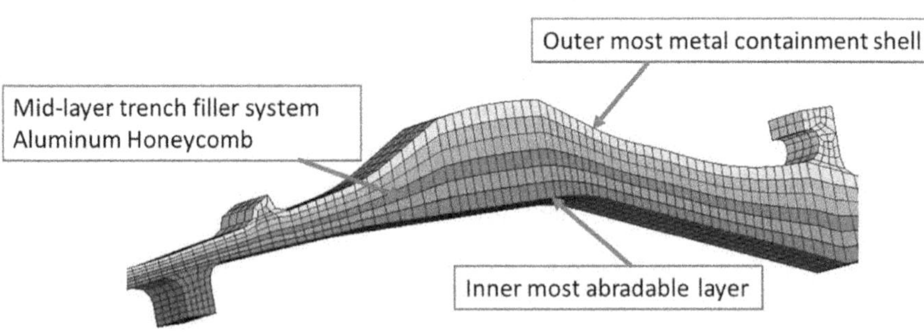

FIGURE 8.20 Cross-sectional view of a generic multilayered hardwall containment system using a preferred flow-through meshing procedure.

main containment metal shell. One should resequence the node renumbering scheme with one PART-ID# at a time.

c. The above process will automatically create a duplicate set of nodes at the mating surfaces, which ensures that there will be no interference or initial spurious penetration due to a small mismatch of nodal coordinates. This step unconditionally guarantees that all mating surface nodes on both the master and salve sides are exactly in the same physical location, and nothing has unknowingly changed due to round-off or truncation errors, commonly observed in the mesh type encountered in the meshing scheme shown in Figure 8.19.

   In LS-DYNA® simulation, this is the recommended way to generate the mesh in order for the interface type *CONTACT_TIED_ SURFACE_TO_ SURFACE to function properly and without any numerical issue of two surfaces passing through each other during the transient analysis run.

d. The following note applies to the Softwall containment system with multiple wraps of Aramid fabric like Kevlar 49 blanket as the final barrier to contain and fully stop the movement of the release blade in the radially outward direction. Since Kevlar cannot take any bending and it can be loaded only in membrane-type tension, each layer of Kevlar blanket must be modeled separately with only one element through the thickness for each individual wrap. Note that if you model more than one element through the thickness, it will automatically start taking bending load, which is not correct and will result in wrong numerical results.

   Each individual wrap should be assigned a separate and its own unique PID #, which is very helpful in postprocessing of the results. Additionally, each individual wrap of the Kevlar blanket should interact with the adjacent wraps by freely sliding without any friction and any surfaces passing through each other. This condition can easily be satisfied by selecting the following contact interface:

   *CONTACT_AUTOMATIC_ SURFACE_TO_SURFACE.

8.7.2 Material Input Data Correction for Multilayered Structure in Bending

The reader will recall from Section 2.3 the limitations of using eight-noded brick elements in the LS-DYNA® meshing scheme. In the explicit time marching forward integration scheme, the eight-noded bricks are the most primitive elements, which use a single-point integration at the element centroid only. In effect, they carry only membrane stresses and do not offer any resistance to bending. Due to his reason of limitation, any mechanical or structural components that are likely to be loaded in bending must be modeled with a minimum of two elements through the thickness. To capture the bending deformation correctly, the recommended value is three elements through the thickness for any typical structural components being represented by eight-noded constant strain brick elements. It should be noted that although the three-element through the thickness model is adequate to capture the deformed shape reasonably well, the accuracy of stress value is still not very precise.

The computed stress values from such models are always underpredicting the actual stresses measured from the strain gage data (see Figure 6.5). This computed stress value essentially after reaching plastic yielding σ_Y and its consistently underpredicting of stresses results in a serious error in "Progressive Failure" value of the structure from the LS-DYNA® analysis. Due to this reason, a small minor CORRECTION is needed in all the material input data related to yield stress σ_Y value for different elastic–plastic material models like *MAT_003, *MAT_019, *MAT_024, and *MAT_098.

This correction can be characterized as "Mesh discretization correction for bending deformation with constant strain brick elements". Here, the "Correction Factor" is determined as follows. If n is the number of constant strain solid brick elements through the thickness, then we can write

$$(\sigma)_{\text{Outermost-element-mid-surface}} = \frac{(n-1)}{n}(\sigma)_{\text{Surface}} \tag{8.35}$$

$$(\sigma)_{\text{Surface}} = \frac{n}{(n-1)}(\sigma)_{\text{LS-DYNA-computed}} \tag{8.36}$$

As a sample example, we apply the above equation for two different models: (i) a three-element through thickness model and (ii) a six-element through thickness model (see Figure 8.21).

Based on the above equations (8.35) and (8.36), we can write the yield stress equation as follows. In the LS-DYNA® input, with a bending type of loading the peak stress is a function of the number of elements used to simulate the thickness "h "of a particular component. Suppose the number of elements is "n" in the thickness direction, then

$$\sigma_Y = (\sigma_Y)_{\text{LS-DYNA FEA model}} = \frac{(n-1)}{n}(\sigma_Y)_{\text{Material Test data}}. \tag{8.37}$$

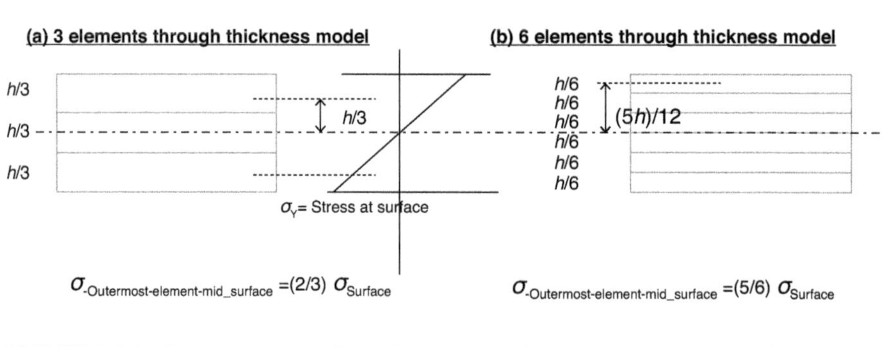

FIGURE 8.21 Sample example of bending stresses with constant strain solid brick elements in (a) a three-element through thickness model and (b) a six-element through thickness model.

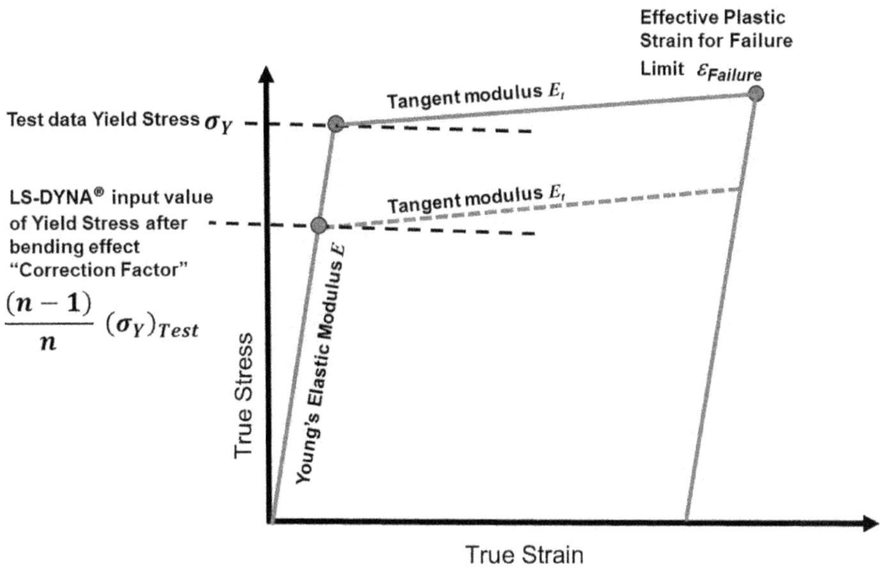

FIGURE 8.22 Schematic bilinear stress–strain diagram for elastic–plastic materials in LS-DYNA® input values of σ_Y and ***UTS*** with bending effect "Correction Factor $(n-1)/n$." (see Figure 8.18 for comparison.)

Figure 8.22 schematically shows the effect of "Correction factor" on the stress–strain diagram corresponding to the input values of yield stress to be used in the LS-DYNA® material data. It is worth pointing out that these corrections related to lowering down the yield stress σ_Y and UTS values are implemented only on the stress side and never on the strain side.

8.7.3 Generic 180-Degrees Sector FBO Containment Model

For the reader's benefit, we have applied all the corrections and lessons learned as outlined earlier in this chapter and made a typical three-blade FBO model shown in Figure 8.23. The model shown here is a 180° sector model of the fan blade containment structure and contains three fan blades. The first one is called the "Release Blade", and the other two blades are called Trail Blade-1 and Trail Blade-2. They are rotated at a constant speed, and a steady-state solution is developed in ANSYS® as the stress initialization is imposed. We also apply the preload on the bolts connecting the inlet and forward flange of the fan containment case.

In the circumferential direction, the two ends of the containment structure at 3 O'clock and 9 O'clock plane should have roller supports as shown in Figure 8.23. This type of boundary condition allows the containment structure to deform in a breathing mode of the case ovalization of radially inward or outward movement in 3–9 O'clock plane due to the main impact by the release blade on the fan case being around 12 O'clock position.

Depending upon the direction of rotation of the fan rotor, the positioning of the three blades under consideration, either at 3 O'clock and 9 O'clock initial positions,

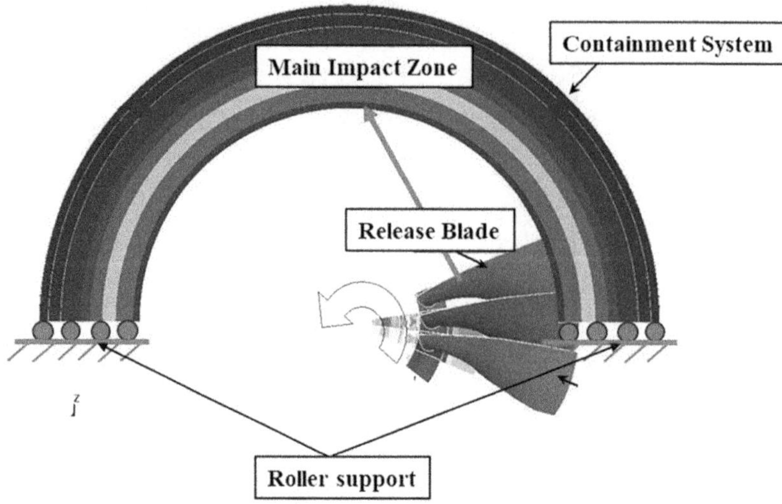

FIGURE 8.23 A typical LS-DYNA® model of 180° sector of the fan containment system with three blades to study its containment capability under the release of one full blade during the FBO event.

ensures that the "main impact zone" of the release blade damage is always around the 12 O'clock position, which is sufficiently away from any end-condition effect.

The reader should be cautious and fully aware of the limitations of 180° circumferential sector of the containment system modeling approach. Although this modeling approach is a quick and efficient way to evaluate the blade containment capability of such a multilayered containment system, it is by no means a replacement for a 360° containment model with fully bladed fan rotor modeling methodology, discussed in detail in Section 4.3. That is still a better modeling approach with higher fidelity to generate all the required dynamic loads to make the design decisions about other components or the mount systems located at far ends of the engine or the rig hardware.

8.8 OIL SLOSHING AND IMPACT OF INFLAMMABLE LIQUIDS INSIDE A CONTAINER

Oil sloshing and the leakage of inflammable liquids inside an oil tank or rupture of fuel lines during the FBO event is a serious fire-hazard situation and has become a matter of concern for regulatory agencies like FAA. Currently, as the LS-DYNA® application is being extended toward full engine models [see Reference 1], there has been significant interest to apply the learnings of oil sloshing analysis from other studies and applications in transportation industries [see Reference 10] to the aerospace and aviation industry as well.

In newer LS-DYNA® applications, researchers have modeled oil with SPH (Smooth Particle Hydrodynamic simulations) elements as well as ALE (Arbitrary Lagrangian Eulerian cell formulations) elements. These elements might be better suited to capture the detailed physics precisely, but have always been extremely challenging to produce credible results in very large models comprising millions of nodes and elements along with thousands of interfaces with potential interactions with surrounding structural components. Here, in the interest of computational time, modeling effort involved, and other availability of testing resources, we will discuss a more conventional robust approach of oil modeling methodology.

Inside an oil tank, the oil elements should be meshed with eight-noded brick elements with an aspect ratio of the elements close to 1 as much as possible. So far, the interface with the surrounding wall of the oil tank structural elements is concerned, the oil elements should be free to move, slide, or separate freely with respect to inside surface segments of the tank walls.

The oil elements are represented by *MAT_009: *MAT_NULL with EOS for compressibility of oil. The typical input parameters for this material model are ρ mass density, $PC \leq 0$ pressure cutoff (in units of load/unit area, i.e., the tension limit), and μ dynamic viscosity. The null material has no yield strength and behaves in a fluid-like manner. When used with solids, this material must be used with an equation of state. Pressure cutoff (Parameter PC) is negative in tension. The equation of state (also called EOS) is essentially a discrete set of digitized tabular data, typically at ten data points, showing the pressure–volume strain characteristics data for the liquid, oil, or any other fluid being considered in the analytical model. For reference and comparison purposes of the numerical magnitude of these values, typical oil properties are as follows:

ρ = Density = 0.029 Lbs/in^3 = 0.8027 gm/cc
μ = Dynamic viscosity = 0.000014 Lbs-sec/in^2 = 0.09653 Pascal-second (Pa-s)

When the oil tank or any other accessory is included in the LS-DYNA® model, it must be meshed in sufficient detail to capture the full dynamic response of these additional components being connected to the main containment system. Figure 8.24 shows one such accessory link's LS-DYNA® FEA mesh to function properly and provide the actual load path during the transient dynamic response of the entire system.

If the rupture of the fuel lines needs to be considered, then they must be included in the LS-DYNA® FEA mesh explicitly. In that context, although eight-noded brick elements are the preferred choice of element types for most of the engine or rig model, the fuel lines may be represented by one-dimensional beam elements in limited places with proper clip supports in correct geometrical locations. It also has to have the failure "ON" option invoked, so that if the threshold value of the dynamic strain exceeds the failure limit either of the fuel line tubes or the clip support; those elements must be ERODED, in order the progressive failure model to properly capture the full dynamics of the entire engine as well as its minor components like the fuel line tubes and its support clip. In this way using the modeling procedure outlined in this section, the reader can develop a model, which will be capable of predicting any possible failure of fuel lines or any other leakage of inflammable liquids like oil from an oil tank during the FBO event or the rig-test outcome in advance prior to the real test.

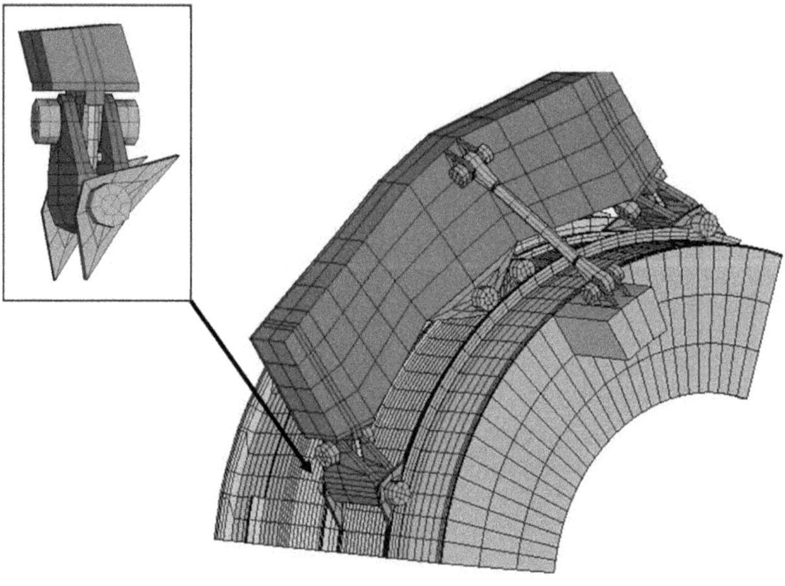

FIGURE 8.24 Typical detailed LS-DYNA® meshing of Links and accessory's connections.

Here, we have discussed in detail as how to capture the entire full dynamic response of the entire engine in the LS-DYNA® simulation on the physics-based standalone analytical model, prior to the actual FBO test. This step is needed as to have full confidence in analytical predictions without making the calibration or validation run of the analytical model and still being able to produce the pretest results accurately, which we had discussed earlier in this chapter about a similar test as well as the pre-test prediction [see Reference 1].

8.9 SUMMARY AND CONCLUSIONS

This chapter applies all the learning about the fundamentals of impact dynamics to solve real-life design problems encountered in various aerospace applications essentially in the propulsion system area. We started with the basic theories of impact situations with conservation of energy and momentum equations and applications of impulse-momentum theorem. The governing equations of motion to describe the stress-wave propagation in deformable bodies were used to develop the mathematical techniques such as (i) implicit and (ii) explicit time marching forward integration methods. We also focused on the numerical accuracy aspects of both techniques. And with the superiority of explicit technique and using the tremendous advancements in numerical simulations and as an industry-standard of commercially available code like LS-DYNA®, we have shown how to tackle and solve true real-life design challenges in the impact dynamics area of numerical simulations.

REFERENCES

1. S. K. Sinha, G. Czarnecki and R. L. Hinrichsen (2013) "Dynamic Analysis of Damage to Aircraft Propulsion System Impacted by Exploding Missile", *AIAA Journal of Aircraft*, Vol. 50, No. 5, pp. 1526–1532. https://doi.org/10.2514/1.C032055
2. T. Nicholas (1980) "Dynamic Tensile Testing of Structural Materials Using a Split-Hopkinson Bar Apparatus", Wright–Patterson AFB, Rept. No. AFWAL-TR-80-4053. https://apps.dtic.mil/sti/pdfs/ADA092832.pdf
3. D. W. Hyde "ConWep: Conventional Weapons Effects Program", ERDC Rep. No. TM-5–855-1, The Engineer Research and Development Center (ERDC), U.S. Army Corps of Engineers Distributed Research and Development Command, Vicksburg, MS, pp. 1–339. https://www.scribd.com/document/863151732/ConWep-Manual
4. R. P. Czachor (2005) "Unique Challenges for Bolted Joint Design in High-Bypass Turbofan Engines." *Journal of Engineering for Gas Turbines and Power*, Vol. 127, No. 2, 240–248. https://doi.org/10.1115/1.1806453
5. HexWeb™ Honeycomb Attributes and Properties (1999) published by Hexcel Corporation, Pleasanton, California. (Total 40 pages). https://www.pccomposites.com/wp-content/uploads/2015/07/PCHC4-4TY4_TDS.pdf
6. LS- DYNA®, KEYWORD USER' S MANUAL, VOLUME II, Material Models, LS-DYNA R11, 12 October 2018 (r:10572), Livermore Software Technology Corporation (LSTC).
7. LS- DYNA®, KEYWORD USER' S MANUAL, VOLUME I, LS-DYNA R11, 18 October 2018 (r:10580), Livermore Software Technology Corporation (LSTC).
8. D. Leseur (1999) "Experimental Investigations of Material Models for Ti-6A1–4V and 2024-T3 Aluminum", No. UCRL-ID-134691. Lawrence Livermore National Laboratory (LLNL), Livermore, CA (United States), 1999. FAA Report No. DOT/FAA/AR-0025, Office of Aviation Research, Washington, DC . https://ntlrepository.blob.core.windows.net/lib/17000/17400/17488/PB2001101864.pdf
9. Y. Huang, J. Ji, and K.-M. Lee (2019) "An Improved Material Constitutive Model Considering Temperature-Dependent Dynamic Recrystallization for Numerical Analysis of Ti-6Al-4V Alloy Machining", *The International Journal of Advanced Manufacturing Technology*, Vol. 97, No. 9, pp. 3655–3670. https://par.nsf.gov/servlets/purl/10095033
10. W. Siraye, A. Wakjira and R. B. Nallamothu (2023) "An Analysis of Fuel Oil Sloshing in Partially Filled Cargo Tanker Trucks Under Cornering Conditions Using Various Baffle Systems", *Journal of Engineering*, No. 1, p. 9941864. https://doi.org/10.1155/2023/9941864

Index

D

E

F

G

H

I

J

K

L

T

U

V

W

Z

For Product Safety Concerns and Information please contact our EU representative GPSR@taylorandfrancis.com
Taylor & Francis Verlag GmbH, Kaufingerstraße 24, 80331 München, Germany

www.ingramcontent.com/pod-product-compliance
Ingram Content Group UK Ltd.
Pitfield, Milton Keynes, MK11 3LW, UK
UKHW022216190926
475702UK00003B/383
9781041262756